Smooth Compactifications of Locally Symmetric Varieties

Second Edition

The new edition of this celebrated and long-unavailable book preserves much of the content and structure of the original, which is still unrivalled in its presentation of a universal method for the resolution of a class of singularities in algebraic geometry. At the same time, the book has been completely re-typeset, errors have been eliminated, proofs have been streamlined, the notation has been made consistent and uniform, and an index and a guide to recent literature have been added.

The authors begin by reviewing, in Chapter I, key results in the theory of toroidal embeddings and by explaining examples that illustrate the theory. Chapter II develops the theory of open self-adjoint homogeneous cones and their polyhedral reduction theory. Chapter III is devoted to basic facts on hermitian symmetric domains and culminates in the construction of toroidal compactifications of their quotients by an arithmetic group. The final chapter considers several applications of the general results.

The book brings together ideas from algebraic geometry, differential geometry, representation theory and number theory, and will continue to prove of value for researchers and graduate students in these areas.

Smooth Compactifications of Locally Symmetric Varieties

Second Edition

AVNER ASH
Boston College

DAVID MUMFORD
Brown University

MICHAEL RAPOPORT
University of Bonn

YUNG-SHENG TAI
Haverford College

With the collaboration of

PETER SCHOLZE
University of Bonn

CAMBRIDGE
UNIVERSITY PRESS

CAMBRIDGE
UNIVERSITY PRESS

University Printing House, Cambridge CB2 8BS, United Kingdom

One Liberty Plaza, 20th Floor, New York, NY 10006, USA

477 Williamstown Road, Port Melbourne, VIC 3207, Australia

314-321, 3rd Floor, Plot 3, Splendor Forum, Jasola District Centre, New Delhi - 110025, India

79 Anson Road, #06-04/06, Singapore 079906

Cambridge University Press is part of the University of Cambridge.

It furthers the University's mission by disseminating knowledge in the pursuit of
education, learning and research at the highest international levels of excellence.

www.cambridge.org
Information on this title: www.cambridge.org/9780521739559

A catalogue record for this publication is available from the British Library

ISBN 978-0-521-73955-9 Paperback

Contents

Preface to the second edition

When CUP approached us with the proposal of a second edition to our book, we first consulted graduate students and younger colleagues to test this idea on them. Their enthusiastic response convinced us of the soundness of the proposition.

In order to keep this project within realistic bounds, we did not rewrite the book, but rather TEX-ed the original text and corrected mistakes that have come to our attention. We also smoothed somewhat the presentation and homogenized the notation. Finally, in order to increase its usability, we added an index.

So, all in all, this is still essentially the same book. In particular, the text of this new edition does not reflect the developments in the field in the last 30 years. To compensate for this, we added a guide to the more recent literature at the end of the book.

In this whole project we were assisted by Peter Scholze, who read the whole manuscript, corrected many mistakes, and helped us with the proof-reading. We thank him heartily. We also thank Y. May, who assisted us in TEX-problems.

We also thank all those who pointed out mistakes in the first edition and often indicated to us how to correct them: we are thus grateful to C.-L. Chai, E. Looijenga, R. Pink, Y. Namikawa, and I. Satake. Finally, we thank the staff of CUP, and particularly D. Tranah, for their expert cooperation.

Avner Ash,
David Mumford,
Michael Rapoport,
Yung-sheng Tai.

Preface to the first edition

Let D be a bounded symmetric domain and let Γ be a neat (see Ch. III, §7) arithmetic subgroup of $\mathrm{Aut}\,(D)^o$. The goal of this monograph is the construction of a family of non-singular† compactifications $\overline{D/\Gamma}$ of D/Γ. This theory was announced and described in rough outline in [2]. Very similar ideas were developed independently by Satake in [3]. Both of us were following the path indicated by the work of Igusa when $\Gamma = \mathrm{Sp}\,(2n, \mathbb{Z})$ and by Hirzebruch when $\Gamma = \mathrm{SL}(2, \mathcal{O})$, where $\mathcal{O} = $ integers in a real quadratic field.

Here is an outline of the monograph. Since this work builds heavily on [1] (referred to as TE I below), we review quickly some of these results and add some comments particular to the complex case in Ch. I, §§1–3. Then, in Ch. I, §§4,5, we describe smooth compactifications of two surfaces D/Γ, in order to illustrate the general theory which follows (actually, in §4, D is not bounded – it is $\Delta \times \mathbb{C}$ – so this is not strictly a special case). Chapter II, by A. Ash, is devoted to self-adjoint homogeneous cones. The main result is a comparison of Siegel sets and polyhedral subcones inside such homogeneous cones. These results are essential for the construction of $\overline{D/\Gamma}$. The construction itself is taken up in Ch. III. The final results require considerable notation to state, but may be found in Ch. III, §§5, 7. The principal technical contribution here is M. Rapoport's calculation of the Satake topology on $D^* = D \bigcup (\text{rat. boundary comp.})$ in terms of the presentation of D as a Siegel domain of third kind, which is crucial to proving that our $\overline{D/\Gamma}$ is Hausdorff. The final chapter by Y. S. Tai adds two important results. Firstly, he applies the construction to prove that D/Γ is a variety "of general type" in Kodaira's classification when Γ is small enough. Secondly, although our general $\overline{D/\Gamma}$ is only an analytic compactification of D/Γ, he shows that many of these $\overline{D/\Gamma}$'s are indeed projective varieties.

One of the main obstacles in our research was that none of us were symmetric space specialists when we began, and, of course, roots are the name of the game throughout. For our sake as well as the reader's, we thought it useful to include a considerable amount of expository material in the hope of making

---

† We are mainly concerned with a larger class of compactifications with "toroidal" singularities on the boundary. Within this class, there are plenty of non-singular compactifications, but these do not play any special role in our study.

the monograph more self-contained. We were greatly aided by similar expository projects of P. Deligne and I. Satake, who graciously lent us their notes.
In general, the expository sections emphasize the geometric aspects somewhat
more than the references, and, in particular, develop the ideas in the form in
which we need them. Experts can skim rapidly through Ch. II, §§1–3 (note,
however, the very crucial tie-up between Pierce decompositions and split tori
which appears to be new) and Ch. III, §§2–4 (note here the key role played by
the intermediate open set $D(F): D \subset D(F) \subset \check{D}$, $D(F) = U(F)_{\mathbb{C}} \cdot D$, in the
construction of the Siegel Domain realization).

I hope that the space $\overline{D/\Gamma}$ here constructed will have other applications in
the theory of automorphic forms, e.g., to calculating invariants of the field
$\mathbb{C}(D/\Gamma)$ and the dimension of the spaces of automorphic forms. Besides these
applications, the theory can hopefully be pushed further in three essential directions: (i) at least for $D = \mathrm{Sp}(2n, \mathbb{R})/K$, extend it to a construction of a
scheme $\overline{D/\Gamma}$ over $\mathbb{Z}$; (ii) to extend Hirzebruch's proportionality theorems to
the non compact case; (iii) in view of the fact that the results describe con-
cretely the degeneration of Hodge structures of a very special type – find an
extension of them, combining the ideas of Ch. III, §7 with Schmid's results on
families of Hodge structures over $\mathring{\Delta}$, to describe asymptotically all families of
Hodge structures on $(\mathring{\Delta})^k$.

David Mumford

Authorship of the various chapters

Chapter I: David Mumford

Chapter II: Avner Ash

Chapter III: Michael Rapoport and David Mumford

Chapter IV: Yung-sheng Tai

References

[1] G. Kempf, F. Knudsen, D. Mumford and B. Saint-Donat, *Toroidal Embeddings I,* Lecture Notes in Mathematics 339. Berlin: Springer 1972.†

[2] D. Mumford, A new approach to compactifying locally symmetric varieties, in *Proceedings of the Tata Institute Colloquium, Jan. 1973*, Oxford Univ.
Press, 1975.

[3] I. Satake, On the arithmetic of tube domains, *Bull. Amer. Math. Soc.*, **79**
(1973), 1076-1094.

† This is referred as 'TE I' throughout the monograph.

I

Basics on torus embeddings; examples

1 Torus embeddings over the complex numbers

We wish to review here quickly some results of TE I† and to give a more explicit description of the complex varieties obtained via certain real spaces of half the dimension.

Let T be an algebraic torus, i.e., $T \cong \mathbb{G}_m^n$ for some n, and let

$$M = \mathrm{Hom}\,(T, \mathbb{G}_m)\,,$$

the character group of T, and

$$N = \mathrm{Hom}\,(\mathbb{G}_m, T)\,,$$

the group of 'one-parameter' subgroups of T (in the algebraic sense).

Then $M \cong \mathbb{Z}^n$ and $N \cong \mathbb{Z}^n$, and there is a natural non-degenerate pairing $\langle \cdot, \cdot \rangle : M \times N \longrightarrow \mathbb{Z}$ of determinant 1. All this is valid over any field k. When $k = \mathbb{C}$, however, T can be described analytically as $\widetilde{T}/\pi$, where $\widetilde{T}$ is a complex vector space and π is a discrete subgroup, generating $\widetilde{T}$ over $\mathbb{C}$ and isomorphic to $\mathbb{Z}^n$. Here $\widetilde{T}$ is the universal covering space of T and π is $\pi_1(T)$ acting on $\widetilde{T}$ via translations. Note, however, that for all $a \in \pi$ the map

$$\widetilde{\phi}_a : \mathbb{C} \longrightarrow \widetilde{T}$$
$$\lambda \longmapsto \lambda \cdot a$$

induces a map

$$\phi_a : \mathbb{C}/\mathbb{Z} \longrightarrow \widetilde{T}/\pi = T\,,$$

and that $\mathbb{C}/\mathbb{Z} \cong \mathbb{G}_m$ canonically via $\lambda \longmapsto e^{2\pi i \lambda}$. Thus every $a \in \pi$ induces $\phi_a \in N$, and this is easily checked to be an isomorphism between π and N. Thus π is just N up to a canonical identification. Since $\widetilde{T} = \pi \otimes \mathbb{C}$, it follows that we have canonical maps:

† Recall this reference from p. x.

1

(i) $N \cong$ the usual topological π_1 of T;

(ii) $N \otimes \mathbb{C} \cong$ the universal covering space of T;

(iii) $(N \otimes \mathbb{C})/N \cong T$.

We abbreviate $N \otimes \mathbb{C}$ by $N_\mathbb{C}$ and $N \otimes \mathbb{R}$ by $N_\mathbb{R}$.

Next, in the isomorphism $N_\mathbb{C}/N \cong T$, consider the subgroup corresponding to $N_\mathbb{R}/N$: this a compact real torus, and is the maximal compact subgroup of T. We denote it by T_c (short for "compact torus"). Moreover, $N_\mathbb{R} \subset N_\mathbb{C}$ has a natural complement, viz. $iN_\mathbb{R}$, and, by quotienting, $iN_\mathbb{R}$ injects into $N_\mathbb{C}/N$. In other words, we get a canonical decomposition

$$N_\mathbb{C}/N \cong (N_\mathbb{R}/N) \times (iN_\mathbb{R}) ,$$

and hence (dividing by i in the second factor)

$$T \cong T_c \times N_\mathbb{R} .$$

We denote the projection $T \longrightarrow N_\mathbb{R}$ by "ord," which is then defined by

$$\mathrm{ord}(x + iy \bmod N) = y , \text{ for all } x, y \in N_\mathbb{R} .$$

If $\alpha \in M$, and $\mathfrak{X}^\alpha : T \longrightarrow \mathbb{C}^*$ is the corresponding function (as in TE I, it is useful to think of M as an additive group, and hence to adopt exponential notation for the characters regarded as functions on T), we obtain the formula

$$\mathfrak{X}^\alpha(x + iy \bmod N) = e^{2\pi i(\langle \alpha, x \rangle + i\langle \alpha, y \rangle)} , \text{ for all } x, y \in N_\mathbb{R} ;$$

hence

$$|\mathfrak{X}^\alpha(z)| = e^{-2\pi \langle \alpha, \mathrm{ord}\, z \rangle} , \text{ for all } z \in T .$$

Next, in TE I, Ch. I, §1, we define embeddings of T in normal affine varieties X_σ, with the action of T on itself extending to an action of T on X_σ, whenever $\sigma \subset N_\mathbb{R}$ is a closed rational polyhedral cone not containing a line. Recall that

$$X_\sigma = \mathrm{Spec}\,\mathbb{C}[\ldots, \mathfrak{X}^\alpha, \ldots]_{\alpha \in M \cap \check{\sigma}} ;$$

here $\check{\sigma} \subset M_\mathbb{R}$ is the dual cone to σ, so $M \cap \check{\sigma}$ is a sub-semigroup of M. In order to study convergence in the classical topology and other details on X_σ, it will be convenient to introduce here the topological space (in the classical, not Zariski, topology) obtained by dividing X_σ by T_c. This will look like $N_\mathbb{R}$ with points at infinity added. Let us first construct these embeddings, which we call N_σ, of $N_\mathbb{R}$ and then show there is a map $\mathrm{ord} : X_\sigma \longrightarrow N_\sigma$ inducing a homeomorphism $X_\sigma/T_c \xrightarrow{\sim} N_\sigma$.

The simplest way to define N_σ is via a basis $\alpha_1, \ldots, \alpha_m$ of the semigroup $\check\sigma \cap M$. Then define

$$i : N_\mathbb{R} \longrightarrow \mathbb{R}^m_{>0},$$
$$x \longmapsto (e^{-2\pi\langle\alpha_1,x\rangle}, \ldots, e^{-2\pi\langle\alpha_m,x\rangle}),$$

and let

$$N_\sigma = \text{closure of } iN_\mathbb{R} \text{ in } \mathbb{R}^m_{\geq 0}.$$

It is very easy to see that this space is independent of the choice of basis (check that if you add to the α_i one more α, then N_σ does not change). If we let $N_\mathbb{R}$ act on $\mathbb{R}^m$ by

$$x \cdot (y_1, \ldots, y_m) = (e^{-2\pi\langle\alpha_1,x\rangle} y_1, \ldots, e^{-2\pi\langle\alpha_m,x\rangle} y_m),$$

then N_σ is the closure of the orbit of $(1, 1, \ldots, 1)$. In particular, $N_\mathbb{R}$ acts on N_σ, extending its action on itself by translation. Exactly as in the theory of torus embeddings (see TE I, Ch. I, §1, Theorem 2), we can decompose N_σ into $N_\mathbb{R}$-orbits; these will correspond bijectively to the faces of σ, and each one will contain a unique point $(y_1, \ldots, y_m)$ with $y_i = 0$ or 1 for all i. Explicitly, for every face τ of σ, the corresponding orbit is:

$$O(\tau) = \left\{ (y_1, \ldots, y_m) \in N_\sigma \;\middle|\; \begin{array}{l} y_i = 0 \text{ if } \alpha_i > 0 \text{ on Int } \tau \\ y_i \neq 0 \text{ if } \alpha_i \equiv 0 \text{ on Int } \tau \end{array} \right\}$$
$$= N_\mathbb{R}\text{-orbit of } \varepsilon_\tau = (\varepsilon_1, \ldots, \varepsilon_m),$$

where

$$\varepsilon_i = \begin{cases} 0 & \text{if } \alpha_i > 0 \text{ on Int } \tau \\ 1 & \text{if } \alpha_i \equiv 0 \text{ on Int } \tau. \end{cases}$$

This can be proven following TE I, substituting the following lemma for the use of $k[[t]]$.

Lemma 1.1 *If* $\{x_k\}$ *is a sequence in* $N_\mathbb{R}$ *and* $S \subset \{1, \ldots, m\}$ *satisfies*

$$\lim_{k \longrightarrow \infty} \alpha_i(x_k) = \lambda_i, \; i \in S,$$
$$\lim_{k \longrightarrow \infty} \alpha_i(x_k) = \infty, \; i \notin S,$$

then

(a) *there is some* $y \in N_\mathbb{R}$ *with* $\alpha_i(y) = 0$ *for* $i \in S$; $\alpha_i(y) > 0$ *for* $i \notin S$;

(b) *there is some* $z \in N_\mathbb{R}$ *with* $\alpha_i(z) = \lambda_i$ *for* $i \in S$.

Proof Left to reader. ☐

Now if we map X_σ into $\mathbb{R}^m_{\geq 0}$ as follows:

$$
\begin{array}{ccc}
T & \xrightarrow{\ \text{ord}\ } & N_\mathbb{R} \\
\cap\downarrow & & \downarrow \\
X_\sigma & & N_\sigma \\
& \searrow f \qquad \curvearrowright & \\
& \mathbb{R}^m_{\geq 0} &
\end{array}
$$

$$
f(x) = \left(|\mathfrak{X}^{\alpha_1}(x)|, \dots, |\mathfrak{X}^{\alpha_m}(x)| \right),
$$

we get a commutative diagram. Since T is dense in X_σ, it follows that f defines a map

$$
\text{ord} : X_\sigma \longrightarrow N_\sigma
$$

and that $\text{ord}(gx) = \text{ord}(x)$ for all $g \in T_c$. Conversely, if $\text{ord}(x_1) = \text{ord}(x_2)$, it follows that $|\mathfrak{X}^\alpha(x_1)| = |\mathfrak{X}^\alpha(x_2)|$ for all $\alpha \in \check{\sigma} \cap M$, from which it follows readily that $x_1 = gx_2$ for some $g \in T_c$. Note that if $\mathbb{O}^\tau \subset X_\sigma$ is the orbit corresponding to τ, then $\text{ord}^{-1}(O(\tau)) = \mathbb{O}^\tau$.

For some purposes, it is convenient to have a coordinate-invariant way of describing N_σ as $N_\mathbb{R}$ plus a set of ideal points at infinity. To describe N_σ this way, for every face τ of σ, let

$$
L(\tau) = \text{smallest linear space containing } \tau.
$$

Then $L(\tau)$ is the stabilizer of ε_τ, so we get:

$$
N_\mathbb{R}/L(\tau) \xrightarrow{\ \sim\ } O(\tau)
$$

$$
x \longmapsto x \cdot \varepsilon_\tau.
$$

Let $x + \infty \cdot \tau \in N_\sigma$ denote $x \cdot \varepsilon_\tau$ (where $x_1 + \infty \cdot \tau = x_2 + \infty \cdot \tau$ if and only if $x_1 - x_2 \in L(\tau)$). The reason for this notation is as follows: decompose $N_\mathbb{R} = N'_\mathbb{R} \oplus L(\tau)$, choose any sequence $x_n = y_n + z_n \in N_\mathbb{R} = N'_\mathbb{R} \oplus L(\tau)$, and choose any $y \in N'_\mathbb{R}$. Then one sees easily that

$$
\left[\lim_{n \longrightarrow \infty} x_n = y + \infty \cdot \tau \text{ in } N_\sigma \right] \Longleftrightarrow \left[\begin{array}{l} \lim_{n \longrightarrow \infty} y_n = y \text{ and, for every} \\ w \in L(\tau), z_n \in \tau + w \text{ if } n \gg 0. \end{array} \right]
$$

Heuristically, we have added a lower-dimensional vector space isomorphic to $N_\mathbb{R}/L(\tau)$ of ideal points $x + \infty \cdot \tau$ obtained by starting at x and moving out to infinity in the direction determined by the cone τ.

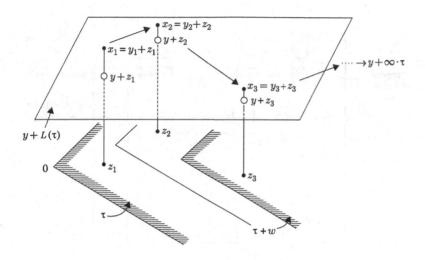

Our convergence condition may be rephrased by saying that a fundamental system of neighborhoods of $y + \infty \cdot \tau$ in $N_{\mathbb{R}}$ is given by

$$U_{\varepsilon,w}^0(y + \infty \cdot \tau) = y + w + B_\varepsilon + \tau \,,$$

for any $w \in L(\tau)$ and any $\varepsilon > 0$, where B_ε denotes the ε-ball around 0 (take any metric on $N_{\mathbb{R}}$). More generally, with this notation, a fundamental system of neighborhoods of $y + \infty \cdot \tau$ in N_σ is given by

$$U_{\varepsilon,w}(y + \infty \cdot \tau) = U_{\varepsilon,w}^0(y + \infty \cdot \tau) \cup \bigcup_{\tau' \text{ face of } \tau} (y + w + B_\varepsilon + \tau + \infty \cdot \tau') \,.$$

For instance, if $N_{\mathbb{R}} = \mathbb{R}^2$ and σ is the positive quadrant, we get the following picture:

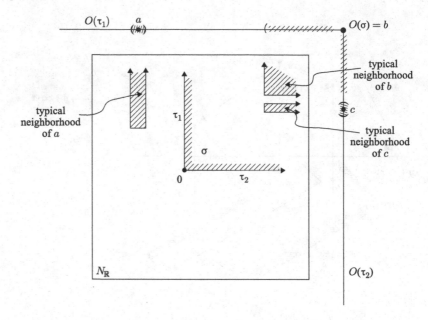

Next recall that in TE I, Ch. I, §2, we glue the affine varieties X_σ together: whenever $\{\sigma_\alpha\}$ is a *rational partial polyhedral decomposition of* $N_{\mathbb{R}}$, meaning

(i) if σ is a face of σ_α, then $\sigma = \sigma_\beta$, for some β;

(ii) for all α, β, the cone $\sigma_\alpha \cap \sigma_\beta$ is a face of σ_α and σ_β,

then we can glue the X_{σ_α} together, obtaining a scheme $X_{\{\sigma_\alpha\}}$. In TE I, we asked that $\{\sigma_\alpha\}$ be a finite set, so that $X_{\{\sigma_\alpha\}}$ was a variety. This is in fact totally irrelevant: for any set $\{\sigma_\alpha\}$ as above, we get an $X_{\{\sigma_\alpha\}}$ as before, except that it may require an infinite number of affines to cover it. Now $X_{\{\sigma_\alpha\}}$ is always a separated normal irreducible scheme, *locally* of finite type over $\mathbb{C}$ and containing T as an open dense subset. In exactly the same way, we glue the N_{σ_α} together into a topological space $N_{\{\sigma_\alpha\}}$, which is $N_{\mathbb{R}}$ plus a large number of ideal vector spaces situated at infinity in many different directions. Moreover, we glue the ord maps together into one map:

$$\mathrm{ord} : X_{\{\sigma_\alpha\}} \longrightarrow N_{\{\sigma_\alpha\}} \ .$$

For instance, $X_{\{\sigma_\alpha\}}$, as a set, is the disjoint union of T-orbits $\mathbb{O}^{\sigma_\alpha}$, one for each α; likewise $N_{\{\sigma_\alpha\}}$ as a set is the disjoint union of $N_{\mathbb{R}}$-orbits $O(\sigma_\alpha)$, one for each α, and $\mathrm{ord}^{-1}(O(\sigma_\alpha)) = \mathbb{O}^{\sigma_\alpha}$.

2 The functor of a torus embedding

In order to make some of our later constructions of compactifications D/Γ purely algebraic and valid for schemes over any ground fields, it will be useful to learn what functor a torus embedding represents. This also gives us another view of what torus embeddings are. First some notations and definitions.

(1) If S is a scheme and X is a set, X_S denotes the constant sheaf on S with stalk X.

(2) Every semigroup or sheaf of semigroups will have an identity element e or identity section e.

(3) If A_1, A_2 are semigroups, a homomorphism $\phi : A_1 \longrightarrow A_2$ is called *strict* if $\phi(e_1) = e_2$ and $\phi(x)$ invertible implies x invertible. If A_1, A_2 are sheaves of semigroups on S, we require that, for every $s \in S$, the map on stalks $\phi_s : A_{1,s} \longrightarrow A_{2,s}$ is strict.

(4) If S is a scheme, then $\mathscr{O}_S^{(\times)}$ will be the semigroup sheaf $(\mathscr{O}_S, \text{mult.})$.

The result is:

Theorem 2.1 *Let T be a torus over k and $T \subset X_{\{\sigma_\alpha\}}$ a torus embedding, where $\sigma_\alpha \subset N(T)_{\mathbb{R}}$ are polyhedral cones. For any k-scheme S, let $F_{\{\sigma_\alpha\}}(S)$ be the set of pairs (Σ, π) consisting of a sub-semigroup sheaf $\Sigma \subset M(T)_S$ and a strict homomorphism $\pi : \Sigma \longrightarrow \mathscr{O}_S^{(\times)}$ such that, for all $s \in S$, we have $\Sigma_s = \check{\sigma}_\alpha \cap M(T)$ for some α. Then there are canonical isomorphisms, functorial in k-schemes S:*

$$\text{Hom}_k(S, X_{\{\sigma_\alpha\}}) \cong F_{\{\sigma_\alpha\}}(S) .$$

Proof We first show how to associate a pair (Σ, π) to a morphism $f : S \longrightarrow X_{\{\sigma_\alpha\}}$. Define:

$$U_\alpha = f^{-1}(X_{\sigma_\alpha}) ,$$
$$\Sigma = \text{the union of the subsheaves } (\check{\sigma}_\alpha \cap M(T))_{U_\alpha} \text{ of } M(T)_S .$$

Note that, for all $s \in S$, if $f(s) \in \mathbb{O}^\alpha$, then

$$s \in U_\beta \iff f(s) \in X_{\sigma_\beta}$$
$$\iff \mathbb{O}^\alpha \subset X_{\sigma_\beta}$$
$$\iff \sigma_\alpha \text{ is a face of } \sigma_\beta$$
$$\iff \check{\sigma}_\beta \cap M(T) \subseteq \check{\sigma}_\alpha \cap M(T) ;$$

hence the stalk of Σ at s is the union of the subsets $\check{\sigma}_\beta \cap M(T)$ of $M(T)$ for all σ_β with face σ_α, i.e., just $\check{\sigma}_\alpha \cap M(T)$. Hence if $r \in \Sigma_s$, then $r \in \check{\sigma}_\alpha$, so

$\mathfrak{X}^r$ is defined on X_{σ_α} and $f^*(\mathfrak{X}^r)$ is defined at s. Therefore we can define $\pi : \Sigma \longrightarrow \mathcal{O}_S^{(\times)}$ by

$$\pi(r) = f^*(\mathfrak{X}^r) .$$

Note that

$$\pi(r) \text{ invertible in } \mathcal{O}_{S,s} \Longleftrightarrow \pi(r)(s) \neq 0$$
$$\Longleftrightarrow \mathfrak{X}^r(f(s)) \neq 0$$
$$\Longleftrightarrow \mathfrak{X}^r \not\equiv 0 \text{ on } \mathbb{O}^\alpha$$
$$\Longleftrightarrow r \equiv 0 \text{ on } \sigma_\alpha$$
$$\Longleftrightarrow -r \in \check{\sigma}_\alpha \cap M(T)$$
$$\Longleftrightarrow r \text{ invertible in } \Sigma_s ,$$

hence π is a strict homomorphism.

Next, let us start with (Σ, π) and define a morphism f. Define open sets U_α by

$$U_\alpha = \{s \in S \mid \check{\sigma}_\alpha \cap M(T) \subset \Sigma_s\} .$$

These form an open covering of S such that if σ_α is a face of σ_β, then $U_\beta \subset U_\alpha$. Next define

$$f_\alpha : U_\alpha \longrightarrow X_{\sigma_\alpha} = \operatorname{Spec} k[\ldots, \mathfrak{X}^r, \ldots]_{r \in \check{\sigma}_\alpha \cap M(T)}$$

via $f_\alpha^*(\mathfrak{X}^r) = \pi(r)$ for all $r \in \check{\sigma}_\alpha \cap M(T)$: this is correct since such an r is in $\Gamma(U_\alpha, \Sigma)$ and since $\pi(r_1 + r_2) = \pi(r_1) \cdot \pi(r_2)$. Now, for any α and β, let $\sigma_\gamma = \sigma_\alpha \cap \sigma_\beta$, which is a face of σ_α and σ_β. Then

$$U_\alpha \cap U_\beta = \{s \in S \mid \check{\sigma}_\alpha \cap M(T) \subset \Sigma_s \text{ and } \check{\sigma}_\beta \cap M(T) \subset \Sigma_s\} .$$

But if $\Sigma_s = \check{\sigma}_\delta \cap M(T)$, then

$$\left[\begin{array}{c} \Sigma_s \supset \check{\sigma}_\alpha \cap M(T) \text{ and} \\ \Sigma_s \supset \check{\sigma}_\beta \cap M(T) \end{array} \right] \Longleftrightarrow \check{\sigma}_\delta \supset \check{\sigma}_\alpha \text{ and } \check{\sigma}_\delta \supset \check{\sigma}_\beta$$
$$\Longleftrightarrow \sigma_\delta \subset \sigma_\alpha \text{ and } \sigma_\delta \subset \sigma_\beta$$
$$\Longleftrightarrow \sigma_\delta \subset \sigma_\gamma$$
$$\Longleftrightarrow \Sigma_s \subset \check{\sigma}_\gamma \cap M(T) ,$$

so $U_\alpha \cap U_\beta = U_\gamma$. Finally, it is clear from the definition that $f_\alpha = \operatorname{res} f_\beta$ whenever $U_\alpha \subset U_\beta$. Therefore the f_α patch together to form a morphism $f : S \longrightarrow X_{\{\sigma_\alpha\}}$.

It is now straightforward to check that these two procedures – associating a (Σ, π) to an f and associating an f to a (Σ, π) – are inverse to each other: we leave this to the reader. $\qquad\qquad\square$

For instance, we find:

$$X_{\{\sigma_\alpha\}}(k) \cong \{(\alpha,\pi) \mid \pi : \check{\sigma}_\alpha \cap M(T) \longrightarrow k^{(\times)} \text{ strict homomorphism}\} \ .$$

If $k = \mathbb{C}$, one can easily prove also that

$$N_{\{\sigma_\alpha\}} \cong \{(\alpha,\rho) \mid \rho : \check{\sigma}_\alpha \cap M(T) \longrightarrow \mathbb{R}_{\geq 0}^{(\times)} \text{ strict homomorphism}\}$$

$$\cong \{(\alpha,\sigma) \mid \sigma : \check{\sigma}_\alpha \cap M(T) \longrightarrow \mathbb{R} \cup \{\infty\} \text{ strict homomorphism}\} \ ,$$

where $\mathbb{R} \cup \{\infty\}$ is a semigroup via $+$. Here

$$\text{ord} : X_{\{\sigma_\alpha\}} \longrightarrow N_{\{\sigma_\alpha\}}$$

is given by

$$\rho(x) = |\pi(x)| \ ,$$
$$\sigma(x) = -\log\rho(x) \ .$$

3 Toroidal embeddings over the complex numbers

We wish to review here quickly some results of TE I, Ch. II, indicating ways to interpret them over $\mathbb{C}$, and generalizing them slightly. A pair

$$U \subset X,$$

where U is a Zariski-open subset of a normal variety X, was called a *toroidal embedding* if, for all $x \in X$, we have that (X,U) is formally isomorphic at x to (X_σ,T) at some $t \in X_\sigma$ (for some torus embedding $T \subset X_\sigma$). Equivalently, this means that there is an étale correspondence between X and X_σ, relating x and t, with U and T corresponding open sets. Over $\mathbb{C}$, a pair

$$U \subset X \ ,$$

where X is an analytic space and U is open in the complex topology, will be called a *toroidal embedding* if, for all $x \in X$, there exists a small neighborhood $W_x \subset X$ of x such that $(W_x, W_x \cap U)$ is isomorphic to $(V_t, V_t \cap T)$ for some neighborhood $V_t \subset X_\sigma$ of some $t \in X_\sigma$ (for some torus embedding $T \subset X_\sigma$). When X, U are varieties, this coincides with the previous definition. Now, this implies immediately that W_x has a canonical stratification $\{Y_{\alpha,x}\}$ into non-singular locally closed analytic strata with $\overline{Y}_{\alpha,x}$ normal: let E_i be the irreducible components of $W_x \setminus W_x \cap U$, and let the $Y_{\alpha,x}$ be the sets

$$\bigcap_{i\in I} E_i \setminus \bigcup_{i\notin I} E_i \ .$$

We shrink W_x if necessary, so that these $Y_{\alpha,x}$ are connected. As x varies, these strata patch up on overlaps, so we can uniquely stratify the whole of X into

$\{Y_\alpha\}$, where the Y_α are connected, locally closed, non-singular analytic strata, and where $Y_\alpha \cap W_x$ is a union of the $Y_{\beta,x}$. However, it may happen that

$$Y_\alpha \cap W_x \supset \text{more than one } Y_{\beta,x}.$$

This means that there is a path in X starting and ending in W_x and lying all in one stratum, but linking two distinct local strata:

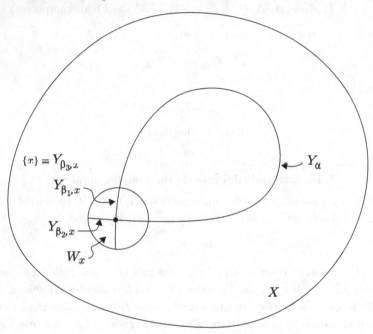

Since this will mean that $\overline{Y}_\alpha$ has more than one branch through x, it is equivalent to $\overline{Y}_\alpha$ being non-normal. As in TE I, p. 57, we say that (X,U) has or has not *self-intersection* according to whether $Y_\alpha \cap W_x$ can be more than one local stratum, or $Y_\alpha \cap W_x$ is always one local stratum. In TE I, we stuck with (X,U)'s without self-intersection. However, there is a class of toroidal embeddings with self-intersection that are almost as nice and that arise in the examples we will treat. Suppose $Y_{\beta_1,x}$ and $Y_{\beta_2,x}$ are part of the same global stratum Y_α. Locally at x there is a unique stratum $Y_{\beta_3,x}$ such that

$$\overline{Y}_{\beta_3,x} = \overline{Y}_{\beta_1,x} \cap \overline{Y}_{\beta_2,x}.$$

Let $Y_{\beta_3,x}$ define a global stratum Y_γ. We say that (X,U) is *without monodromy* if Y_γ has a neighborhood W such that $Y_{\beta_1,x}$ and $Y_{\beta_2,x}$ lie in different components of $Y_\alpha \cap W$. To visualize this, note that, for every path in Y_γ beginning and ending at x, we can uniquely propagate the germ of analytic space $\overline{Y}_{\beta_1,x}$ along this path. If this germ can be taken to $\overline{Y}_{\beta_2,x}$ by such a path, then, for every neighborhood W of Y_γ, we may connect $Y_{\beta_1,x}$ and $Y_{\beta_2,x}$ within $Y_\alpha \cap W$. If not, then, in some small enough W, they cannot be connected:

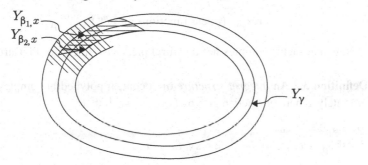

A toroidal embedding with monodromy

If (X,U) is without monodromy, then every stratum Y_α has a small complex neighborhood, which we call $\mathrm{Star}_0(Y_\alpha)$, in which all the local strata $Y_{\beta,x}$ (where $x \in Y_\alpha, \overline{Y}_{\beta,x} \supset Y_\alpha$) remain distinct; i.e., $\mathrm{Star}_0(Y_\alpha)$ is a union of semi-local strata $Y_\beta^{(\alpha)}$ such that $\overline{Y}_\beta^{(\alpha)}$ is normal or, equivalently, $Y_\beta^{(\alpha)} \cap W_x$ is one local stratum. We may even assume that there is a stratum-preserving *homeomorphism*

$$\mathrm{Star}_0(Y_\alpha) \approx X_\sigma \times Y_\alpha ,$$

where $T \subset X_\sigma$ is a true embedding. Of course, in the whole space X, we may have $Y_{\beta_1}^{(\alpha)}$ and $Y_{\beta_2}^{(\alpha)}$ as part of the same stratum Y_γ.

The main point of Ch. II, §1 of TE I was to associate to each toroidal embedding without self-intersection (X,U) a *conical polyhedral complex with integral structure*. If we generalize slightly our definition of such a complex, we can do this for any analytic toroidal embedding without monodromy too. The following definition may be compared with the definition in TE I, p. 69.

Definition 3.1 A *conical polyhedral complex* Σ is a topological space $|\Sigma|$, plus a stratification $\{S_\alpha\}$ of $|\Sigma|$ (i.e., a partition of $|\Sigma|$ into disjoint locally closed pieces S_α such that each $\overline{S}_\alpha$ is a union of finitely many S_β's), plus, for each α, a finite-dimensional vector space V_α of real-valued continuous functions on S_α such that:

(a) if $n_\alpha = \dim(V_\alpha)$ and $F_1,\ldots,F_{n_\alpha}$ is a basis of V_α, then

$$(f_i) : S_\alpha \longrightarrow \mathbb{R}^{n_\alpha}$$

is a homeomorphism of S_α with an open convex polyhedral cone $C_\alpha \subset \mathbb{R}^{n_\alpha}$;

(b) $(f_i)^{-1}$ extends to a continuous surjective map

$$(f_i)^{-1} : \overline{C}_\alpha \longrightarrow \overline{S}_\alpha$$

mapping the open faces $C_\alpha^{(\beta)}$ of $\overline{C}_\alpha$ homeomorphically to the strata S_β in $\overline{S}_\alpha$, and inducing isomorphisms

$$\mathrm{res}_{C_\alpha^{(\beta)}}\,(\text{linear functions on } \mathbb{R}^{n_\alpha}) \xrightarrow{\sim} V_\beta\;.$$

Note that such a complex has a natural piecewise-linear (or PL) structure.

Definition 3.2 An *integral structure* on a conical polyhedral complex is a set of finitely generated abelian groups $L_\alpha \subset V_\alpha$ such that

(i) $L_\alpha \otimes \mathbb{R} \cong V_\alpha$;
(ii) if S_β is a face of S_α, then $\mathrm{res}_{S_\beta} L_\alpha = L_\beta$.

The changes from TE I are: (a) that the collection $\{S_\alpha\}$ is not supposed to be finite; and (b) we allow two faces of the same polyhedron C_α in $\mathbb{R}^n$ to be identified in X. We sketch how to associate a complex $\Sigma = (|\Sigma|, \{S_\alpha\}, \{V_\alpha\})$ to a toroidal embedding (X, U) without monodromy in the following.

(a) For all strata Y of (X, U), let

$$M^Y = \text{group of Cartier divisors on } \mathrm{Star}_0(Y), \text{supported on}$$
$$\mathrm{Star}_0(Y) \setminus U \cap \mathrm{Star}_0(Y)\,,$$
$$M_+^Y = \text{sub-semigroup of effective divisors}\,,$$
$$N_{\mathbb{R}}^Y = \mathrm{Hom}\,(M^Y, \mathbb{R})\,,$$
$$\sigma^Y = \{x \in N_{\mathbb{R}}^Y \mid \langle D, x \rangle \geq 0, \text{ for all } D \in M_+^Y\}\,.$$

(b) For all strata Z_0 in $\mathrm{Star}_0(Y)$, let Z be the stratum of X containing Z_0; then we get a map

$$\alpha_{Z_0} : M^Y \longrightarrow M^Z$$

by restricting a divisor on $\mathrm{Star}_0(Y)$ to the component of $\mathrm{Star}_0(Y) \cap \mathrm{Star}_0(Z)$ containing Z_0, and then extending it to $\mathrm{Star}_0(Z)$. This induces an isomorphism

$$\beta_{Z_0} : \sigma^Z \xrightarrow{\sim} \text{ a face of } \sigma^Y\;.$$

(c) Define

$$|\Sigma| = \bigcup_Y \sigma^Y \Big/ \left(\begin{array}{c} \text{equivalence relation generated} \\ \text{by the maps } \beta_{Z_0} \end{array} \right) \cong \bigsqcup_Y \mathrm{Int}\,\sigma^Y\,,$$
$$S_\alpha = \text{image of } \mathrm{Int}\,(\sigma^{Y_\alpha})\,,$$
$$V_\alpha = \text{the functions } M^{Y_\alpha} \otimes \mathbb{R} \text{ on } \sigma^{Y_\alpha}\,,$$
$$L_\alpha = \text{the functions } M^{Y_\alpha} \text{ on } \sigma^{Y_\alpha}\,.$$

This is an immediate generalization of the construction of TE I, pp. 59–72. The map "ord" also has an analytic version. Define†

$$\text{R.S.}^U(X) = \{\phi : \Delta \longrightarrow X \text{ holomorphic such that } \phi(\mathring{\Delta}) \subset U\},$$

where Δ is the open unit disc and $\mathring{\Delta} = \Delta \setminus \{0\}$. Define

$$\text{ord} : \text{R.S.}^U(X) \longrightarrow |\Sigma|$$

as follows. Let $\phi \in \text{R.S.}^U(X)$ with $\phi(0) \in \text{Stratum } Y$. Then, on some smaller disc Δ', we have $\phi(\Delta') \subset \text{Star}_0(Y)$; hence, for all divisors $D \in M^Y$, the pullback $\phi^* D$ is a divisor on Δ' with a definite multiplicity $\text{ord}_0(\phi^* D)$ at 0. Define $\text{ord}(\phi)$ in σ^Y by

$$\langle D, \text{ord}(\phi) \rangle = \text{ord}_0(\phi^* D),$$

and define $\text{ord}(\phi)$ in $|\Sigma|$ as the image of this.

We may also give a purely topological definition of ord via monodromy. In fact, suppose we choose a nice neighborhood $\text{Star}_0(Y)$ so that

$$\text{Star}_0(Y) \approx X_\tau \times Y.$$

Note that in this case we get isomorphisms

$$M^Y \cong M(T);$$

hence

$$N_{\mathbb{R}}^Y \cong N(T)_{\mathbb{R}}$$

and

$$\sigma^Y \cong \tau.$$

In particular this shows that

$$\begin{aligned}
\pi_1(\text{Star}_0(Y) \cap U) &\cong \pi_1(T \times Y) \\
&\cong \pi_1(F) \times \pi_1(Y) \\
&\cong N(T) \times \pi_1(Y) \\
&\cong N^Y \times \pi_1(\text{Star}_0(Y));
\end{aligned}$$

hence

$$\text{Ker}\,[\pi_1(\text{Star}_0(Y) \cap U) \longrightarrow \pi_1(\text{Star}_0(Y))] \xrightarrow[\zeta]{\sim} N^Y.$$

Using this isomorphism, we find:

---

† For the justification of the notation "R. S." (short for "Riemann surface" as used by Zariski), see TE I, p. 64.

Proposition 3.3 *Let $\phi \in \text{R.S.}^U(X)$ and assume $\phi(0) \in$ stratum Y. For c small, we have*

$$\text{res}\,\phi : \Delta_c = \{z \mid |z| \le c\} \longrightarrow \text{Star}_0(Y) \,;$$

hence res ϕ *induces*

$$\phi_* : \pi_1(\partial\Delta_c) \longrightarrow \text{Ker}\,[\pi_1(\text{Star}_0(Y) \cap U) \longrightarrow \pi_1(\text{Star}_0(Y))]\,.$$

If 1 *is the canonical generator of the left-hand side, then*

$$\zeta(\phi_*(1)) = \text{ord}(\phi)\,.$$

Proof By the way that $\zeta : \pi_1(T) \xrightarrow{\sim} N(T)$ is defined, it follows that, for every loop λ in T, and every character $\mathfrak{X}^\alpha$,

$$2\pi\langle\alpha, \zeta(\lambda)\rangle = \text{change in arg}\,\mathfrak{X}^\alpha \text{ around the loop } \lambda\,.$$

Now let $T \subset X_\sigma$ be a torus embedding, $t \in X_\sigma$, let V be a neighborhood of t, and let $u \cdot \mathfrak{X}^\alpha$ be a function on V, where u is a unit on V; moreover, let λ be a loop in V arising by restricting to $\partial\Delta_c$ a holomorphic map $\phi : \Delta_c \longrightarrow V$. Then

$$2\pi\langle\alpha, \zeta(\lambda)\rangle = \text{change in arg}\,(u \cdot \mathfrak{X}^\alpha) \text{ around the loop } \lambda\,.$$

Next, go over to a toroidal embedding $U \subset X$, let $x \in X$ and let V be a neighborhood of x, and let δ be a meromorphic function on V with no zeroes or poles on $V \cap U$. Moreover, let λ be a loop in V arising by restricting to $\partial\Delta_c$ a holomorphic map $\phi : \Delta_c \longrightarrow V$. Then the principal divisor (δ) is an element of M^Y, and

$$\begin{aligned}
2\pi\langle(\delta), \zeta(\lambda)\rangle &= \text{change in arg}\,\delta \text{ around the loop } \lambda \\
&= \text{change in arg}\,(\delta \circ \phi) \text{ on } |z| = c \\
&= 2\pi \cdot (\text{order of zero or pole of } \delta \circ \phi \text{ at } 0) \\
&= 2\pi\langle(\delta), \text{ord}(\phi)\rangle\,.
\end{aligned}$$

Since $\lambda = \phi_*(1)$, this proves what we want. $\square$

4 Compactification of the universal elliptic curve

We now take up perhaps the simplest example of our theory. We deal first with this example over the complex numbers.

Fix an integer $k \ge 3$ once and for all. Let

$$\Gamma = \left\{ \begin{pmatrix} a & b \\ c & d \end{pmatrix} \,\middle|\, a \equiv d \equiv 1 \bmod k, b \equiv c \equiv 0 \bmod k, ad - bc = 1 \right\}$$

act on the upper half plane $\mathfrak{H}$ by

$$\omega \longmapsto \frac{a\omega + b}{c\omega + d}.$$

Let

$$\Gamma^A = \Gamma \ltimes \mathbb{Z}^2$$

(semidirect product, where $\mathbb{Z}^2$ is normal with Γ acting on $\mathbb{Z}^2$ by

$$(m,n) \longmapsto (am + cn, bm + dn);$$

"A" stands for "affine"). Then Γ^A acts on $\mathbb{C} \times \mathfrak{H}$ by

$$(z, \omega) \longmapsto \left(\frac{z + m\omega + n}{c\omega + d}, \frac{a\omega + b}{c\omega + d} \right).$$

Then $M = \mathfrak{H}/\Gamma$ is the moduli space for elliptic curves with level-k structure and $X = (\mathbb{C} \times \mathfrak{H})/\Gamma^A$ is the universal level-k elliptic curve over M, via the canonical projection $p : X \longrightarrow M$ (see, for example, Lang [2]). The problem is that M, and hence X, are not compact and we seek compactifications:

$$
\begin{array}{ccc}
X & \lhook\joinrel\longrightarrow & \tilde{X} \\
\downarrow & & \downarrow \\
M & \lhook\joinrel\longrightarrow & \tilde{M}.
\end{array}
$$

The usual procedure for M is to note that, since it is one-dimensional, there is a unique non-singular complete algebraic curve $\tilde{M}$ such that $M = \tilde{M} \setminus \{\text{finite set}\}$. Then, from the theory of algebraic surfaces, one can also find a canonical $\tilde{X}$: the unique so-called non-singular relatively minimal model over $\tilde{M}$. These methods do not however generalize to higher-dimensional cases, and we seek to describe $\tilde{M}$ and $\tilde{X}$ by a more direct "scissors and glue" construction involving torus embeddings.

We deal first with the cusp $i\infty \in \partial\mathfrak{H}$. Consider the subgroup

$$\Gamma_1 = \left\{ \begin{pmatrix} 1 & b \\ 0 & 1 \end{pmatrix} \mid b \equiv 0 \bmod k \right\}$$

and factor $\pi : \mathfrak{H} \longrightarrow \mathfrak{H}/\Gamma = M$ via

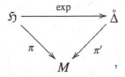

where q is the coordinate on $\overset{\circ}{\Delta}$ and exp is defined by

$$q = e^{2\pi i \omega / k}.$$

This makes $\overset{\circ}{\Delta}$ isomorphic to $\mathfrak{H}/\Gamma_1$, hence π factors via exp. Moreover, define

$$\mathfrak{H}_d = \{\omega \mid \operatorname{Im} \omega \geq d\} \, ,$$
$$\overset{\circ}{\Delta}_d = \{q \mid 0 < |q| \leq e^{-2\pi d/k}\} \, .$$

Then $\mathfrak{H}_d = \exp^{-1}(\overset{\circ}{\Delta}_d)$ and $\overset{\circ}{\Delta}_d \cong \mathfrak{H}_d/\Gamma_1$. The following lemma is easy to check.

Lemma 4.1 *There exists d_0 such that, for all $\omega \in \mathfrak{H}$, $\gamma \in \Gamma$,*

$$\omega \text{ and } \gamma\omega \in \mathfrak{H}_{d_0} \Longrightarrow \gamma \in \Gamma_1 \, .$$

□

Therefore res π' maps $\overset{\circ}{\Delta}_{d_0}$ *injectively* to M:

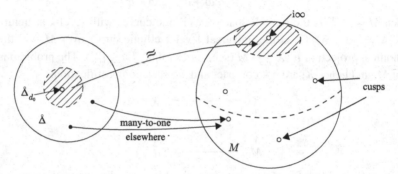

Moreover, as $d \longrightarrow \infty$, it is well known that the sets $\pi(\mathfrak{H}_d) \subset M$ are a fundamental system of neighborhoods of the cusp i∞. Therefore, we find that we can glue via this map by taking M plus

$$\Delta_{d_0} = \{q \mid |q| \leq e^{-2\pi d_0/k}\}$$

and identifying them via res π' on $\overset{\circ}{\Delta}_{d_0}$.

Next, every rational point $p/q \in \partial\mathfrak{H}$ also defines a cusp of M, except that p/q and $\gamma(p/q)$ for $\gamma \in \Gamma$ define the same cusp. Now, a fundamental system of neighborhoods of p/q in $\widetilde{M}$ are the sets $\pi(W_d(p/q))$, where $W_d(p/q)$ is the closed disc in $\mathfrak{H}$ of radius d, tangent to the real axis at p/q (a so-called *horocycle*) and, if d is small enough, the Γ-equivalence of points of $W_d(p/q)$ becomes $\Gamma_1(p/q)$-equivalence, where

$$\Gamma_1(p/q) = \{\gamma \in \Gamma \mid \gamma(p/q) = p/q\}$$
$$= \left\{ \begin{pmatrix} 1+pq & -p^2 \\ q^2 & 1-pq \end{pmatrix}^n \mid n \in \mathbb{Z} \right\} \, .$$

So we can mimic the above construction to obtain $\widetilde{M}$ by glueing. But, even more simply, we can use the fact that $\mathrm{SL}(2,\mathbb{Z})$ acts transitively on the set of rational points plus ∞, hence $SL(2, \mathbb{Z})/\Gamma$ acts on M and permutes transitively all its cusps. Thus, if we know how to fill in one, we can fill in the others by acting by $SL(2, \mathbb{Z})/\Gamma$.

Now look upstairs at $\mathbb{C} \times \mathfrak{H}$. Define

$$\Gamma_1^A = \left(\begin{array}{c} \text{subgroup of } \Gamma^A \text{ generated by} \\ (z, \omega) \longmapsto (z+1, \omega) \text{ and} \\ (z, \omega) \longmapsto (z, \omega + k) \end{array} \right) \cong \mathbb{Z}^2 \, ;$$

$$\Gamma_2^A = \left(\begin{array}{c} \text{subgroup of } \Gamma^A \text{ generated by } \Gamma_1^A \text{ and} \\ \alpha : (z, \omega) \longmapsto (z + \omega, \omega) \end{array} \right) .$$

Factor $\pi : \mathbb{C} \times \mathfrak{H} \longrightarrow (\mathbb{C} \times \mathfrak{H})/\Gamma^A = X$ via:

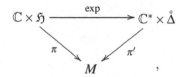

where x is the coordinate on $\mathbb{C}^*$, and q that on $\mathring{\Delta}$, and where exp is defined by

$$x = e^{2\pi i z} \, ,$$
$$q = e^{2\pi i \omega / k} \, .$$

This makes $\mathbb{C}^* \times \mathring{\Delta}$ isomorphic to $(\mathbb{C} \times \mathfrak{H})/\Gamma_1^A$. Now, Γ_1^A is a normal subgroup of Γ_2^A and $\Gamma_2^A/\Gamma_1^A \cong \mathbb{Z}$, with generator α, and Γ_2^A/Γ_1^A acts on $\mathbb{C}^* \times \mathring{\Delta}$. The previous lemma now gives us:

Corollary 4.2 *There exists d_0 such that, for all $(z, \omega) \in \mathbb{C} \times \mathfrak{H}$ and all $\gamma \in \Gamma^A$,*

$$(z, \omega) \text{ and } \gamma(z, \omega) \in \mathbb{C} \times \mathfrak{H}_{d_0} \Longrightarrow \gamma \in \Gamma_2^A \, .$$

$\square$

Therefore,

$$\operatorname{res} \pi' : (\mathbb{C}^* \times \mathring{\Delta}_{d_0})/\{\alpha^n\} \longrightarrow X$$

is injective. To compactify X over $i\infty \in \tilde{M}$, it suffices to enlarge $\mathbb{C}^* \times \mathring{\Delta}_{d_0}$ to an analytic manifold Y over Δ_{d_0}, equivariantly with respect to the action of α and so that, mod α, we get a manifold proper over Δ_{d_0}:

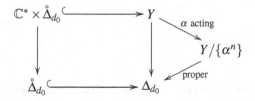

Here is where tori come in: think of $\mathbb{C}^* \times \overset{\circ}{\Delta}$ as an open subset of the two-dimensional torus $\mathbb{C}^* \times \mathbb{C}^*$ (with coordinates x, q). Thus α acts on the whole torus by

$$(x,q) \longmapsto (q^k x, q) \ .$$

We shall construct a torus embedding $\mathbb{C}^* \times \mathbb{C}^* \subset X_{\{\sigma_\alpha\}}$ and the sought-for analytic manifold Y will be the closure of $\mathbb{C}^* \times \overset{\circ}{\Delta}_{d_0}$ in $X_{\{\sigma_\alpha\}}$. In fact, identify $N(\mathbb{C}^* \times \mathbb{C}^*)$ with $\mathbb{Z} \times \mathbb{Z}$ and note that α acts on $N(\mathbb{C}^* \times \mathbb{C}^*)$ by

$$(a,b) \longmapsto (a+kb,b) \ .$$

We choose $\{\sigma_\alpha\}$ to be the following infinite chain σ_n, $n \in \mathbb{Z}$:

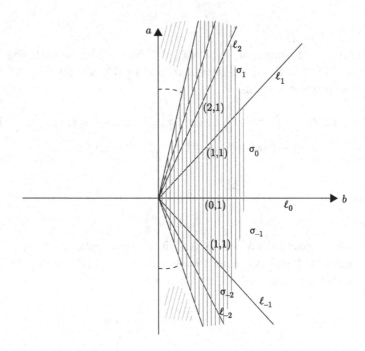

Note that α carries σ_n to σ_{n+k}, so that, mod α, there are only finitely many

σ. The corresponding $X_{\{\sigma_n\}}$ may be pictured as follows:

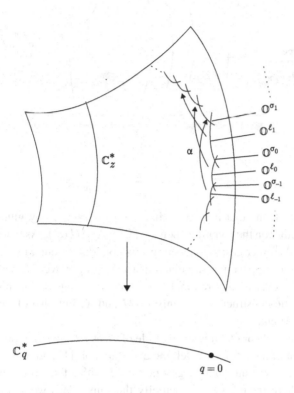

Clearly α acts on $X_{\{\sigma_n\}}$. Since each σ_i is generated by a basis of $\mathbb{Z} \times \mathbb{Z}$, it follows that $X_{\{\sigma_n\}}$ is a manifold, i.e., smooth. Moreover, a whole neighborhood of the boundary $X_{\{\sigma_n\}} \setminus (\mathbb{C}^* \times \mathbb{C}^*)$ is contained in $\mathbb{C}^* \times \mathring{\Delta}_{d_0}$, so define Y to be

$$Y = \text{interior of closure of } \mathbb{C}^* \times \mathring{\Delta}_{d_0} \text{ in } X_{\{\sigma_n\}}$$
$$= (\mathbb{C}^* \times \mathring{\Delta}_{d_0}) \cup \left(X_{\{\sigma_n\}} \setminus (\mathbb{C}^* \times \mathbb{C}^*)\right) .$$

What happens when we divide by α? Clearly α does not act discontinuously on the whole of $\mathbb{C}^* \times \mathbb{C}^*$ (i.e., if $|q| = 1$) so we cannot form $(\mathbb{C}^* \times \mathbb{C}^*)/\{\alpha^n\}$. However, it can be checked to act discontinuously on Y, and the quotient looks like this:

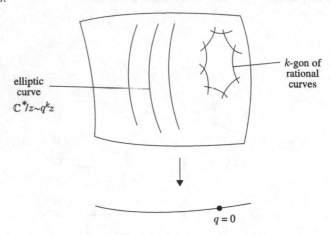

We can now define that part of $\widetilde{X}$ that lies over $i\infty \in \widetilde{M}$ by glueing X and $Y/\{\alpha^n\}$ together on the common open set $(\mathbb{C}^* \times \mathring{\Delta}_{d_0})/\{\alpha^n\}$. As before, we can take care of the other cusps by pushing this boundary around by $\mathrm{SL}(2,\mathbb{Z})/\Gamma$. This gives us a compact non-singular surface $\widetilde{X}$, proper over $\widetilde{M}$, with its fibers elliptic curves over M and rational k-gons over the cusps. In this way, we find an analytic construction not only of M and X, but also of their natural completions $\widetilde{M}$ and $\widetilde{X}$.

We want to study briefly this same circle of ideas from another point of view, similar to that of the articles of Deligne and Rapoport [1] and of Mumford [3]. Suppose we are over an arbitrary ground field L with a fixed primitive kth root of unity ζ (here $(\mathrm{char}\,L, k) = 1$; actually the same ideas work over suitable ground rings too). Then we can define $\widetilde{M}$ and $\widetilde{X}$ over L as the schemes which represent a certain functor. Restricting ourselves to the corresponding formal schemes at the cusps, we can then see directly that $\widetilde{M}$ and $\widetilde{X}$ are quotients of the formal neighborhood of the locus at infinity in the same torus embeddings introduced above.

We need some definitions.

(a) If E is an elliptic curve with given origin e, a *level-k-structure* on E is a pair of points $x, y \in E$ of order k such that $e_k(x,y) = \zeta$ (here e_k is Weil's pairing); equivalently, if $\mathscr{L}$ is a line bundle of degree k, and $\phi, \psi : \mathscr{L} \longrightarrow \mathscr{L}$ are automorphisms lifting translation by x, resp. y, then

$$\phi \circ \psi \circ \phi^{-1} \circ \psi^{-1} = \text{multiplication by } \zeta.$$

(b) If E is a k-gon of rational curves with given group law on $E_0 \subset E$ (where E_0 are the smooth points of E), a *level-k-structure* on E is a pair of points $x, y \in E_0$ of order k such that $e_k(x,y) = \zeta$. Here, to define e_k, let $\mathscr{L}$ be a line bundle of degree 1 on each component of E: then there exist $\phi, \psi : \mathscr{L} \longrightarrow$ $\mathscr{L}$ lifting translation by $x, y : E \longrightarrow E$. Then

$$\phi \circ \psi \circ \phi^{-1} \circ \psi^{-1} = \text{multiplication by } \zeta .$$

Then consider the following functor on schemes over L:

$$\widetilde{\mathfrak{M}}(S) = \begin{cases} \text{set of 4-tuples } (p, \sigma, u, v) \text{ where:} \\ \text{(a) } p : Y \longrightarrow S \text{ is a proper flat morphism, all fibers of} \\ \quad \text{which are elliptic curves or } k\text{-gons of rational curves;} \\ \text{(b) } \sigma : Y_0 \times_S Y \longrightarrow Y \text{ is a morphism (here } Y_0 \text{ is the} \\ \quad \text{open set in } Y \text{ where } p \text{ is smooth), making } Y_0 \text{ a group} \\ \quad \text{scheme over } S \text{ and making } Y_0 \text{ act on } Y;\dagger \\ \text{(c) } u, v : S \longrightarrow Y_0 \text{ are sections of order } k \text{ inducing a} \\ \quad \text{level-}k\text{-structure on each fiber.} \end{cases}$$

Let $\mathfrak{M} \subset \widetilde{\mathfrak{M}}$ be the subfunctor of (p, σ, u, v), where p is smooth. Let

$$\widetilde{\mathfrak{X}}(S) = \begin{cases} \text{set of 5-tuples } (p, \sigma, u, v, w), \text{ with } p, \sigma, u, v \text{ as before} \\ \text{and } w : S \longrightarrow Y \text{ any section of } Y \text{ over } S \end{cases},$$

$\mathfrak{X} = $ subfunctor of $\widetilde{\mathfrak{X}}$ of (p, σ, u, v, w) for which p is smooth.

Then the following theorem is well-known (see, e.g., [1]).

Theorem 4.3
(a) $\widetilde{\mathfrak{M}}$ *is represented by a curve* $\widetilde{M}$ *smooth and proper over* L;
(b) $\mathfrak{M}$ *is represented by* $\widetilde{M} \setminus C$, *where* C *is a finite set of "cusps"*;
(c) *if* $(p, \sigma, \alpha, \beta) \in \widetilde{\mathfrak{M}}(\widetilde{M})$ *corresponds to the identity map* $\widetilde{M} \longrightarrow \widetilde{M}$, *and* p *is the morphism from* $\widetilde{X}$ *to* $\widetilde{M}$, *then* $\widetilde{X}$ *represents* $\widetilde{\mathfrak{X}}$, *and* $X = p^{-1}(M)$ *represents* $\mathfrak{X}$;
(d) *if* $L = \mathbb{C}$, *then* $\widetilde{M}$ *and* $\widetilde{X}$ *are the varieties constructed above.*

Fix a cusp $c \in C$, i.e., a 4-tuple

$$\overline{p} : \overline{Y} \longrightarrow \text{Spec} L,$$
$$\overline{\sigma} : \overline{Y}_0 \times_{\text{Spec} L} \overline{Y} \longrightarrow \overline{Y},$$
$$\overline{u}, \overline{v} : \text{Spec} L \longrightarrow \overline{Y} ,$$

where $\overline{Y}$ is a k-gon. These are all isomorphic modulo the natural action of $\text{SL}_2(\mathbb{Z}/k\mathbb{Z})$ on $\widetilde{\mathfrak{M}}$ taking

$$(p, \sigma, u, v) \longmapsto (p, \sigma, au + bv, cu + dv) .$$

---

† To exclude some stupid examples of σ, one must add the condition that, for every geometric point $s \in S$ such that the fiber $p^{-1}(s)$ is a k-gon, all translations $x \longmapsto \sigma(y, x)$, $y \in p^{-1}(s) \cap Y_0$, induce rotations of the graph of the k-gon $p^{-1}(s)$.

So let us assume $\bar{v}$ is in the identity component of $\overline{Y}_0$, and $\bar{u}$ is in an adjacent one:

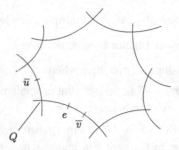

(When we assume this, $(\bar{p}, \bar{\sigma}, \bar{u}, \bar{v})$ is unique up to isomorphism.) Let $\widehat{M}$ (resp. $\widehat{X}$) denote the formal completion of $\widetilde{M}$ (resp. $\widetilde{X}$) at c (resp. along $p^{-1}(c)$).

Now go back to the torus embeddings introduced earlier:

$$
\begin{array}{ccc}
\mathbb{G}_m \times \mathbb{G}_m & \hookrightarrow & X_{\{\sigma_n\}} \\
\scriptstyle p_2 \downarrow & & \downarrow \\
\mathbb{G}_m & \hookrightarrow & \mathbb{A}^1_q \ ,
\end{array}
$$

where σ_n is the set of sectors $\langle (n,1), (n+1,1) \rangle$, $n \in \mathbb{Z}$. Let $\alpha : X_{\{\sigma_n\}} \longrightarrow X_{\{\sigma_n\}}$ be the automorphism $(z, q) \longrightarrow (z q^k, q)$, where z and q are coordinates on the first and second factor $\mathbb{G}_m$. Let $\widehat{X}_{\{\sigma_n\}}$ (resp. $\widehat{\mathbb{A}}^1_q$) be the formal completion of $X_{\{\sigma_n\}}$ (resp. $\mathbb{A}^1_q$) along the complement of the torus. Then we claim that there are canonical isomorphisms:

$$
\begin{array}{ccc}
\widehat{X}_{\{\sigma_n\}}/\{\alpha^n\} & \xrightarrow{\ \sim\ } & \widehat{X} \\
\scriptstyle p_2 \downarrow & & \downarrow \scriptstyle p \\
\widehat{\mathbb{A}}^1_q & \xrightarrow{\ \sim\ } & \widehat{M} \ .
\end{array}
$$

We will show here only how to construct:

(a) a morphism $\widehat{\mathbb{A}}^1_q \longrightarrow \widehat{M}$;

(b) a map of functors

$$
\widehat{M}(S) \longrightarrow \widehat{\mathbb{A}}^1_q(S)
$$

for $S = \operatorname{Spec} R$, where R is an Artin local L-algebra with residue field L.

We let the reader check that these are inverse to each other and that we get corresponding maps on the curves over these bases.

Construction (a) The point is that $\widehat{X}_{\{\sigma_n\}}/\{\alpha^n\}$ itself is flat and proper over $\widehat{\mathbb{A}}^1_q$ and its (one) closed fiber is indeed a k-gon. The action of $\mathbb{G}_m \times \mathbb{G}_m$ on $X_{\{\sigma_n\}}$ is a morphism:

$$(\mathbb{G}_m \times \mathbb{G}_m) \times_{\mathbb{A}^1_q} X_{\{\sigma_n\}} \longrightarrow X_{\{\sigma_n\}} ,$$

and it is readily checked that this extends to

$$\sigma : (X_{\{\sigma_n\}})_0 \times_{\mathbb{A}^1_q} X_{\{\sigma_n\}} \longrightarrow X_{\{\sigma_n\}} ,$$

where $(X_{\{\sigma_n\}})_0$ is the open subset where $X_{\{\sigma_n\}}$ is smooth over $\mathbb{A}^1_q$. This induces

$$\sigma : (\widehat{X}_{\{\sigma_n\}}/\{\alpha^n\})_0 \times_{\mathbb{A}^1_q} (\widehat{X}_{\{\sigma_n\}}/\{\alpha^n\}) \longrightarrow \widehat{X}_{\{\sigma_n\}}/\{\alpha^n\} .$$

Finally, the sections $z = q$ and $z = \zeta$ of $\mathbb{G}_m \times \mathbb{G}_m$ over $\mathbb{G}_m$ define sections

$$u, v : \widehat{\mathbb{A}}^1_q \longrightarrow (\widehat{X}_{\{\sigma_n\}}/\{\alpha^n\})_0 .$$

Construction (b) Now start with an element of $\widehat{M}(S)$, where $S = \operatorname{Spec} R$ as above, i.e., (p, σ, u, v), with $Y \longrightarrow S$, extending the fixed $(\bar{p}, \bar{\sigma}, \bar{u}, \bar{v})$ defining the cusp. Then $Y_{\mathrm{red}} = \overline{Y}$ is a k-gon of rational curves, so the universal cover Y^* of Y is locally of finite type over S and Y^*_{red} is just an infinite string of copies of $\mathbb{P}^1$. If we fix an identity $e : S \longrightarrow Y^*$ over $e : S \longrightarrow Y$, then we get a unique lifting of $\sigma : Y_0 \times Y \longrightarrow Y$ to

$$\sigma^* : Y_0^* \times Y^* \longrightarrow Y^*$$

such that

(a) $\sigma^*(e, e) = e$,
(b) $\sigma^*(x, y) = \sigma^*(y, x)$ if $x, y \in Y_0^*$.

Next, let $\overline{Y}_0(e)$ and $Y_0(e)$ denote the identity components of $\overline{Y}_0$ and Y_0. Then there are two isomorphisms

$$\overline{Y}_0(e) \cong \mathbb{G}_{m,L} ,$$

and we can fix one of them by requiring that the point 0 in the closure of $\mathbb{G}_{m,L}$ corresponds to the intersection point Q of the closures of $\overline{Y}_0(e)$ and $\overline{Y}_0(\bar{u})$ in $\overline{Y}$. According to a result of Grothendieck [4], exp. IX, §3, tori are "rigid", so this isomorphism lifts to a unique S-isomorphism

$$Y_0(e) \cong \mathbb{G}_{m,S} .$$

But $Y_0^*(e) \cong Y_0(e)$, so we get an action of $\mathbb{G}_{m,S}$ on Y^*; now we begin to see why torus embeddings are involved. Let U^- be the open subset of Y^* consisting of "half the string" starting at $Y_0^*(e)$ and in the direction of the limit point 0 of $\mathbb{G}_{m,S}$:

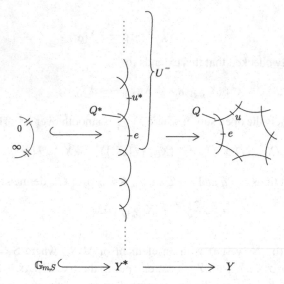

$$\mathbb{G}_{m,S} \hookrightarrow Y^* \longrightarrow Y$$

Also lift u to $u^* : S \longrightarrow Y^*$ so as to lie in the component adjacent to $Y_0^*(e)$. Then:

Lemma 4.4 *There exists a unique* $\mathfrak{X} \in \Gamma(U^-, \mathscr{O}_{Y^*})$ *such that* $\mathfrak{X}(e) = 1$, *and such that, under the action* $\sigma_0 : \mathbb{G}_{m,S} \times_S U^- \longrightarrow U^-$ *of* $\mathbb{G}_{m,S}$ *on* U^-,

$$\sigma_0^*(\mathfrak{X}) = z \cdot \mathfrak{X},$$

where z *is the coordinate on* $\mathbb{G}_{m,S}$.

Proof Decompose $\Gamma(U^-, \mathscr{O}_{Y^*})$ into eigenspaces under the action of $\mathbb{G}_{m,S}$:

$$\Gamma(U^-, \mathscr{O}_{Y^*}) = \bigoplus_{n=-\infty}^{+\infty} W_n,$$

where $\sigma_0^*(f) = z^n \cdot f$ if $f \in W_n$. Since U^- is flat over R, it follows that W_n is a flat R-module. But

$$\Gamma(U^-, \mathscr{O}_{\overline{Y}^*}) = \bigoplus_{n=-\infty}^{+\infty} W_n \otimes_R L,$$

and since $\overline{Y}^*$ is just a string of copies of $\mathbb{P}^1$, one sees immediately that

$$W_1 \otimes_R L = L \cdot \overline{\mathfrak{X}},$$

where the function $\overline{\mathfrak{X}}$ is z on $\overline{Y}_0^*(e)$ and 0 on all the other components. Therefore W_1 is a free rank-1 R-module, and the condition $\mathfrak{X}(e) = 1$ picks out a
unique element. $\square$

Now let $\sigma_1(x) = \sigma^*(u^*, x)$, giving us an automorphism $\sigma_1 : Y^* \longrightarrow Y^*$, and a
morphism $\sigma_1 : U^- \longrightarrow U^-$. Then $\sigma_1^*(\mathfrak{X})$ is another element of $\Gamma(U^-, \mathcal{O}_{Y^*})$ in
W_1 (notation as in the proof of the preceding lemma), so

$$\sigma_1^*(\mathfrak{X}) = q \cdot \mathfrak{X} \, , \text{ for some } q \in R \, .$$

This q is the sought-for period! It defines a homomorphism

$$L[[q]] \longrightarrow R \, ,$$

and hence an element of $\widehat{A}_q^1(S)$.

5 Hirzebruch's theory of the Hilbert modular group

We now investigate a second beautiful example in order to motivate and illustrate further all the theory which follows.

Let $K = \mathbb{Q}(\sqrt{d})$ be a real quadratic number field and associate to it the
following objects:

$$\mathcal{O} = \text{ring of integers in } K,$$
$$\mathfrak{a} = \text{a fixed ideal in } \mathcal{O},$$
$$u_0 = \text{a generator of the group of units } u \in \mathcal{O}, \text{ such that}$$
$$u \equiv 1 \bmod \mathfrak{a} \text{ (or of this group mod } \pm 1).$$

We do not regard K as a subfield of $\mathbb{R}$, but rather as an abstract field extension
of $\mathbb{Q}$, with two embeddings:

$$\phi_1, \phi_2 : K \longrightarrow \mathbb{R} \, ,$$

neither being more important than the other. Let

$$\Gamma = \left\{ \begin{pmatrix} a & b \\ c & d \end{pmatrix} \in \mathrm{SL}(2, \mathcal{O}) \,\Big|\, \begin{pmatrix} a & b \\ c & d \end{pmatrix} \equiv \begin{pmatrix} 1 & 0 \\ 0 & 1 \end{pmatrix} \bmod \mathfrak{a} \right\} \, .$$

Consider the following embedding:

$$\mathrm{SL}(2, K) \longrightarrow \mathrm{SL}(2, \mathbb{R} \times \mathrm{SL}(2, \mathbb{R}))$$
$$\begin{pmatrix} a & b \\ c & d \end{pmatrix} \longmapsto \left(\begin{pmatrix} \phi_1(a) & \phi_1(b) \\ \phi_1(c) & \phi_1(d) \end{pmatrix}, \begin{pmatrix} \phi_2(a) & \phi_2(b) \\ \phi_2(c) & \phi_2(d) \end{pmatrix} \right) \, .$$

Then there is a unique $\mathbb{Q}$-structure on $\mathrm{SL}(2, \mathbb{R}) \times \mathrm{SL}(2, \mathbb{R})$ with $\mathrm{SL}(2, K)$ as its
$\mathbb{Q}$-rational points and Γ is an arithmetic subgroup for this $\mathbb{Q}$-structure. We let

$SL(2,\mathbb{R}) \times SL(2,\mathbb{R})$ act componentwise on $\mathfrak{H} \times \mathfrak{H}$ and seek to compactify the so-called *Hilbert modular surface*:

$$\mathfrak{F}_{K,\mathfrak{a}} = (\mathfrak{H} \times \mathfrak{H})/\Gamma .$$

It is known (see Shimizu [5]) that $\mathfrak{F}_{K,\mathfrak{a}}$ can be embedded in a compact normal analytic surface $\overline{\mathfrak{F}}_{K,\mathfrak{a}}$ by adding only a finite number of points, called *cusps* $F_1, \dots, F_N$:

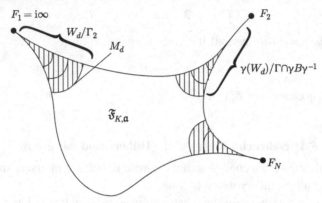

Topologically, starting with the simplest cusp "i∞", here is the picture: let

$$W_d = \{(z_1, z_2) \mid \operatorname{Im} z_1 \cdot \operatorname{Im} z_2 \ge d\} \subset \mathfrak{H} \times \mathfrak{H} ;$$

$$\Gamma_1 = \left\{ \begin{pmatrix} 1 & b \\ 0 & 1 \end{pmatrix} \mid b \in \mathfrak{a} \right\},$$

and let

$$\Gamma_2 = \left\{ \begin{pmatrix} u_0^n & b \\ 0 & u_0^{-n} \end{pmatrix} \mid n \in \mathbb{Z}, b \in \mathfrak{a} \right\}$$

$$= \Gamma \cap \left\{ \begin{pmatrix} a & b \\ 0 & d \end{pmatrix} \right\} / \Gamma \cap \{\pm 1\} .$$

Then $\Gamma_2 \cdot W_d = W_d$ (use the fact that $\phi_1(u_0) \cdot \phi_2(u_0) = \operatorname{Norm}(u_0) = \pm 1$: see (ii) below), and it turns out that if $d \gg 0$, then Γ-equivalence on $\mathfrak{H} \times \mathfrak{H}$ reduces on W_d to $\pm\Gamma_2$-equivalence, i.e.,

$$z_1, z_2 \in W_d, z_1 = \gamma z_2 \text{ for some } \gamma \in \Gamma \Longrightarrow \gamma \in \pm\Gamma_2 .$$

Since ± 1 acts trivially on $\mathfrak{H} \times \mathfrak{H}$, we get $W_d/\Gamma_2 \subset (\mathfrak{H} \times \mathfrak{H})/\Gamma = \mathfrak{F}_{K,\mathfrak{a}}$. But W_d/Γ_2 is easy to visualize:

(i) Γ_1 acts by $(z_1, z_2) \mapsto (z_1 + \phi_1(b), z_2 + \phi_2(b))$ and

$$\Phi(\mathfrak{a}) = \{(\phi_1(b), \phi_2(b)) \mid b \in \mathfrak{a}\}$$

is a lattice in $\mathbb{R} \times \mathbb{R}$;

(ii) Γ_1 is a normal subgroup of Γ_2 and $\Gamma_2/\Gamma_1 \cong \mathbb{Z}$ with generator γ_0 equal to the image of $\begin{pmatrix} u_0 & 0 \\ 0 & u_0^{-1} \end{pmatrix}$. Now Γ_1 leaves invariant $\operatorname{Im} z_1$ and $\operatorname{Im} z_2$, and Γ_2/Γ_1 acts on these by

$$\gamma_0^*(\operatorname{Im} z_1) = \phi_1(u_0)^2 \cdot \operatorname{Im} z_1,$$
$$\gamma_0^*(\operatorname{Im} z_2) = \phi_2(u_0)^2 \cdot \operatorname{Im} z_2 = \phi_1(u_0)^{-2} \cdot \operatorname{Im} z_2 \ ;$$

(iii) hence $\operatorname{Im} z_1 \cdot \operatorname{Im} z_2$ is invariant under Γ_2.

Think of $1/(\operatorname{Im} z_1 \cdot \operatorname{Im} z_2)$ as measuring the distance to the cusp. For every fixed e, if $z_i = x_i + i y_i$, we get a diagram

$$M_e = \{(z_1, z_2) \mid \operatorname{Im} z_1 \cdot \operatorname{Im} z_2 = e\}/\Gamma_2$$

$$\Big\downarrow {\scriptstyle \text{fiber } \{(x_1, x_2) \in \mathbb{R}^2\}/\Phi(\mathfrak{a})}$$

$$\{(y_1, y_2) \in \mathbb{R}^2_{>0} \mid y_1 \cdot y_2 = e\}/\{\gamma_0''\} \underset{\text{homeo}}{\approx} S^1 \ ,$$

from which it follows that M_e is a compact 3-manifold which is an $S^1 \times S^1$ bundle over S^1. As e varies, the manifolds M_e are all homeomorphic and we get

$$(z_1, z_2) \quad \in \quad W_d/\Gamma_2$$

$$\Big\downarrow \qquad\qquad \Big\downarrow {\scriptstyle \text{fibers } M_e}$$

$$\operatorname{Im} z_1 \cdot \operatorname{Im} z_2 \quad \in \quad [d, \infty) \ ,$$

i.e.,

$$W_d/\Gamma_2 \underset{\text{homeo}}{\approx} M_d \times [d\infty) \ .$$

The compactification $\overline{\mathfrak{F}}_{k,\mathfrak{a}}$ in the subset W_d/Γ_2 simply results, topologically, by embedding $M_d \times [d, \infty)$ in the cone over M_d, i.e., the one-point compactification of W_d/Γ_2. Thus the subsets $M_{d'}/\Gamma_2$, as $d' \longrightarrow \infty$, are a fundamental system of neighborhoods of the cusp $F_1 = i\infty$.

There may be other cusps too: for every $\gamma \in \mathrm{SL}(2, K)$, consider the subset $\gamma(W_d) \subset \mathfrak{H} \times \mathfrak{H}$. If $B \subset \mathrm{SL}(2, K)$ is the subgroup of matrices $\begin{pmatrix} a & b \\ 0 & a^{-1} \end{pmatrix}$, then $\gamma(W_d)$ is left invariant by the group $\Gamma \cap \gamma B \gamma^{-1}$, and if $d \gg 0$, then, as before,

$$\gamma(W_d)/\Gamma \cap \gamma B \gamma^{-1} \subset (\mathfrak{H} \times \mathfrak{H})/\Gamma \ .$$

Moreover, $\Gamma \cap \gamma B \gamma^{-1} \cong \gamma^{-1}\Gamma\gamma \cap B$, which is an extension

$$1 \longrightarrow \left\{ b \in K \mid \begin{pmatrix} 1 & b \\ 0 & 1 \end{pmatrix} \in \gamma^{-1}\Gamma\gamma \right\} \longrightarrow \gamma^{-1}\Gamma\gamma \cap B \longrightarrow$$

$$\longrightarrow \left\{ u \in K^* \mid \begin{pmatrix} u & b \\ 0 & u^{-1} \end{pmatrix} \in \gamma^{-1}\Gamma\gamma \text{ for some } b \right\} \longrightarrow 1$$

$$\underset{\text{\rotatebox{90}{$\cong$}}}{}$$

$$\mathbb{Z} \text{ or } \mathbb{Z} \times (\mathbb{Z}/2\mathbb{Z})$$

(the 2-torsion comes from $u = \pm 1$, and if this occurs it acts trivially on $\mathfrak{H} \times \mathfrak{H}$, so we ignore it), from which it follows, as before, that

$$\gamma(W_d)/\Gamma \cap \gamma B \gamma^{-1} \underset{\text{homeo}}{\approx} M(\gamma) \times [d, \infty)$$

for some compact 3-manifold $M(\gamma)$ which is an $S^1 \times S^1$-bundle over S^1. Again, we make a one-point compactification.

Actually, there are only finitely many cusps. In fact

(a) if $\gamma' = \gamma \cdot \delta$, with $\delta = \begin{pmatrix} a & b \\ 0 & a^{-1} \end{pmatrix} \in B$, then $\gamma'(W_d) = \gamma(W_{d'})$, where $d' = d \cdot \text{Norm}(a)^2$, and $\Gamma \cap \gamma' B \gamma'^{-1} = \Gamma \cap \gamma B \gamma^{-1}$, so we have the same cusp;

(b) if $\gamma' = \varepsilon \cdot \gamma$ with $\varepsilon \in \Gamma$, then the images of $\gamma(W_d)$ and $\gamma'(W_d)$ in $(\mathfrak{H} \times \mathfrak{H})/\Gamma$ are the same, hence

$$\begin{array}{c} \gamma(W_d)/\Gamma \cap \gamma B \gamma^{-1} \\ \| \qquad\qquad \searrow \\ \gamma'(W_d)/\Gamma \cap \gamma' B \gamma'^{-1} \end{array} \hookrightarrow (\mathfrak{H} \times \mathfrak{H})/\Gamma$$

Hence the cusp depends only on the double coset $\Gamma\gamma B \in \Gamma \backslash \text{SL}(2, K)/B$. It is easy to check that if $\Gamma\gamma B \neq \Gamma\gamma' B$, then, for $d \gg 0$, the images of $\gamma(W_d)$ and $\gamma'(W_d)$ in $(\mathfrak{H} \times \mathfrak{H})/\Gamma$ are disjoint; hence if we want $\overline{\mathfrak{F}}_{K,\mathfrak{a}}$ to be normal (and we do), we must have different cusps here. Thus

$$\#\text{cusps} = \#(\Gamma \backslash \text{SL}(2, K)/B),$$

which is finite by a classical theorem (in fact, one can check that if $\mathfrak{a} = \mathcal{O}$, then

$$\Gamma \backslash \text{SL}(2, K)/B \cong \text{ideal class group of } K) .$$

Now, how do we put an analytic structure on $\overline{\mathfrak{F}}_{K,\mathfrak{a}}$? We shall only do this at the end and instead, by following the suggestions made by the above topological construction, plus our knowledge of toroidal embeddings, define directly a blown-up non-singular compactification $\widetilde{\mathfrak{F}}_{K,\mathfrak{a}}$ of $\mathfrak{F}_{K,\mathfrak{a}}$. Again we start with the cusp $i\infty$. The idea is first to factor the canonical map $\mathfrak{H} \times \mathfrak{H} \longrightarrow \mathfrak{F}_{K,\mathfrak{a}}$ as follows:

$$\mathfrak{H} \times \mathfrak{H} \longrightarrow (\mathfrak{H} \times \mathfrak{H})/\Gamma_1 \longrightarrow (\mathfrak{H} \times \mathfrak{H})/\Gamma_2 \longrightarrow \mathfrak{F}_{K,\mathfrak{a}} .$$

We may embed $(\mathfrak{H} \times \mathfrak{H})/\Gamma_1$ in a torus as follows: Γ_1 acts on $\mathfrak{H} \times \mathfrak{H}$ by translations by the lattice $\Phi(\mathfrak{a})$ in $\mathbb{R} \times \mathbb{R}$:

$$(z_1, z_2) \longmapsto (z_1 + \phi_1(b), z_2 + \phi_2(b)) , \quad b \in \mathfrak{a} .$$

Let T be the torus given by $\mathbb{C} \times \mathbb{C}$ modulo the same group of translations:

$$T = \mathbb{C} \times \mathbb{C}/\Phi(\mathfrak{a}) ,$$

so that

$$N(T) = \Phi(\mathfrak{a}) ,$$
$$N(T)_{\mathbb{R}} = \mathbb{R} \times \mathbb{R} ,$$

and we get the exact sequence:

$$0 \longrightarrow T_c \longrightarrow T \xrightarrow{\text{ord}} N_{\mathbb{R}}(T) \longrightarrow 0$$
$$ \| \| \phantom{T \xrightarrow{\text{ord}}} \|$$
$$\mathbb{R} \times \mathbb{R}/\Phi(\mathfrak{a}) \quad \mathbb{C} \times \mathbb{C}/\Phi(\mathfrak{a}) \longrightarrow \mathbb{R} \times \mathbb{R}$$

$$(z_1, z_2) \longmapsto (\operatorname{Im} z_1, \operatorname{Im} z_2)$$

It follows that:

$$(\mathfrak{H} \times \mathfrak{H})/\Gamma_1 \cong \operatorname{ord}^{-1}(\mathbb{R}_{>0} \times \mathbb{R}_{>0}) ,$$
$$W_d/\Gamma_1 \cong \operatorname{ord}^{-1}(\underbrace{\{(y_1, y_2) \mid y_1 y_2 \geq d, y_i > 0\}}_{\text{call this } V_d}) .$$

The first set is open in T, the second set is closed in the first. Next, $\Gamma_2/\Gamma_1 \cong \{\gamma_0^n\}$, and γ_0 acts on $\mathfrak{H} \times \mathfrak{H}$ by

$$\gamma_0(z_1, z_2) = (\phi_1(u_0)^2 z_1, \phi_1(u_0)^{-2} z_2) .$$

Let $v_0 = \phi_1(u_0)^2$. Now the action of γ_0 on $\mathfrak{H} \times \mathfrak{H}$ extends to $\mathbb{C} \times \mathbb{C}$, hence to T and to the open subset $(\mathfrak{H} \times \mathfrak{H})/\Gamma_1 \subset T$. In particular, γ_0 acts on $N_{\mathbb{R}}(T)$ by

$$\gamma_0(x_1, x_2) = (v_0 x_1, v_0^{-1} x_2) ,$$

an action which preserves the positive quadrant and the (irrational) lattice $\Phi(\mathfrak{a})$. We have thus arrived at the following situation: T is a two-dimensional torus and $\gamma_0 : T \longrightarrow T$ is a hyperbolic automorphism of infinite order (in fact, up to replacing γ_0 by a power or a root, we have obtained the most general hyperbolic automorphism of $\mathbb{G}_m^2$).

At this point, the idea is to enlarge T – and its open subsets $(\mathfrak{H} \times \mathfrak{H})/\Gamma_1$ and W_d/Γ_1 – by adding some analytic boundary $\mathcal{E}$, so that γ_0 still acts on $T \cup \mathcal{E}$, and then to divide by $\{\gamma_0^n\}$ so that $E = \mathscr{E}/\{\gamma_0^n\}$ is the boundary that can be added to $(\mathfrak{H} \times \mathfrak{H})/\Gamma$ 'in the direction $i\infty$':

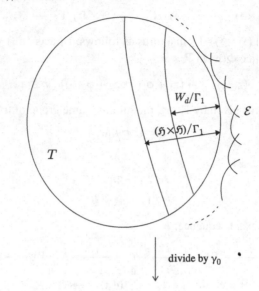

divide by γ_0

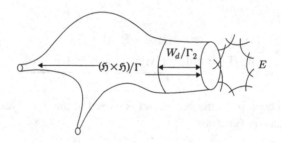

To enlarge T, we use the theory of torus embeddings.

(i) We seek a decomposition of the positive quadrant $\mathbb{R}_{\geq 0} \times \mathbb{R}_{\geq 0}$ into rational sectors $\{\sigma_\alpha\}$.

(ii) These should satisfy: $\gamma_0 \sigma_\alpha = $ some σ_β; $\bigcup \sigma_\alpha = \mathbb{R}_{\geq 0} \times \mathbb{R}_{\geq 0}$; $\sigma_\alpha \cap \sigma_\beta = (0)$ or a common edge. Note that since $(0) \times \mathbb{R}_{\geq 0}$ and $\mathbb{R}_{\geq 0} \times (0)$ are *irrational* half-lines, all σ_α are in the *interior* of the positive quadrant, and we will need an infinite number of σ_α. However, we require:

> *modulo the action of γ_0, there are only finitely many σ_α .*

(iii) In this case, if we denote one of the σ_α by σ_0, we can number them uniquely $\ldots, \sigma_{-2}, \sigma_{-1}, \sigma_0, \sigma_1, \sigma_2, \ldots$ so that $\sigma_i \cap \sigma_j$ is a common edge if and only if $i = j \pm 1$, and $\gamma_0 \sigma_i = \sigma_{i+d}$ for all i and some fixed d. Let $\ell_i = \sigma_i \cap \sigma_{i+1}$. It looks like this:

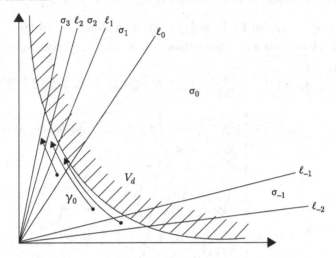

(iv) Given such $\{\sigma_\alpha\}$, we obtain a torus embedding

$$T \subset X_{\{\sigma_\alpha\}},$$

where $X_{\{\sigma_\alpha\}}$ is a scheme locally of finite type over $\mathbb{C}$ (in fact, locally a normal complex variety), with the action of γ_0 extending to $X_{\{\sigma_\alpha\}}$. Moreover, $X_{\{\sigma_\alpha\}}$ is the union of T and an infinite chain of non-singular rational curves E_i, one for each half-line ℓ_i, meeting at points P_i, one for each sector σ_i:

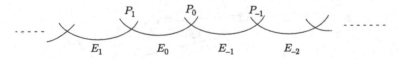

(v) Next enlarge $(\mathfrak{H} \times \mathfrak{H})/\Gamma_1$ by setting

$$\widetilde{(\mathfrak{H} \times \mathfrak{H})/\Gamma_1} = (\mathfrak{H} \times \mathfrak{H})/\Gamma_1 \cup \bigcup_{i=-\infty}^{+\infty} E_i \,.$$

In fact, in the corresponding embedding

$$N_{\mathbb{R}}(T) \subset N_{\{\sigma_i\}},$$

it is clear that

$$\widetilde{\mathbb{R}_{>0} \times \mathbb{R}_{>0}} = \mathbb{R}_{>0} \times \mathbb{R}_{>0} \cup \underbrace{N_{\{\sigma_i\}} \setminus N_{\mathbb{R}}(T)}_{\text{chain of boundary segments}}$$

is the interior of the closure of $\mathbb{R}_{>0} \times \mathbb{R}_{>0}$ in $N_{\{\sigma_i\}}$. Therefore taking ord^{-1}, $(\widetilde{\mathfrak{H} \times \mathfrak{H}})/\Gamma_1$ is the interior of the closure of $(\mathfrak{H} \times \mathfrak{H})/\Gamma_1$ in $X_{\{\sigma_\alpha\}}$.

(vi) It is easy to see that $\Gamma_2/\Gamma_1 = \{\gamma_0^n\}$ acts discontinuously on $(\widetilde{\mathfrak{H} \times \mathfrak{H}})/\Gamma_1$. In fact, check first that it acts discontinuously on $\mathbb{R}_{>0} \times \mathbb{R}_{>0}$ with fundamental domain $\Omega = \left(\overline{\bigcup_{i=0}^{d-1} \sigma_i} \right)$:

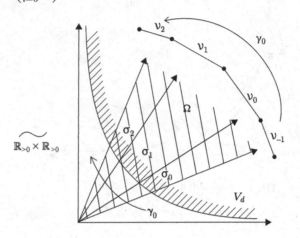

with quotient looking topologically like this:

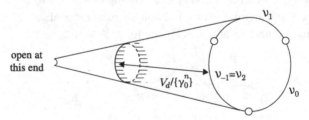

Note that $\Omega \cap V_d$ is compact modulo $\mathbb{R}_{>0}$, hence $V_d/\{\gamma_0^n\}$ is compact modulo $\mathbb{R}_{>0}$. Therefore $\mathrm{ord}^{-1}(\Omega)$ is a fundamental domain in $(\widetilde{\mathfrak{H} \times \mathfrak{H}})/\Gamma_1$ and we get an analytic space

$$((\widetilde{\mathfrak{H} \times \mathfrak{H}})/\Gamma_1)/(\Gamma_2/\Gamma_1) ,$$

consisting of the open piece $(\mathfrak{H} \times \mathfrak{H})/\Gamma_2$ and a closed analytic set

$$E = \left(\bigcup_{i=-\infty}^{+\infty} E_i \right)/\{\gamma_0^n\},$$

which is a d-sided polygon of rational curves $\overline{E}_0 \cup \cdots \cup \overline{E}_{d-1}$ (with $\overline{E}_i$ the image of E_i):

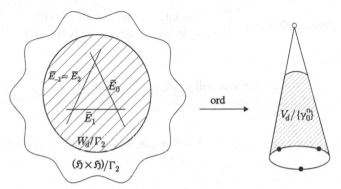

(Here $\overline{E}_i \longmapsto v_i$ and $\overline{E}_i \cap \overline{E}_{i+1} \longmapsto v_i \cap v_{i+1}$ under the ord map.) Moreover, in here, $W_d / \Gamma_2 \cup E$ is compact (since ord is proper). Thus, as above, since $W_d / \Gamma_2 \subset \mathfrak{F}_{K,\mathfrak{a}}$, we can form $\mathfrak{F}_{K,\mathfrak{a}} \cup E$ and make it into an analytic space by the above analytic structures on the two subsets $\mathfrak{F}_{K,\mathfrak{a}}$ and $W_d / \Gamma_2 \cup E$.

To recover our previous compactification of the cusp i∞, we shall blow down E to a point. This can be accomplished by checking that E has a fundamental system of *strongly pseudoconvex* neighborhoods. But in fact, V_d is a *strongly convex* subset of $\mathbb{R}_{>0} \times \mathbb{R}_{>0}$ in the following sense: for all $x \in \partial V_d$, the subset V_d is defined near x by an equation $\varphi_x \geq 0$, where, for all $t \neq 0$ in the tangent space to ∂V_d at x, defined by $d\varphi_x(t) = 0$, we have

$$d^2 \varphi_x(t,t) < 0 .$$

It is an easy lemma that, if a closed subset $W \subset \mathbb{C}^n$ is defined at a boundary point $z \in \partial W$ by

$$\varphi_z(\operatorname{Re} z_1, \dots, \operatorname{Re} z_n) \geq 0 ,$$

then W is strongly pseudoconvex at z if and only if the closed set $V \subset \mathbb{R}^n$ given by $\varphi_z(x_1, \dots, x_n) \geq 0$ is strongly convex. Since $\log|z_i| = \operatorname{Re} \log z_i$, this implies that W_d is strongly pseudoconvex. Therefore

$$\bigcup_{i=0}^{d-1} \overline{E}_i \subset ((\widetilde{\mathfrak{H} \times \mathfrak{H}})/\Gamma_1)/(\Gamma_2/\Gamma_1)$$

is an exceptional set and can be blown down to a point.

But the advantage of our boundary is that we can make it non-singular. In fact, let e_i be the smallest integral point on the half-line ℓ_i, i.e., $e_i \in \ell_i \cap \Phi(\mathfrak{a})$. Then from TE I, we know:

$X_{\{\sigma_i\}}$ non-singular $\Longleftrightarrow$ e_i and e_{i+1} generate the lattice $\Phi(\mathfrak{a})$ for all i .

But it is easy to check that, for any two $e, e' \in \Phi(\mathfrak{a})$,

$$e, e' \text{ generate } \Phi(\mathfrak{a}) \Longleftrightarrow \left[\begin{array}{c} \text{the intersection of } \Phi(\mathfrak{a}) \text{ and the} \\ \text{triangle } \overline{0, e, e'} \text{ contains only } 0, e \text{ and } e' \end{array} \right].$$

Now introduce

$$\Sigma = \text{convex hull of } \mathbb{R}_{>0} \times \mathbb{R}_{>0} \cap \Phi(\mathfrak{a}) .$$

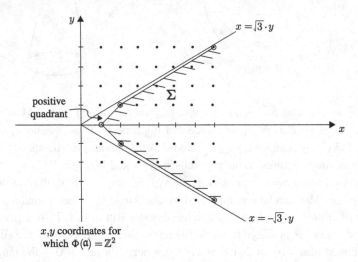

x,y coordinates for
which $\Phi(\mathfrak{a}) = \mathbb{Z}^2$

It follows that if $X_{\{\sigma_i\}}$ is non-singular, then the e_i must include all lattice points on $\partial\Sigma$. On the other hand, if we take the e_i to be precisely the points $\Phi(\mathfrak{a}) \cap \partial\Sigma$, and let $\ell_i = \mathbb{R}_{\geq 0} e_i$, $\sigma_i = \langle e_{i-1}, e_i \rangle$, then the corresponding $X_{\{\sigma_i\}}$ is non-singular. So this is the minimal non-singular choice of $X_{\{\sigma_i\}}$ and leads to the minimal desingularization of the old boundary $(\mathfrak{H} \times \mathfrak{H})/\Gamma_2 \cup (\text{one point})$. One can, by the way, easily compute from the theory of TE I the intersection matrix $(\overline{E}_i \cdot \overline{E}_j)$ for the curves $\overline{E}_i$ on $((\widetilde{\mathfrak{H} \times \mathfrak{H}})/\Gamma_1)/(\Gamma_2/\Gamma_1)$ by the following recipe:

(a) number the $\overline{E}_i$ cyclically, i.e., $\overline{E}_{i+d} = \overline{E}_i$;
(b) then, if $i \neq j$,

$$(\overline{E}_i \cdot \overline{E}_j) = \begin{cases} 0 & \text{if } |i - j| > 1 \\ 1 & \text{if } |i - j| = 1; \end{cases}$$

(c) $(\overline{E}_i^2) = -d_i$, where d_i is the index in $\Phi(\mathfrak{a})$ of the subgroup generated by e_{i-1} and e_{i+1}; alternatively,

$$d_i = 2 \cdot \text{area}(\overline{0, e_{i-1}, e_{i+1}}) ,$$

if the area is normalized so that $\text{area}(\mathbb{R} \times \mathbb{R}/\Phi(\mathfrak{a})) = 1$.

Also, it is well-known that the vertices of Σ are nothing but the *consecutive best rational approximations* to the irrational lines $(0) \times \mathbb{R}$ and $\mathbb{R} \times (0)$, generated by the continued fraction expansion of the slope of these lines (expressed via a basis of $\Phi(\mathfrak{a})$).

In the same way, we can handle the other cusps and glue in polygons of rational curves there. For every cusp F corresponding to $\Gamma \gamma B$, the isomorphism

$$(\mathfrak{H} \times \mathfrak{H})/\Gamma \xrightarrow{\sim} (\mathfrak{H} \times \mathfrak{H})/\gamma^{-1}\Gamma\gamma,$$
$$x \longmapsto \gamma^{-1}x,$$

carries F to the standard cusp i∞, so it suffices to repeat the above construction, replacing Γ by $\gamma^{-1}\Gamma\gamma$ and then to carry back the resulting boundary piece E to $(\mathfrak{H} \times \mathfrak{H})/\Gamma$ via the above isomorphism.

References

[1] P. Deligne and M. Rapoport, Les schémas de modules de courbes elliptiques, in *Modular Functions of One Variable, II (Proc. Internat. Summer School, Univ. Antwerp, Antwerp, 1972)*. Lecture Notes in Mathematics 349. Berlin: Springer, 1973, pp. 143–316.

[2] S. Lang, *Elliptic Functions*. Reading, MA: Addison-Wesley Publishing Co., Inc., 1973. With an appendix by J. Tate.

[3] D. Mumford, An analytic construction of degenerating abelian varieties over complete rings, *Compositio Math.* **24** (1972), 239–272.

[4] M. Demazure et A. Grothendieck (eds.), Séminaire de Géométrie Algébrique du Bois Marie 1962/64–Schémas en groupes (SGA 3)–Vol.2. Lecture Notes in Mathematics 152. New York: Springer-Verlag, 1970.

[5] H. Shimizu, On discontinuous groups operating on the product of the upper half planes, *Ann. of Math.* **77** (2) (1963), 33–71.

II

Polyhedral reduction theory in self-adjoint cones

Minkowski was the first to demonstrate the existence of a polyhedral fundamental domain for the action of $GL(n, \mathbb{Z})$ on the self-adjoint cone of all positive-definite quadratic forms in n variables with real coefficients. His work was generalized by Weyl and others to many other cases. Recently, A. Borel has produced a theory of coarse fundamental domains (called Siegel sets) for any arithmetic subgroup Γ of a reductive algebraic group with $\mathbb{Q}$-structure. Using this we have gone back to exact polyhedral fundamental domains, showing their existence in a unified approach for any arithmetic subgroup acting on a self-adjoint homogeneous cone, C. More precisely, a set of closed polyhedral cones $\{\sigma_\alpha\}$, such that $\sigma_\alpha \subset \overline{C}$ and σ_α is spanned by a set of rational vertices for every α, is called a Γ-*admissible polyhedral decomposition* of C when the following hold:

(1) a face of a σ_α is a σ_β;
(2) $\sigma_\alpha \cap \sigma_\beta$ is a common face of σ_α and σ_β;
(3) $\gamma \sigma_\alpha$ is a σ_β, for all $\gamma \in \Gamma$;
(4) modulo Γ, there are only a finite number of σ_α;
(5) $C = \bigcup_\alpha (\sigma_\alpha \cap C)$.

The result that such decompositions exist is proven at the end of Section 4.

Section 1 and part of Section 2 follows closely some notes of Deligne (unpublished). The first two sections describe the work of Vinberg and Koecher on self-adjoint homogeneous cones and their connection with Jordan algebras. In Section 1 the cones appear by themselves, while in Section 2, after some background on Jordan algebras, the link between cone and algebra is explained in detail. Several small facts for later use also find proof in Section 2.

Section 3 analyzes the structure of the cone, especially with respect to the way its boundary breaks up into the disjoint union of lower-dimensional self-adjoint homogeneous cones, called boundary components. The Peirce decomposition is the main tool used here. We derive a correspondence between maximal split tori in the automorphism group of the cone and maximal strictly commutative rational subalgebras of the Jordan algebra. An easy computation then yields an explicit description of the rational root-space structure of the group of automorphisms.

We use the results of Section 3 to prove in Section 4 the main reduction theorem, roughly that Siegel sets and rational polyhedral cones are "cofinal" with respect to inclusion. In Section 5 we define "cores" and "co-cores" with two purposes: (1) the theorem near the beginning will be used later in comparing the topologies of various compactifications of the locally symmetric varieties D/Γ; and (2) the construction of explicit polyhedral fundamental domains in the spirit of Voronoi. If L is a lattice in V giving the $\mathbb{Q}$-structure on V, then a typical core is the closed convex hull of $C \cap L$. A typical co-core is the closed convex hull of $(\overline{C} \cap L) \setminus \{0\}$. By taking the cones over the faces of a co-core, one obtains a Γ-admissible polyhedral decomposition of the cone. Everything in this section uses the existence of some polyhedral fundamental domain, proved in Section 4.

1 Homogeneous self-adjoint cones

1.1

Let V be a finite-dimensional real vector space. We call $C \subset V \setminus \{0\}$ a *cone* if C is open and if $\mathbb{R}_{>0}C = C$. We say that C is *non-degenerate* if $\overline{C}$ does not contain an entire straight line. Another expression for this property is given as follows. Let $\overline{C}^* \subset V^*$ be the set of linear forms, ≥ 0 on C. The dual cone C^* is the interior of $\overline{C}^* \setminus \{0\}$. Then a convex cone C is non-degenerate if and only if $C^* \neq \emptyset$. One always has $C \subset C^{**}$ with equality if and only if C is convex non-degenerate (or $C = V \setminus \{0\}$). We say that C is *self-dual* (or *self-adjoint*) if there exists a positive-definite form on V such that the resulting isomorphism between V and V^* transforms C into C^*.

1.2

Let $C \subset V$ be a convex non-degenerate cone, with $C^* \subset V^*$ its dual. Let $G = \operatorname{Aut}(C,V)$ be the group of linear transformations of V which preserve C. We are going to define the *characteristic function* of C. Fix dual Haar measures dx, dx^* on V and V^*.

Proposition 1.1 *The expression*

$$\varphi(x) = \int_{C^*} e^{-\langle x, x^* \rangle} \, dx^*$$

defines a real-valued function on C.

Proof Fix a point $x \in C$. The following condition defines a Haar measure $\mathrm{d}x_1^*$ on the hyperplane $H_\alpha = \{x^* \mid \langle x, x^* \rangle = \alpha\} \subset V^*$: For any continuous function f of compact support on V^* we have

$$\int_{V^*} f(x^*)\mathrm{d}x^* = \int_{-\infty}^{\infty} \mathrm{d}\alpha \int_{H_\alpha} f(x_1^*)\mathrm{d}x_1^* .$$

We can thus write

$$\varphi(x) = \int_0^{\infty} e^{-\alpha}\mathrm{d}\alpha \int_{H_\alpha \cap C^*} \mathrm{d}x_1^* .$$

But, as is easily seen, $H_\alpha \cap \overline{C^*}$ is compact; thus the volume $v(\alpha)$ of $H_\alpha \cap C^*$ is finite. Since H_α is obtained from H_1 by a homothety with coefficient α,

$$v(\alpha) = \alpha^{n-1}v(1) .$$

So $\varphi(x) = v(1) \int_0^{\infty} e^{-\alpha}\alpha^{n-1}\,\mathrm{d}\alpha < \infty.$ $\qquad\square$

Remark 1.2 The quantity $\varphi(x)$ is canonical up to multiplication by a constant (depending on the choice of the Haar measures).

Lemma 1.3 *We have* $\varphi(gx) = \frac{1}{\det(g)}\varphi(x)$. *In particular, the measure*

$$\mu = \varphi(x)\mathrm{d}x$$

on C is G-invariant.

Proof Indeed,

$$\varphi(gx) = \int_{C^*} e^{-\langle gx, x^* \rangle}\,\mathrm{d}x^* = \int_{C^*} e^{-\langle x, {}^t g x^* \rangle}\,\mathrm{d}x^* .$$

Now make a change of variables. $\qquad\square$

In particular, for $\lambda \in \mathbb{R}_{>0}$, we have

$$\varphi(\lambda x) = \frac{1}{\lambda^n}\varphi(x) ,$$

where $n = \dim V$.

Recall that a continuous function f on a convex subset M of affine space is (strictly) convex if, for all $x_1, x_2 \in M$, for any point x on the interval joining x_1 and x_2 and dividing it into the ratio $p : q$ (where $p, q > 0$ and $p + q = 1$), we have $f(x) < pf(x_1) + qf(x_2)$. If M is open and $f \in C^2$, a sufficient condition for f to be convex is that the quadratic form $d^2 f$ be positive-definite at all points of M.

Proposition 1.4

  (i) $\log \varphi$ *is convex;*
  (ii) φ *is convex.*

Proof We have

$$d\log \varphi = \frac{d\varphi}{\varphi}, \quad d^2 \log \varphi = \frac{d^2\varphi}{\varphi} - \left(\frac{d\varphi}{\varphi}\right)^2 .$$

From this it follows that it suffices to show that $d^2 \log \varphi$ is positive-definite. To prove this, we calculate, for $x \in C$ and $a \in V = T_x(C)$:

$$(d\varphi(x))(a) = -\int\limits_{C^*} e^{-\langle x,x^*\rangle} \langle a,x^*\rangle \, dx^*$$

$$(d^2\varphi(x))(a) = \int\limits_{C^*} e^{-\langle x,x^*\rangle} \langle a,x^*\rangle^2 \, dx^* .$$

Put $F(x^*) = e^{-\frac{1}{2}\langle x,x^*\rangle}$, $G(x^*) = e^{-\frac{1}{2}\langle x,x^*\rangle}\langle a,x^*\rangle$. Then, for $a \neq 0$,

$$(d^2 \log \varphi(x))(a) = \frac{1}{(\varphi(x))^2}\left[\int\limits_{C^*} F^2 \, dx^* \int\limits_{C^*} G^2 \, dx^* - \left(\int\limits_{C^*} FG \, dx^*\right)^2\right] > 0,$$

because F and G are not proportional functions. $\square$

As a consequence of this proposition we have that

$$g = \frac{d^2 \log \varphi(x)}{d^2 x} \, dx^2$$

defines a G-invariant Riemannian metric on C.

Example 1.5 Let $V = \mathbb{R}$, $C = \mathbb{R}_{>0} \subset \mathbb{R}$. Then $\varphi(x) = \frac{1}{x}$ and the Riemannian metric g on $\mathbb{R}_{>0}$ is $g = \frac{dx^2}{x^2}$.

Proposition 1.6 φ *goes to infinity upon approach of a point of the boundary of* C.

Proof Indeed, if x_k, $k = 1, 2, \ldots$, converges to $x_0 \in \partial C$, then $f_k(x') = e^{-\langle x_k, x'\rangle}$ converges to $f_0(x') = e^{-\langle x_0, x'\rangle}$ uniformly on any bounded set in V^*. So, it suffices to show that

$$\int\limits_{C^*} f_0(x^*)\mathrm{d}x^*$$

is a divergent integral.

Let $x_0^* \in \overline{C^*}$ with $x_0^* \neq 0$, $\langle x_0, x_0^* \rangle = 0$. Take a small ball K lying entirely in C^* and consider the set $L = K + \mathbb{R}_{\geq 0} x_0^* \subset C^*$. Let $c = \min\limits_{x^* \in K} f_0(x^*)$. Then $c > 0$ and $c = \min\limits_{x^* \in L} f_0(x^*)$. So we have

$$\int\limits_{C^*} f_0(x^*)\mathrm{d}x^* \geq \int\limits_{L} f_0(x^*)\mathrm{d}x^* \geq c \int\limits_{L} \mathrm{d}x^* \ .$$

The final expression is indeed infinite since L has infinite volume. $\qquad\square$

1.3

Proposition 1.7 *Let C be a convex non-degenerate cone in V and let $G = \mathrm{Aut}(C,V)$. Then G is a Lie group, the stabilizer of $e \in C$ in G is compact and maximally compact if C is homogeneous.*

Proof The first two assertions follow from the existence of the G-invariant Riemannian metric g on C. The final one follows from the fact that any compact subgroup of G has a fixed point in C, as one proves by the usual averaging method. $\qquad\square$

Proposition 1.8 *Let $C \subset V$ be an open set in V, such that $C^* \neq \emptyset$. Let $G \subseteq \mathrm{Aut}(C,V)$. If there is $e \in C$ such that $G \cdot e$ is open in V, then $G \cdot e = C$ and C is a convex homogeneous cone under G.*

Corollary 1.9 *Let $C \subset V$ be a convex homogeneous cone under G. Then $C^* \subset V^*$ is a convex homogeneous cone under the dual group G^*.*

Proof (*of Proposition* 1.8) First, $G \cdot e \subset C \subset C^{**}$, and C^{**} still satisfies the hypotheses of the proposition. So we can suppose that C is non-degenerate convex. Let $r > 0$ such that the ball with radius r around e (with respect to the Riemannian metric g) is contained in $G \cdot e$. For every sequence of points x_i such that $d(x_{i-1}, x_i) < r$ and $x_0 = e$ we get $x_n \in G \cdot e$ (because g is G-invariant). Since C is connected, $C = G \cdot e$. $\qquad\square$

Proposition 1.10 *Let $\mathscr{G}$ be a reductive connected algebraic group and let (V,ρ) be a representation of $\mathscr{G}$ all defined over $\mathbb{R}$. Let $e \in V$ and $G = \mathscr{G}(\mathbb{R})^o$.*

Suppose that the stabilizer in G of e is maximal compact and that G · e is open. Then G · e is a homogeneous self-adjoint cone. Conversely, the automorphism group of a self-adjoint cone is reductive.

Proof Let K be the stabilizer of e. Let σ be the Cartan involution and let B be a positive-definite symmetric bilinear form on V such that $\rho(\sigma(g)) = {}^t\rho(g)^{-1}$.

The corresponding Cartan decompositions are given by

$$\mathrm{Lie}\,(G) = \mathfrak{k} \oplus \mathfrak{p}\,,\ G = K \cdot \exp(\mathfrak{p}) = K \cdot P\,.$$

Let $g_1, g_2 \in G$ and let $\sigma(g_2)^{-1}g_1 = p^2k$, with $k \in K$, $p \in P$. Then

$$\langle g_1 e, g_2 e \rangle = \langle \sigma(g_2)^{-1}g_1 e, e \rangle = \langle p^2 e, e \rangle = \langle pe, pe \rangle > 0\,.$$

Thus, identifying V with V^* by B, we have

$$G \cdot e \subset (G \cdot e)^*\,.$$

But $(G \cdot e)^*$ is stable under G: for $x \in (G \cdot e)^*$, $h \in G$, and $g \in G$,

$$\langle ge, hx \rangle = \langle \sigma(h)ge, x \rangle > 0\,.$$

Then Proposition 1.8 applied to $(G \cdot e)^*$ shows that $G \cdot e = (G \cdot e)^*$. Conversely, if C is self-adjoint, then $\mathrm{Aut}\,(C, V)$ is stable under $g \longmapsto {}^tg$ (transpose with respect to the positive-definite metric on V making C self-adjoint), and consequently G is reductive. $\qquad\square$

Remark 1.11 It may appear at present that the concept of a self-adjoint homogeneous cone is very wide. This is not so. Any convex cone C splits in a unique way into the direct sum of indecomposable convex cones C_i. Then C is homogeneous (respectively self-adjoint) if and only if all the C_i are.

Now the indecomposable self-adjoint homogeneous cones have been completely classified. They are:

(1) the cone of positive-definite symmetric matrices;
(2) the cone of positive-definite hermitian complex matrices;
(3) the cone of positive-definite hermitian quaternion matrices;
(4) the spherical cone in $\mathbb{R}^{n+1}$ of $(x_0, \ldots, x_n)$ with

$$x_0 > \sqrt{x_1^2 + \ldots + x_n^2}\,;$$

(5) the 27-dimensional cone of positive-definite hermitian octavic matrices of third order.

Here (1)–(3) are classical, (4) is semi-classical, and (5) is exceptional.

2 Jordan algebras

Jordan algebras arose in quantum mechanics from the desire to make operators commute, at the expense of their associativity. Their usefulness for our purposes comes from the one-to-one correspondence between self-adjoint homogeneous cones and formally real Jordan algebras. For comprehensive references, see [5], [7], [8], and [13].

Definition 2.1 A *Jordan algebra A* over a field k (of char $k \neq 2$) is a finite-dimensional (non-associative) algebra with unit element p such that, for all $a, b \in A$

(i) $a \cdot b = b \cdot a$;
(ii) $a^2 \cdot (b \cdot a) = (a^2 \cdot b) \cdot a$, where $a^2 = a \cdot a$.

Although not strictly necessary, we will assume for convenience that char $k \neq 2, 3$. In fact, we will only need the case char $k = 0$ in applications.

Start with a Jordan algebra A over k. For $a \in A$, we denote by L_a multiplication by a:

$$L_a(x) = x \cdot a .$$

Set

$$S(a,b,c,d) = (a \cdot b) \cdot (c \cdot d) + (a \cdot c) \cdot (b \cdot d) + (a \cdot d) \cdot (b \cdot c) .$$

By (i) this is symmetric in a, b, c, d. Set

$$T(a;b,c,d) = (a \cdot (b \cdot c)) \cdot d + (a \cdot (c \cdot d)) \cdot b + (a \cdot (d \cdot b)) \cdot c .$$

Again using (i), one sees that T is symmetric in b, c, d. Then

$$S(a,b,c,d) = T(a;b,c,d) .$$

Indeed, for $b = c = d$ this identity reduces to (ii); the general case follows from this by polarization, since 6 is invertible in k. In particular, T is symmetric. This can be expressed as

$$T(a;b,c,d) = T(b;a,c,d) ,$$

which can be reformulated by saying that $D_{a,b} = [L_a, L_b]$ is a derivation, i.e.,

$$D_{a,b}(c \cdot d) = c \cdot D_{a,b}(d) + d \cdot D_{a,b}(c) .$$

Theorem 2.2 (Vinberg) *Let A be an algebra with unit over a field k of characteristic $\neq 2, 3$. Let $t : A \longrightarrow k$ be a linear form, the "trace", such that*

(a) *A is commutative and $D_{a,b}$ is a derivation;*
(b) $t([L_a, L_b](c)) = 0$;

(c) $t(a \cdot b)$ *is a non-degenerate bilinear form.*

   Then A is a Jordan algebra.

Proof It suffices to prove $S = T$. Set

$$\{a,b,c,d;e\} = t((S(a,b,c,d) - T(a;b,c,d)) \cdot e) .$$

We have just seen that assumption (a) is equivalent to T being symmetric. Hence $\{a,b,c,d;e\}$ is symmetric in a,b,c,d.

   Observe that

$$S(a,b,c,d) - T(a;b,c,d) = (D_{a,b\cdot c} + D_{b,c\cdot a} + D_{c,a\cdot b})(d) . \qquad (2.1)$$

It follows with (a) and (b) that $\{a,b,c,d;e\}$ is anti-symmetric in d and e: indeed,

$$\{a,b,c,d;e\} + \{a,b,c,e;d\} = t((D_{a,b\cdot c} + D_{b,c\cdot a} + D_{c,a\cdot b})(d \cdot e)) = 0 .$$

But this implies that the expression $\{\ \}$ is identically 0: indeed, $\{a,b,c,d;e\} = -\{a,b,c,e;d\}$; but, whereas the L.H.S. is *symmetric* in c and d, the R.H.S. is *anti-symmetric* in c and d.

   Since the element e is arbitrary, assumption (c) now implies the assertion. $\qquad\square$

Definition 2.3 If A is a Jordan algebra, $x,y \in A$, then x and y are said to *commute strictly (or strongly)* if L_x and L_y commute. A subalgebra $B \subset A$ is *strongly associative* if any two elements $x,y \in B$ strongly commute. (This makes sense because it implies that, for all $x,y,z \in B$,

$$x(zy) = L_x L_y(z) = L_y L_x(z) = (xz)y .)$$

Proposition 2.4 *For all* $x \in A$, *define* x^n *inductively by* $x^n = x \cdot x^{n-1}$. *Then:*

(a) x^i, x^j *strongly commute for all* $i,j \geq 0$;
(b) $x^i \cdot x^j = x^{i+j}$;
(c) $p,x,x^2,\ldots$ *span a strongly associative subalgebra* $k[x] \subset A$.

Proof Interpret $S = T$ by taking d as a variable. It then says:

$$L_{a(bc)} = L_{ab} \cdot L_c + L_{ac} \cdot L_b + L_{bc} \cdot L_a - L_b \cdot L_a \cdot L_c - L_c \cdot L_a \cdot L_b . \qquad (2.2)$$

In particular, if $a = b = x$, $c = x^k$, this says:

$$L_{x^{k+2}} = 2L_{x^{k+1}} \cdot L_x + L_{x^2} \cdot L_{x^k} - L_{x^k} \cdot L_x^2 - L_x^2 \cdot L_{x^k} , \qquad (2.3)$$

hence all L_{x^k} belong to the subalgebra of $\mathrm{Hom}_k(A,A)$ generated by L_x, L_{x^2}. But condition (ii) in the definition of a Jordan algebra says that L_x, L_{x^2} commute. So all L_{x^k} commute. Finally

$$x^i \cdot x^j = L_{x^i}(L_{x^{j-1}}(x)) = L_{x^{j-1}}(L_{x^i}(x)) = x^{j-1} \cdot x^{i+1},$$

which, by a simple induction, proves (b), and then (c) is obvious. $\qquad\square$

A useful identity is:

Lemma 2.5 *For all* $x, y \in A$, x *and* y^2 *strictly commute if and only if* $x \cdot y$ *and* y *strictly commute.*

Proof In the equation $S = T$, take $b = c = y$, $a = x$ to find $2[L_{x \cdot y}, L_y] = [L_x, L_{y^2}]$. $\qquad\square$

Definition 2.6 For any $x \in A$, we say that $y \in A$ is the *inverse* of x if x and y strictly commute and $x \cdot y = p$.

Note that an inverse is unique: if y_1, y_2 were inverses

$$\begin{aligned} y_1 &= y_1 \cdot (x \cdot y_2) \\ &= x \cdot (y_1 \cdot y_2) \ (x, y_1 \text{ strongly commute}) \\ &= y_2 \cdot (x \cdot y_1) \ (x, y_2 \text{ strongly commute}) \\ &= y_2 \,. \end{aligned}$$

Lemma 2.7 *If* x *has an inverse* x^{-1}, *then* $x^{-1} \in k[x]$.

Proof In fact by Lemma 2.5, x^{-1} and x^2 strictly commute. Since L_x^k, for $k \geq 3$, are polynomials in L_x and L_x^2, in fact x^{-1} and x^k strictly commute for all $k \geq 1$. Then if x were a zero divisor in $k[x]$, i.e., $x \cdot f(x) = 0$, for $f(x) \neq 0$, one would get

$$0 = x^{-1} \cdot (x \cdot f(x)) = f(x) \cdot (x \cdot x^{-1}) = f(x),$$

which is a contradiction. But $k[x]$ is finite-dimensional, so every non-zero divisor has an inverse in $k[x]$. $\qquad\square$

Corollary 2.8 *If* x *has an inverse, then* x^2 *has an inverse, and* $(x^2)^{-1} = (x^{-1})^2$.

We now take up the Peirce decomposition of a Jordan algebra with respect to an idempotent $\varepsilon \in A$. By the recursion formula (2.3), L_ε satisfies the identity $\varphi(L_\varepsilon) = 0$, where

$$\varphi(T) = 2T^3 - 3T^2 + T.$$

Define

$$\varphi_0(T) = 2T^2 - 3T + 1 \,,$$

$$\varphi_{\frac{1}{2}}(T) = -4T^2 + 4T \,,$$

$$\varphi_1(T) = 2T^2 - T \,.$$

Then $\varphi_0 + \varphi_{\frac{1}{2}} + \varphi_1 = 1$, and φ divides $(T - \lambda)\varphi_\lambda(T)$ for $\lambda = 0, \frac{1}{2}, 1$. If $x \in A$, we get $x = x_0 + x_{\frac{1}{2}} + x_1$ with $x_\lambda = \varphi_\lambda(L_\varepsilon)x$, and $\varepsilon \cdot x_\lambda = \lambda x_\lambda$.

Writing $A_\lambda = \varphi_\lambda(L_\varepsilon)A$, we have

$$A = A_0 \oplus A_{\frac{1}{2}} \oplus A_1$$

and A_λ is the λ-eigenspace for L_ε. We write $A_\lambda(\varepsilon)$ for A_λ, if necessary to avoid confusion. This is called the *Peirce decomposition* of A. Note that Lemma 2.5 with $x \in A_0$, $y = \varepsilon$, implies that ε and A_0 strongly commute; hence L_x, for all $x \in A_0$, preserves the eigenspaces of ε. Similarly, Lemma 2.5 with $x \in A_1$, $y = p - \varepsilon$, implies that ε and A_1 strongly commute; hence L_x, for all $x \in A_1$, preserves the eigenspaces of ε.

In particular, if $x \in A_0$, $y \in A_1$, then $x \cdot y \in A_0 \cap A_1 = (0)$. In fact, recall that $S = T$ and hence, by (2.1),

$$D_{a,b \cdot c} + D_{b,c \cdot a} + D_{c,a \cdot b} = 0 \,.$$

Take $a = x$, $b = y$, $c = \varepsilon$ to see that x and y strongly commute. If one takes $a = \varepsilon$ and $b, c \in A_{\frac{1}{2}}$, and evaluates at $d = \varepsilon$, we see that $\varepsilon \cdot (\varepsilon \cdot (b \cdot c)) = \varepsilon \cdot (b \cdot c)$, i.e., $\varphi_{\frac{1}{2}}(L_\varepsilon)(b \cdot c) = 0$. Thus $b \cdot c \in A_0 + A_1$. We summarize this in the following multiplication table:

$\mu \backslash \lambda$	0	1/2	1
0	0	1/2	product zero, strongly commute
1/2	1/2	0 + 1	1/2
1	product zero, strongly commute	1/2	1

Here if $x \in A_\mu$, $y \in A_\lambda$, then $x \cdot y \in A_\nu$, where ν is graphed as a function of μ, λ in the multiplication table above.

Corollary 2.9 *With respect to the inner product* $\langle x, y \rangle = \mathrm{Tr}(L_{x \cdot y})$, *the spaces* A_0, $A_{1/2}$, *and* A_1 *are perpendicular.*

We call a set of idempotents $\varepsilon_1, \ldots, \varepsilon_n$ *mutually orthogonal* if $\varepsilon_i \cdot \varepsilon_j = \delta_{ij} \cdot \varepsilon_i$, for $1 \le i, j \le n$. By Lemma 2.5 with $x = \varepsilon_i$, $y = \varepsilon_j$, the ε_i all strictly commute, and generate a strongly associative subalgebra $W = \bigoplus_{i=1}^n k\varepsilon_i$.

We say $\{\varepsilon_1, \ldots, \varepsilon_n\}$ is a *complete* set of mutually orthogonal idempotents if $\varepsilon_1 + \cdots + \varepsilon_n = p$. In this case, we can have a Peirce decomposition for all the ε_i at once because the L_{ε_i} commute with each other. We write $A = \bigoplus A_v$, where the A_v are simultaneous eigenspaces for $L_{\varepsilon_1}, \ldots, L_{\varepsilon_n}$. Because $L_{\varepsilon_1} + \cdots + L_{\varepsilon_n} = \mathrm{id}$, and the various eigenvalues are all 0, $\frac{1}{2}$, or 1, we end up with $A = \bigoplus_{i \leq j} A_{ij}$ where, if $x \in A_{ij}$,

$$\varepsilon_k \cdot x = \tfrac{1}{2}(\delta_{ki} + \delta_{kj})x \, .$$

Also, if $x \in A_{ij}$, $y \subset A_{k\ell}$, and $(i, j) \neq (k, \ell)$, then $\mathrm{Tr}(L_{x \cdot y}) = 0$.

We get the finest decomposition from a *maximal* set of mutually orthogonal idempotents, that is $\{\varepsilon_1, \cdots, \varepsilon_n\}$ such that the n is maximal for sets of orthogonal idempotents. Note that a maximal set is complete: if $f = \varepsilon_1 + \cdots + \varepsilon_n \neq p$, then $\{\varepsilon_1, \ldots, \varepsilon_n, p - f\}$ is a larger set of mutually orthogonal idempotents.

Definition 2.10 If $k = \mathbb{R}$, then a Jordan algebra A is said to be *formally real* if

$$x^2 + y^2 = 0 \Longrightarrow x = y = 0 \, .$$

If B is a subalgebra of A, it too is formally real. An associative Jordan algebra is formally real if and only if it is isomorphic to a product of copies of $\mathbb{R}$ with componentwise multiplication.

Definition 2.11 A *positive-definite trace form* $t : A \longrightarrow \mathbb{R}$ on a real Jordan algebra is a linear map such that $t(x^2) > 0$ for all $x \in A$, $x \neq 0$.

Proposition 2.12 *Let A be a real Jordan algebra. The following are equivalent:*

(a) *A is formally real;*
(b) *A has a positive-definite trace form t;*
(c) *$t(x) = \mathrm{Tr}(L_x)$ is a positive-definite trace form (hence $\langle x, y \rangle = t(x \cdot y)$ is a positive-definite inner product).*

Proof That (c) $\Longrightarrow$ (b) $\Longrightarrow$ (a) is clear. Assume (a). Then, for all $x \in A$, $\mathbb{R}[x] \cong \bigoplus_{i=1}^{N} \mathbb{R}\varepsilon_i$, where ε_i are idempotents. If $x = \sum \lambda_i \varepsilon_i$, it follows that $x^2 = \sum \lambda_i^2 \varepsilon_i$, hence

$$\mathrm{Tr}(L_{x^2}) = \sum \lambda_i^2 \mathrm{Tr} L_{\varepsilon_i},$$

and, by the Peirce decomposition, $\mathrm{Tr} L_{\varepsilon_i} > 0$. $\qquad \square$

The formally real Jordan algebras were classified in 1934 by P. Jordan, Wigner, and von Neumann. This gives the classification of self-adjoint homogeneous cones in Section 1, in light of the following one-to-one correspondence between the two classes of objects. See [5], Ch. 11.

Let $C \subset V$ be a homogeneous self-adjoint cone and let $e \in C$, $G = \mathrm{Aut}\,(C,V)$, $\mathfrak{g} = \mathrm{Lie}\,(G)$. Let $K \subset G$ be the stabilizer of e in G and $\mathfrak{k}$ its Lie algebra. Let σ be the corresponding Cartan involution with $\mathfrak{g} = \mathfrak{k} \oplus \mathfrak{p}$ being the Cartan decomposition, i.e.,

$$\sigma = \left\{ \begin{array}{ll} 1 & \text{on } \mathfrak{k}\,, \\ -1 & \text{on } \mathfrak{p}\,. \end{array} \right.$$

Now $\mathfrak{g}$ acts on V by taking the differential at 0 of the G-action, $\mathfrak{k}$ is the stabilizer of e, and the map

$$\pi \longmapsto \pi \cdot e$$

from $\mathfrak{p}$ into V is bijective. We define an algebra structure on $\mathfrak{p}$ by setting

$$(\pi \cdot \pi') \cdot e = \pi \cdot (\pi' \cdot e)\,.$$

Let $p \in \mathfrak{p}$ be such that $p \cdot e = e$.

One verifies that:

(1) $\pi \cdot \pi' = \pi' \cdot \pi$ (indeed $(\pi \cdot \pi' - \pi' \cdot \pi) \cdot e = [\pi, \pi'] \cdot e = 0$ since $[\pi, \pi'] \in \mathfrak{k}$);
(2) clearly p is a unit element;
(3) $[L_\pi, L_{\pi'}] = \mathrm{ad}\,[\pi, \pi']$.

Furthermore, let $\langle \cdot, \cdot \rangle$ be a positive-definite form on V "which makes C self-adjoint" and such that σ is the restriction to $\mathfrak{g} \subset \mathrm{End}\,(V)$ of $u \longmapsto -{}^t u$ (such forms exist, see Proposition 1.10 and its proof). Define the linear form t on $\mathfrak{p}$ by

$$t(\pi) = \langle \pi \cdot e, e \rangle\,.$$

Then

$$\begin{aligned} t([L_a, L_b](c)) &= \langle [[a,b],c] \cdot e, e \rangle \text{ (by (3) above)} \\ &= \langle [a,b] \cdot c \cdot e, e \rangle - \langle c \cdot [a,b] \cdot e, e \rangle \\ &= -\langle c \cdot e, [a,b] \cdot e \rangle - \langle c \cdot [a,b] \cdot e, e \rangle \text{ (since } [a,b] \in \mathfrak{k}) \\ &= 0 \text{ (since } \mathfrak{k} \cdot e = 0)\,. \end{aligned}$$

Also, for $\pi \neq 0$,

$$t(\pi \cdot \pi) = \langle \pi \cdot \pi \cdot e, e \rangle = \langle \pi \cdot e, \pi \cdot e \rangle > 0\,.$$

We thus see that $(\mathfrak{p}, \cdot, t)$ satisfies the hypotheses of Vinberg's theorem (Theorem 2.2 above), which implies that $(\mathfrak{p}, \cdot)$ is a Jordan algebra $(A, \cdot)$.

Going in the converse direction, the cone C can be recaptured in the following way:

$$C = \{ \exp(\rho(a)) \cdot e \mid a \in A \}\,,$$

where $\rho : A = \mathfrak{p} \longrightarrow \mathrm{End}\,(V)$ comes from the representation of $\mathfrak{g}$ on V.

Identifying A with V, $\exp(\rho(a))\cdot e = \exp_J(a)$, where $\exp_J(a) = 1 + a + \frac{a^2}{2} + \cdots$ is the Jordan algebra exponential which may be calculated in the algebra $\mathbb{R}[a]$ generated by a in A. Since $\mathbb{R}[a]$ is an associative formally real Jordan algebra, it is a product of copies of $\mathbb{R}$. In particular, the exponentials in A are also the squares of invertible elements. So, we can also describe the cone as the set of squares of invertible elements.

Finally, starting from a formally real Jordan algebra A, define

$$\mathfrak{g} = \operatorname{Der} A \oplus A$$

(here $\operatorname{Der} A$ means derivations of A for its Jordan multiplication). Now $\mathfrak{g}$ is a Lie algebra if we define

$$[D_1, D_2] = D_1 D_2 - D_2 D_1 \, ,$$
$$[D, x] = D(x) \, ,$$
$$[x, y] = L_x L_y - L_y L_x \, ,$$

for $x, y \in A$ and $D, D_1, D_2 \in \operatorname{Der} A$.

In fact, $\mathfrak{g}$ is a subspace of $\operatorname{Hom}_{\mathbb{R}}(A, A)$ if (D, x) acts on A by $y \longrightarrow Dy + x \cdot y$. One sees immediately that $[\cdot, \cdot]$ on $\mathfrak{g}$ is just the commutator in $\operatorname{Hom}_{\mathbb{R}}(A, A)$. Let $G \subset \operatorname{GL}(A)$ be the corresponding Lie group, and let $K \subset G$ be the subgroup corresponding to $\operatorname{Der} A$. Putting the inner product $\langle x, y \rangle = \operatorname{Tr}(L_{x \cdot y})$ on A, and defining $\sigma : \mathfrak{g} \longrightarrow \mathfrak{g}$ to be $+1$ on $\operatorname{Der} A$, and -1 on A, one sees:

(a) exponentiating a derivation leads to an automorphism, so K acts on A by Jordan automorphisms; in particular

$$\langle kx, y \rangle = \langle x, k^{-1} y \rangle \text{ for all } k \in K \, ,$$

and hence

$$\langle Dx, y \rangle = -\langle x, Dy \rangle \text{ for all } D \in \operatorname{Der} A \, ;$$

(b) equation (2.2) may be written as

$$L_{a(b \cdot c)} - L_{b(a \cdot c)} = [L_a L_b, L_c] + [L_c, L_b L_a] \, ,$$

hence

$$\operatorname{Tr} L_{a(b \cdot c)} = \operatorname{Tr} L_{b(a \cdot c)} \, ,$$

and therefore

$$\langle L_z x, y \rangle = \langle x, L_z y \rangle$$

for all $z \in A$.

Thus $\sigma g = -{}^t g$, for all $g \in \mathfrak{g}$; it follows that $\mathfrak{g}$ is a reductive sub Lie algebra of $\mathrm{Hom}(A,A)$ and σ is a Cartan involution. Finally, since $\mathfrak{g} \cdot p = A$, the orbit $G \cdot p$ is open in A; Proposition 1.8 shows that $C = G \cdot p$ is a homogeneous self-adjoint cone and $G \subset \mathrm{Aut}(C,V)^o$. Here, as above, $G \cdot p = (\exp A \cdot K) \cdot p = \{\exp_J(a) \mid a \in A\}$. Moreover, writing $G' = \mathrm{Aut}(C,V)^o$, we see $\mathrm{Lie}\, G' = \mathfrak{k} + \mathfrak{p}$ as usual. We have three Lie algebras:

$$(\text{Inner Der}\, A) + A \subset \mathrm{Der}\, A + A \subset \mathfrak{k} + \mathfrak{p}$$

(where inner derivations are the derivations $[L_x, L_y]$, for $x, y \in A$). Since $\pi \longmapsto \pi \cdot p$ defines $\mathfrak{p} \xrightarrow{\sim} A$, and since, for any real reductive Lie group G without compact factors, $\mathfrak{k} = [\mathfrak{p}, \mathfrak{p}]$, it follows that all three are equal! This proves:

Theorem 2.13 *Given a real vector space V and a point $p \in V$, there is a one–one correspondence between homogeneous self-adjoint cones $C \subset V$, with $p \in C$, and formally real Jordan algebra structures $(V, \cdot)$ on V, with identity p, given by:*

(a)

$$C = \{\exp(a) \mid a \in V\} = \{a^2 \mid a \text{ invertible in } V\}$$

and this cone is self-adjoint w.r.t. $\langle x, y \rangle = \mathrm{Tr}(L_{x \cdot y})$;

(b)

$$\mathrm{Lie}(\mathrm{Aut}(C,V)) = \mathrm{Der}\, V \oplus \{L_x \mid x \in V\}$$

(orthogonal with respect to the Killing form on $\mathrm{Lie}\,\mathrm{Aut}(C,V)$).

$\square$

This description of C is useful for describing the *symmetry* on C, i.e., picking a base point $p \in C$, $C \cong G/K$, and the Cartan involution $\sigma : G \longrightarrow G$ induces $\overline{\sigma} : C \longrightarrow C$: we claim that, in this Jordan algebra structure on V, $\overline{\sigma}(x) = x^{-1}$. To see this, suppose $x = \exp_J(a)$. Then

$$x = \exp_J a = (\exp L_a)(p) \,.$$

Now $L_a \in \mathfrak{p}$, hence $\exp L_a \in \exp \mathfrak{p} \subset G$, and hence $\sigma(\exp L_a) = (\exp L_a)^{-1} = \exp(-L_a)$. Thus

$$\overline{\sigma}(x) = \sigma(\exp L_a)(p) = \exp(-L_a)(p) = \exp_J(-a) = \exp_J(a)^{-1} = x^{-1} \,.$$

We mention one more fact about this correspondence. To derive a Jordan algebra from a cone $C \subset V$, we had to choose a basepoint p. Suppose we choose a new basepoint $p' = gp$, with $g \in \exp \mathfrak{p}$. We claim that the new Jordan algebra structure on V is isomorphic to the old one. If $\mathfrak{g} = \mathfrak{k} \oplus \mathfrak{p}$ is the Cartan decomposition corresponding to p, the new one for p' is $\mathfrak{g} = \mathrm{Ad}\,g(\mathfrak{k}) \oplus \mathrm{Ad}\,g(\mathfrak{p})$.

Say $u_1, u_2 \in V$ and $u_i = \pi_i p$ with $\pi_i \in \mathfrak{p}$. Then $gu_i = (g\pi_i g^{-1})p'$. Therefore, if $\perp$ denotes the new Jordan multiplication, we have

$$gu_1 \perp gu_2 = (g\pi_1 g^{-1})(g\pi_2 g^{-1})p' = g(\pi_1(\pi_2 p)) = g(u_1 \cdot u_2) \ .$$

Hence g induces an isomorphism $V \longrightarrow V_\perp$.

Technically $V_\perp$ is called the *mutation* of V by p'. See [5], Ch. 5, or [8], pp. 67ff.

3 Boundary components and Peirce decompositions

3.1

Let V be a real vector space defined over $\mathbb{Q}$ and let $C \subset V$ be a self-adjoint homogeneous cone. Let $G = \operatorname{Aut}(C, V)^o$. Fix once and for all a rational point $p \in C$ to be the basepoint, and let K be the stabilizer of p in G. Since K is maximal compact there corresponds to it a Cartan involution σ of $\operatorname{Lie}(G)$, so that $\operatorname{Lie}(G) = \operatorname{Lie}(K) + \mathfrak{p}$, where $\mathfrak{p}$ is the -1-eigenspace of σ and $\operatorname{Lie}(K)$ is the $+1$-eigenspace. There is an inner product $\langle \cdot, \cdot \rangle$ on V with respect to which $\exp \sigma g = {}^t(\exp g)^{-1}$, where ^{t} denotes the adjoint. Thus, elements of K are orthogonal and elements of $\exp \mathfrak{p}$ are self-adjoint w.r.t. this scalar product. As we saw in Section 1, C is self-adjoint for any such inner product.

We assume that the subgroup $G \subset \operatorname{GL}(V)$ is defined over $\mathbb{Q}$. This strong assumption implies that $\operatorname{Lie}(K)$ and the Killing form are defined over $\mathbb{Q}$, hence also $\mathfrak{p}$ and σ. Therefore we may choose the scalar product so that it is defined over $\mathbb{Q}$ too. Then if L is a lattice in V giving the $\mathbb{Q}$-structure, L and the dual lattice L^* are commensurable.

Regard $\operatorname{Lie}(\operatorname{Aut} V)$ as $\operatorname{End}(V)$, so that $\mathfrak{p} \subset \operatorname{End}(V)$. Then we may identify $\mathfrak{p}$ with V by identifying π and πp for any $\pi \in \mathfrak{p}$. Thus $\mathfrak{p}$ acquires a Jordan algebra structure with Jordan multiplication $\cdot$ defined by $(\pi \cdot \pi')p = \pi(\pi' p)$. We can transfer this structure to V. When we think of the Jordan algebra we will use the symbol V, reserving $\mathfrak{p}$ for the Lie algebra guise. Now $\mathfrak{p}$ has a $\mathbb{Q}$-structure as a vector space which is identical with that of V. Therefore, V *as a Jordan algebra* is defined over $\mathbb{Q}$, since $\mathfrak{p}$ acts $\mathbb{Q}$-morphically on V, and the Jordan multiplication is just the pullback of this action via $\mathfrak{p} \xrightarrow{\sim} V$. Finally, V is formally real, that is, $x^2 + y^2 = 0 \Longrightarrow x = y = 0$.

Conversely, given any formally real Jordan algebra V defined over $\mathbb{Q}$, its identity element p will be rational. Putting

$$C = \{x^2 \mid x \in V \text{ is invertible}\},$$

we acquire a situation as described above if we let $\langle x, y \rangle = \operatorname{Tr}(L_{x \cdot y})$.

3.2

Fix any subfield k of $\mathbb{R}$, and take rational to mean k-rational.

Let $\varepsilon \in V$ be a rational idempotent. For any $x \in V$, denote by L_x the map $L_x(y) = x \cdot y$, for $y \in V$. Then L_ε is semi-simple and has eigenvalues 0, $\frac{1}{2}$, and 1. Let $V_i(\varepsilon)$, or V_i when no confusion is possible, denote the i-eigenspace. This gives the so-called *Peirce decomposition*

$$V = V_0 \oplus V_{\frac{1}{2}} \oplus V_1 \ ;$$

see Section 2. In particular,

$$V_0 \cdot V_1 = 0, V_0 \cdot V_{\frac{1}{2}} \subset V_{\frac{1}{2}}, V_1 \cdot V_{\frac{1}{2}} \subset V_{\frac{1}{2}}, V_{\frac{1}{2}} \cdot V_{\frac{1}{2}} \subset V_0 \oplus V_1 \ ,$$

and V_0, V_1 are sub Jordan algebras. Since p is the identity in V, also $p - \varepsilon$ is an idempotent, and $V_0(\varepsilon) = V_1(p - \varepsilon)$. The Peirce decomposition is defined over k since ε is rational.

Thus if $C_i = C_i(\varepsilon)$, $i = 0, 1$, is the cone of squares of invertible elements in V_i, then C_i is a self-adjoint homogeneous cone since V_i inherits formal-reality from V, and everything is compatibly defined over k. We call C_i a *rational boundary component* of C. Note that C and $\{0\}$ are, by definition, boundary components; we call them *improper* boundary components.

Note† that $\overline{C}$, the closure of C, is simply the set of squares in V.

To persuade ourselves that this is indeed a good notion of boundary component, we prove:

Proposition 3.1 *The closure $\overline{C}$ of C is the disjoint union of the real boundary components.*

Proof Say $y \in \overline{C}$ and $y = x^2$ for some $x \in V$. Then x generates a sub Jordan algebra W, consisting of polynomials in x with no constant term. Note that $W \subset \mathbb{R}[x]$ and, since $\mathbb{R}[x]$ is formally real, $\mathbb{R}[x]$ is isomorphic to a product of copies of $\mathbb{R}$ with componentwise multiplication. Hence also any (not necessarily unital) $\mathbb{R}$-subring of $\mathbb{R}[x]$ is isomorphic to a product of copies of $\mathbb{R}$ and, in particular, possesses a unit. Therefore W has a unit e. In V itself, e is an idempotent, and $W \subset V_1(e)$.

Now $e \in W$ means that $e = a_1 x + a_2 x^2 + \cdots + a_n x^n$ for some $a_i \in \mathbb{R}$. Square both sides to get

$$a_1^2 x^2 + 2a_1 a_2 x^3 + \cdots + a_n^2 x^{2n} = e \ ,$$

---

† In fact, let $S = (V \setminus \{0\})$ modulo homotheties, and $\varphi : S \longrightarrow S$ be $\varphi(x) = x^2$. Because V is formally real, φ is well-defined. The image of φ is closed since S is compact. Thus the set of squares Σ in V is closed, so $\Sigma \supset \overline{C}$. On the other hand, the set of invertible elements in V is an open dense set, so $\Sigma \subset \overline{C}$.

and therefore

$$(a_1^2 x + 2a_1 a_2 x^2 + \cdots)x = e \, .$$

Now $a_1^2 x + 2a_1 a_2 x^2 + \cdots$ is in W, so x is *invertible* in $V_1(e)$. Thus $y = x^2$ is the square of an invertible element in $V_1(e) = V_0(p - e)$ and thus is in some boundary component.

To show the boundary components do not overlap, suppose y is an invertible element in both $V_1(\varepsilon_1)$ and $V_1(\varepsilon_2)$. There exists $x_i \in V_1(\varepsilon_i)$ with $x_i \cdot y = \varepsilon_i$, $i = 1, 2$. In general, if y is an invertible element in a Jordan algebra, then its inverse is an element in the subalgebra W generated by y and p. Note that W is associative. Applying this to $V_1(\varepsilon_1)$ and $V_1(\varepsilon_2)$ gives $x_1, x_2 \in W$. Therefore $x_i \cdot y = \varepsilon_i \in W$ for $i = 1, 2$, and so $\varepsilon_1 = \varepsilon_1 \cdot \varepsilon_2 = \varepsilon_2$.   □

Recall that we have an inner product on V, defined by $\langle x, y \rangle = \mathrm{Tr}\,(L_{x \cdot y})$. The Peirce eigenspaces of L_ε are mutually orthogonal with respect to it. Thus in some orthogonal basis, we can write the matrix for L_ε as

$$L_\varepsilon = \begin{pmatrix} 1 & & & & & & & \\ & \ddots & & & & & 0 & \\ & & 1 & & & & & \\ & & & \frac{1}{2} & & & & \\ & & & & \ddots & & & \\ & & & & & \frac{1}{2} & & \\ & & & & & & 0 & \\ 0 & & & & & & & \ddots \\ & & & & & & & & 0 \end{pmatrix},$$

which we will always write using block notation:

$$L_\varepsilon = \begin{pmatrix} 1 & & \\ & \frac{1}{2} & \\ & & 0 \end{pmatrix} \, .$$

Recall the identification of $\mathfrak{p}$ and V and note that, for any element x, $\exp x = p + x + \frac{1}{2}x^2 + \frac{1}{6}x^3 + \cdots$ is the same in the Jordan algebra as in the Lie algebra $\mathfrak{p} \subset \mathrm{End}\,(V)$. Thus

$$L_{\exp t\varepsilon} = \begin{pmatrix} e^t & & \\ & e^{\frac{1}{2}t} & \\ & & 1 \end{pmatrix} \, .$$

Introducing the new parameter $s = e^{\frac{1}{2}t}$, we have

$$
L_{\exp t\varepsilon} = \begin{pmatrix} s^2 & & \\ & s & \\ & & 1 \end{pmatrix} = a(s),
$$

where $a(s)$ is a one-parameter subgroup contained in $\exp \mathfrak{p} \subset G$. Clearly, $a(s)$ is an *algebraic* one-parameter subgroup, defined over k if ε is. Also, $\varepsilon = \varepsilon p = \frac{1}{2}a'(1)p$, where the prime denotes $\frac{d}{ds}$.

In particular, if we choose for ε the identity p, we get $a(s) = h(s^2)$, where $h(u)$ is the one-parameter subgroup of homotheties.

Since $\varepsilon \cdot \varepsilon = \varepsilon$, we have $\varepsilon(\varepsilon p) = \varepsilon p$, and thus $\pi_{\frac{1}{2}} p = 0$, where π_i is the orthogonal projection onto $V_i(\varepsilon)$, for $i = 0, \frac{1}{2}, 1$. Moreover, if $a(0) = \lim_{s \to 0} a(s)$ which exists in $\mathrm{End}(V)$, then $a(0) = \pi_0$ and $V_0(\varepsilon) = a(0)V$.

Remark 3.2 More generally, let $a(s)$ be *any* one-parameter subgroup in G. There exists a unique $n \in \mathbb{Z}$ such that $\lim_{s \to 0} h(s^n)a(s)$ exists and is nonzero. Denote the limit by $a(0)$. We will show later that $a(0)V$ is a sub Jordan algebra of V, and $a(0)C$ is a boundary component, k-rational if $a(s)$ is.

3.3

Corresponding to any one-parameter subgroup a in G there is a parabolic subgroup $P(a) = \{x \in G \mid \lim_{s \to 0} a(s)^{-1}xa(s) \text{ exists}\}$, see [11], §2.2. In particular, let $a(s)$ be the one-parameter subgroup defined over k corresponding, as above, to a rational idempotent ε. On $\mathrm{Lie}(G)$, $a(s)$ acts semi-simply in the adjoint representation, so that $\mathrm{Lie}(G) = \bigoplus_\chi \mathfrak{U}_\chi$, where χ is a character of $a(s)$ and

$$
\mathfrak{U}_\chi = \{u \in \mathrm{Lie}(G) \mid a(s)^{-1}ua(s) = \chi(s)u\}.
$$

Since a is one-dimensional, $\chi(s) = s^m$ for some $m \in \mathbb{Z}$. Let $\mathfrak{U}_m = \mathfrak{U}_\chi$ in this case.

Let $Z(a)$ be the centralizer of a in G and let U be the subgroup of G with Lie algebra equal to $\bigoplus_{m>0} \mathfrak{U}_m$. Then $Z(a)$ normalizes U, and $P(a) = Z(a)U$ is the parabolic subgroup corresponding to $a(s)$. Clearly, $P(a)$ is k-parabolic since $a(s)$ is defined over k.

Write $\mathrm{Norm}(C_0) = \{g \in G \mid gC_0 \subset C_0\}$, where C_0 is any boundary component. First we establish a lemma.

Lemma 3.3 $\pi_0(C) = C_0$.

Proof Since $\pi_0 = \lim\limits_{s \to 0} a(s)$, we see that $\pi_0(x) \in \overline{C}$ for any $x \in C$. Now, by the self-adjoint property of C and C_0, we know that

$$C = \{z \in V \mid \langle z, y^2 \rangle > 0 \text{ for all } y \in V, \ y \neq 0\},$$
$$C_0 = \{z \in V_0 \mid \langle z, y^2 \rangle > 0 \text{ for all } y \in V_0, \ y \neq 0\}.$$

Because π_0 is an orthogonal projection onto V_0, we have $\langle \pi_0(x), y^2 \rangle = \langle x, y^2 \rangle > 0$ for any $y \in V_0$, since $x \in C$. Therefore $\pi_0(x) \in C_0$.

Conversely, suppose $x_0 \in C_0$. Let $\varepsilon' = p - \varepsilon$, so that ε' is the identity in V_0. Now $x_0 \in C_0$ implies there exist $y, z \in V_0(\varepsilon)$ with $y^2 = x_0$ and $y \cdot z = \varepsilon'$. Then $(y + \varepsilon)^2 = y^2 + 2y \cdot \varepsilon + \varepsilon^2 = x_0 + \varepsilon$ and

$$(y + \varepsilon) \cdot (z + \varepsilon) = y \cdot z + \varepsilon = \varepsilon' + \varepsilon = p.$$

Thus $x_0 + \varepsilon \in C$ and $\pi_0(x_0 + \varepsilon) = x_0$. $\qquad\square$

Proposition 3.4 $P(a) = \mathrm{Norm}\,(C_0)$.

Proof Let u be an element of $\mathfrak{U}_m$ for $m > 0$. Then $s^m a(s) u = u a(s)$. Taking the limit as $s \longrightarrow 0$, we have $u\pi_0 = 0$. Thus $uC_0 = u\pi_0(C) = 0$. Hence $\exp u$ normalizes C_0, and in fact restricts to the identity on C_0. If $z \in Z(a)$, then z commutes with $a(s)$ for all s, hence also commutes with π_0, and z normalizes C_0 as well. Since $P(a) = Z(a)U$, we see that $P(a) \subset \mathrm{Norm}\,(C_0)$.

If $\mathrm{Norm}\,(C_0)$ is actually a *larger* parabolic than $P(a)$, then there must be some $m < 0$ with $\mathfrak{U}_m \cap \mathrm{Lie}\,\mathrm{Norm}\,(C_0) \neq (0)$ (since $\mathrm{Ad}\,a(s)$ is semi-simple). Pick $v \neq 0$ in this intersection. We have $a(s)v = s^{-m} v a(s)$, and, letting $s \longrightarrow 0$, we see $\pi_0 v = 0$. Therefore $\pi_0(\exp v) = \pi_0$. Since $\exp v \in \mathrm{Norm}\,(C_0)$, and $p - \varepsilon \in C_0$, we get $(\exp v)(p - \varepsilon) = (p - \varepsilon)$.

On the other hand, we can apply the first paragraph to the one-parameter subgroup $b(s) = a^{-1}(s)h^2(s)$ corresponding to $p - \varepsilon$ and to C_1, the corresponding boundary component in $V_0(p - \varepsilon)$. This shows that $\exp v$ restricts to the identity on C_1. Because $\varepsilon \in C_1$, we get $(\exp v)\varepsilon = \varepsilon$.

Adding these together gives $(\exp v)p = p$, which says $\exp v \in K$. But K is compact and $\mathfrak{U}_m$ is nilpotent so $v = 0$, a contradiction. $\qquad\square$

Proposition 3.5 *For any boundary component C_0 and any $g \in G$, gC_0 is also a boundary component.*

Proof Let $P = \mathrm{Norm}\,(C_0)$. Then $G = KP$, so it suffices to prove that gC_0 is a boundary component if $g \in K$. But $g \in K$ implies that g fixes p, so it is easy to see that $g : V \longrightarrow V$ is an automorphism of Jordan algebras. Thus, if ε is an idempotent, so is $g\varepsilon$, and $gV_0(\varepsilon) = V_0(g\varepsilon)$. $\qquad\square$

3.4

Any semi-simple Jordan algebra W over k has a unique decomposition (up to permutation of factors) $W = W_1 \oplus \cdots \oplus W_r$, where W_i are simple Jordan algebras over k, ideals in W, and $W_i \cdot W_j = 0$ if $i \neq j$. In our set-up, we say that the cone C is *k-reducible* if and only if C can be written as $C = C_1 + C_2$ (or equivalently $\overline{C} = \overline{C}_1 + \overline{C}_2$), where the C_i are open cones in k-rational linear subspaces $V_i \subset V$ such that $V_1 \cap V_2 = (0)$; otherwise C is called *k-irreducible*.

Lemma 3.6 *The cone C decomposes uniquely into $C = C_1 + \cdots + C_n$ with each C_i being k-irreducible in V_i and $V = V_1 \oplus \cdots \oplus V_n$.*

Proof Suppose $\overline{C} = \overline{C}_1 + \overline{C}_2$ and $\overline{C} = \overline{D}_1 + \overline{D}_2$ are two decompositions of C. We claim that $\overline{C}_1 = (\overline{C}_1 \cap \overline{D}_1) + (\overline{C}_1 \cap \overline{D}_2)$. Indeed, let $x \in \overline{C}_1$ and write $x = d_1 + d_2$, with $d_i \in \overline{D}_i$. Write $d_i = e_{i1} + e_{i2}$, $e_{ij} \in \overline{C}_j$. Then $x - (e_{11} + e_{21}) = (e_{12} + e_{22})$ is in the linear span of C_1 and of C_2. Thus $e_{12} + e_{22} = 0$; hence $e_{12} = e_{22} = 0$ and $d_1, d_2 \in \overline{C}_1$.

By a dimension argument, C has at least one decomposition as desired in the lemma, and the claim shows that it is unique up to permutation of the factors.   □

Proposition 3.7 *The cone C is k-irreducible if and only if V is k-simple as a Jordan algebra.*

Proof If V is k-reducible, then C obviously is also. Now suppose C is k-reducible. Write $C = \sum C_i$ as a sum of k-irreducible factors. Since this decomposition is unique, any $g \in G(k)$ acting on C must permute the C_i.

Since $G(k)$ is Zariski-dense in G, we conclude that each $g \in G$ permutes the factors C_i; and because G is connected, it actually takes each C_i into itself, i.e., $G = \prod \text{Aut}(C_i)^o$. Now it is obvious that $V = \prod V_i$ as Jordan algebras and V is not k-simple.   □

3.5

Proposition 3.8 *Suppose that $\varepsilon_1, \ldots, \varepsilon_n$ are a set of k-idempotents which are mutually orthogonal, i.e., ε_i is k-rational for each i and $\varepsilon_i \cdot \varepsilon_j = \delta_{ij} \varepsilon_j$. Then $\sum_{i=1}^{k} \mathbb{R} \varepsilon_i = \text{Lie}(A)$, where A is a k-split torus of rank k and $A \subset \exp \mathfrak{p}$.*

Proof This follows from the following three facts:

(a) In any Jordan algebra, L_u and $L_{u \cdot v}$ commute if and only if L_{u^2} and L_v do. This implies that $\sum \mathbb{R} \varepsilon_i$ is a strictly commutative sub Jordan algebra.

(b) $[L_\pi, L_{\pi'}] = \text{ad}\,[\pi, \pi']$ for $\pi, \pi' \in \mathfrak{p}$.

(c) For any $x \in \text{Lie}\,(K)$, $\text{ad}\,x(\mathfrak{p}) = 0 \Longrightarrow x = 0$. This is because $x \longmapsto \text{ad}\,x$ defines $\text{Lie}\,(K) \overset{\sim}{\longrightarrow} \text{Der}\,V$ as we have seen in Section 2, cf. Theorem 2.13.

Thus $\mathfrak{A} = \sum \mathbb{R} \varepsilon_i$ is a commutative sub Lie algebra of $\mathfrak{p}$. Each $\mathbb{R} \varepsilon_i = \text{Lie}\,a_i(s)$ is an algebraic Lie algebra, where $a_i(s)$ is the one-parameter subgroup corresponding to ε_i, so $\mathfrak{A}$ is algebraic too. Then $A = \exp \mathfrak{A}$ is a connected diagonalizable algebraic group and $\text{Lie}\,A$ is defined over k, so A is a k-torus. Since each $a_i(s)$ is a k-one-parameter subgroup, we see that A is k-split. The rank of A is n, because $\varepsilon_i = \frac{1}{2}a_i'(1)$, and the ε_i are linearly independent.

Note that, denoting $Y(A) = \text{Hom}\,(\mathbb{G}_m, A)$, there is a canonical isomorphism

$$\sum_{i=1}^{k} \mathbb{Q} \varepsilon_i \overset{\sim}{\longrightarrow} Y(A) \otimes \mathbb{Q}$$

taking ε_i to $a_i \in Y(A)$. $\qquad\qquad\qquad\qquad\qquad\qquad\qquad\qquad\qquad\qquad\square$

3.6

Proposition 3.9 *Every maximal* $\mathbb{R}$*-split torus* A*, such that* $A \subset \exp \mathfrak{p}$*, arises as in Proposition 3.8, i.e.,* $A = \exp(\sum_{i=1}^{n} \mathbb{R} \varepsilon_i)$*, where* ε_i *are mutually orthogonal idempotents of* V*.*

Proof First we establish the following lemma:

Lemma 3.10 *If* $\dim V > 1$*, then* V *possesses an idempotent different from* 0 *and* p*.*

Proof Choose $u \in V$ linearly independent from p, and let W be the sub Jordan algebra generated by u. Then W is totally real and strictly commutative, so $W \cong \mathbb{R}^s$, where the Jordan multiplication in $\mathbb{R}^s$ is just componentwise multiplication. If $s > 1$, then W has at least two non-trivial idempotents, and, if $s = 1$, some multiple of u works. $\qquad\qquad\qquad\qquad\qquad\qquad\square$

Since any two maximal $\mathbb{R}$-split tori contained in $\exp \mathfrak{p}$ are conjugate by an element of K which also acts as an automorphism of V, it suffices to take a maximal set $\varepsilon_1, \ldots, \varepsilon_n$ of mutually orthogonal idempotents and check that the resulting $A = \exp(\sum \mathbb{R} \varepsilon_i)$ is maximal. We will prove this by induction on $\dim V$. If $\dim V = 1$, the result is obvious. If $\dim V > 1$, then $n \geq 2$ by the lemma.

Let $V = V_0 \oplus V_{\frac{1}{2}} \oplus V_1$ be the Peirce decomposition of V with respect to ε_1. Then $\{\varepsilon_2, \ldots, \varepsilon_n\}$ are a maximal set of mutually orthogonal idempotents in V_0 and similarly $\{\varepsilon_1\}$ in V_1. For $i = 0$ or 1, let $C_i \subset V_i$ be the boundary components, $G_i = \mathrm{Aut}\,(C_i)^o$, $A_1 = \exp\mathbb{R}\varepsilon_1$, and $A_0 = \exp\sum_{i=2}^{n}\mathbb{R}\varepsilon_i$. By induction, A_i is a maximal $\mathbb{R}$-split torus in G_i. If $A = A_1 \times A_2$ is not maximal, let $A' \supsetneq A$ be a maximal $\mathbb{R}$-split torus with $A' \subset \exp(\mathfrak{p})$. Since A' commutes with a_1, we have $A' \subset Z(a_1) \subset \mathrm{Norm}\,(C_0) \cap \mathrm{Norm}\,(C_1)$, and we get restriction maps $\varphi : A' \longrightarrow G_1 \times G_2$. By the maximality of A_i in G_i, the image $\varphi(A')$ has to be all of $A_1 \times A_2$. So φ has a non-trivial one-parameter subgroup $b(s)$ in its kernel. But then $b(s)$ is the identity on C_i; hence $b(s)(\varepsilon_1) = \varepsilon_1$ and $b(s)(p - \varepsilon_1) = p - \varepsilon_1$, hence $b(s)(p) = p$, and therefore $b(s) \in K$. Since K is compact, this is a contradiction, so $A' = A$. $\square$

3.7

Proposition 3.11 *For every $k \subset \mathbb{R}$, there exist maximal k-split tori A such that $A \subset \exp\mathfrak{p}$ and all such arise as $\sum_{i=1}^{n}\mathbb{R}\varepsilon_i$, where ε_i are mutually orthogonal k-idempotents of V.*

Proof The existence of such A follows from:

Lemma 3.12 *If G is a reductive algebraic group defined over $k \subset \mathbb{R}$, $\sigma : G \longrightarrow G$ is a Cartan involution over $\mathbb{R}$, and $P \subset G$ is a minimal k-parabolic of G, then there is a unique torus $S \subset P$ such that*

(i) $\sigma(x) = x^{-1}$, *for all $x \in S$;*

(ii) *S is a lifting into P of the unique maximal k-split torus T in $P/R_u(P)$.*

If σ is defined over k, then so is S.

Proof By Proposition 1.8 in [4], there exists a *unique* Levi subgroup L of P which is stable under the Cartan involution σ. The lifting S_1 of T into this L has the required properties. Conversely, for any such torus S, we know that $P = Z_G(S) \ltimes R_u(P)$, and S stable by σ implies $Z_G(S)$ stable by σ, i.e., $Z_G(S) = L$. Thus $S = S_1$. By the uniqueness of S, σ defined over k implies S is $\mathrm{Gal}\,(\mathbb{C}/k)$-invariant, hence S is defined over k. $\square$

Going back to the proposition, let $A \subset \exp\mathfrak{p}$ be a maximal k-split torus. Then $\mathfrak{A} = \mathrm{Lie}\,A \subset \mathfrak{p}$ is defined over k, so that $\mathfrak{A}$ generates a sub Jordan algebra $\mathfrak{A}_0 \subset V$ defined over k (identifying $\mathfrak{p}$ and V as usual).

Now A is contained in some maximal $\mathbb{R}$-split torus B. We may assume $B \subset \exp \mathfrak{p}$. Proposition 3.9 implies that $\operatorname{Lie} B$ is a strictly commutative sub Jordan algebra. Therefore $\mathfrak{A}_0$ is as well, since $\mathfrak{A}_0 \subset \operatorname{Lie} B$.

So we have $\mathfrak{A}_0 \cong \sum_{i=1}^{r} \mathbb{R} \varepsilon_i$ (over $\mathbb{R}$) and, if A_0 is the k-torus defined as $\exp(\mathfrak{A}_0)$,

$$\sum \mathbb{Q} \varepsilon_i \cong Y(A_0) \otimes \mathbb{Q} .$$

Now $\operatorname{Gal}(\mathbb{C}/k)$ acts on both $\mathfrak{A}_0$ and on A_0. Acting on $\mathfrak{A}_0$, it consists of Jordan isomorphisms, and hence is given by permutations of the ε_i. The above isomorphism is $\operatorname{Gal}(\mathbb{C}/k)$-equivariant. If

$$\{\varepsilon_{i(1,1)}, \ldots, \varepsilon_{i(1,\ell_1)}\}, \ \{\varepsilon_{i(2,1)}, \ldots, \varepsilon_{i(2,\ell_2)}\}, \ \cdots$$

are the $\operatorname{Gal}(\mathbb{C}/k)$-orbits among the $\{\varepsilon_i\}$, the subspace $(\sum \mathbb{Q} \varepsilon_i)^{\operatorname{Gal}(\mathbb{C}/k)}$ of invariants under $\operatorname{Gal}(\mathbb{C}/k)$ in $\sum \mathbb{Q} \varepsilon_i$ is generated by

$$\varepsilon_1' = \sum_{j=1}^{\ell_1} \varepsilon_{i(1,j)} , \ \varepsilon_2' = \sum_{j=1}^{\ell_2} \varepsilon_{i(2,j)} , \ \cdots .$$

But then A, which is the maximal k-split torus in A_0, is given by

$$A = \exp \left(\left(\sum \mathbb{R} \varepsilon_i \right)^{\operatorname{Gal}(\mathbb{C}/k)} \right) = \exp \left(\sum \mathbb{R} \varepsilon_i' \right) ,$$

and, since ε_i' are mutually orthogonal idempotents, A has the required form. $\qquad \square$

3.8

Choose a maximal set $\{\varepsilon_1, \ldots, \varepsilon_n\}$ of mutually orthogonal k-idempotents of V. Let A be the corresponding maximal k-split torus, so that $\mathfrak{A} = \operatorname{Lie}(A) \cong \sum \mathbb{R} \varepsilon_i$, by canonical k-isomorphism. We propose to compute the root structure of G with respect to A.

To do this, we use a Peirce decomposition for all the ε_i at once. Since L_{ε_i} and L_{ε_j} commute, and because $L_{\varepsilon_1} + \cdots + L_{\varepsilon_n} = \operatorname{id}$, it is easy to see that V decomposes into simultaneous eigenspaces $V = \bigoplus_{r \leq s} V_{rs}$ with eigenvalues given by

$$\varepsilon_t \cdot v = \frac{1}{2}(\delta_{tr} + \delta_{ts})v$$

if $v \in V_{rs}$.

For convenience of notation, write $V_{sr} = V_{rs}$ if $s \geq r$, and $x_{rs} = x_{sr}$ for an element of V_{rs}. For instance, if $x = \sum_{k \leq \ell} x_{k\ell}$, then $x \cdot \varepsilon_m = \frac{1}{2} \sum_{k \neq m} x_{km} + x_{mm}$. Note that if $r \neq s$, $V_{rs} = V_{\frac{1}{2}}(\varepsilon_r) \cap V_{\frac{1}{2}}(\varepsilon_s)$, and $V_{rr} = V_1(\varepsilon_r)$.

Let $\mathfrak{g} = \operatorname{Lie} G$. By Theorem 2.13, we know that $\mathfrak{g} = \operatorname{Der} V \oplus V$. Here $\operatorname{Der} V$

are the derivations of V as a Jordan algebra. Hence $\mathrm{Der}\,V \subset \mathrm{End}\,(V)$, it is defined over k, and $\mathrm{Der}\,V \cong \mathrm{Lie}\,K$ over k. If $D, D' \in \mathrm{Der}\,V$ and $x, y \in V$, the Lie bracket in $\mathfrak{g}$ is given by

$$[D, x] = Dx\,,$$
$$[D, D'] = DD' - D'D\,,$$
$$[x, y] = L_x L_y - L_y L_x\,.$$

To find the root structure, we are looking for elements $0 \neq (D, x) \in \mathfrak{g}$ such that $[\varepsilon_i, (D, x)] = \lambda_i (D, x)$, for $i = 1, \ldots, n$, and some $\lambda_i \in \mathbb{C}$. Thus we require $[\varepsilon_i, (D, x)] = ([L_{\varepsilon_i}, L_x], -D\varepsilon_i) = \lambda_i (D, x)$.

That is,

$$D\varepsilon_i = -\lambda_i x\,,$$
$$[L_x, L_{\varepsilon_i}] = -\lambda_i D\,.$$

Applying the lower equation to ε_j and using the top gives

$$y \cdot (\varepsilon_i \cdot \varepsilon_j) - \varepsilon_i \cdot (x \cdot \varepsilon_j) = -\lambda_i D\varepsilon_j = \lambda_i \lambda_j x\,.$$

Write $x = \sum_{k \leq \ell} x_{k\ell}$, with $x_{k\ell} \in V_{k\ell}$. We can deduce the following.

In the case $i = j$,

$$\frac{1}{2}\sum_{i \neq \ell} x_{i\ell} + x_{ii} - \frac{1}{4}\sum_{i \neq \ell} x_{i\ell} - x_{ii} = \lambda_i^2 x\,.$$

That is,

$$\frac{1}{4}\sum_{i \neq \ell} x_{i\ell} = \lambda_i^2 x\,.$$

Hence,

(i) either $x \in \bigoplus_{\ell \neq i} V_{i\ell}$ and $\lambda_i = \pm\frac{1}{2}$;
(ii) or $x_{i\ell} = 0$ for $\ell \neq i$ and $\lambda_i = 0$.

In particular, $\lambda_i \in \{0, \pm\frac{1}{2}\}$ for all i.

In the case $i \neq j$,

$$-\frac{1}{4}x_{ij} = \lambda_i \lambda_j x\,.$$

Hence,

(i) either $x \in V_{ij}$ and $-\lambda_j = \lambda_i = \pm\frac{1}{2}$;
(ii) or $x_{ij} = 0$ and $\lambda_j = 0$ or $\lambda_i = 0$.

Putting this information together shows that at most two λ_k can be non-zero. There are three possibilities:

(1) $\lambda_i = \frac{1}{2}, \lambda_j = -\frac{1}{2}, x \in V_{ij}$, and all other λ_k are 0, where $1 \le i \le n, 1 \le j \le n$ and $i \ne j$;

(2) all $\lambda_k = 0$ and $x \in \bigoplus V_{rr}$;

(3) $\lambda_i = \pm\frac{1}{2}$, all other λ_k are 0, where $1 \le i \le n$.

Possibility (3) is actually impossible. Since $\lambda_\ell = 0$ for $\ell \ne i$, we know $x_{i\ell} = 0$ for $i \ne \ell$, but $\lambda_i = \pm\frac{1}{2}$ implies $x \in \bigoplus_{\ell \ne i} V_{i\ell}$. So $x = 0$, and then $\lambda_i D = -[L_x, L_{\varepsilon_i}]$ makes $D = 0$ also.

Possibility (2) just gives elements that centralize $\mathfrak{A}$.

Possibility (1) is in fact realized, as long as $V_{ij} \ne 0$. To check this, take, without loss of generality, $i = 1$, $j = 2$, pick $x \in V_{12}$, and let $D = -2[L_x, L_{\varepsilon_1}]$.

Claim

(a) $[L_x, L_{\varepsilon_2}] = -[L_x, L_{\varepsilon_1}]$ *and* $[L_x, L_{\varepsilon_j}] = 0$ *for* $j \ge 3$;

(b) $D\varepsilon_1 = \frac{1}{2}x$;

(c) $D\varepsilon_2 = -\frac{1}{2}x$;

(d) $D\varepsilon_j = 0$ *for* $j \ge 3$;

(e) (D, x) *is in a root space for* $\mathfrak{g}$.

Proof Since $x \in V_{12} = V_{\frac{1}{2}}(\varepsilon_1) \cap V_{\frac{1}{2}}(\varepsilon_2) \cap \bigcap_{j \ge 3} V_0(\varepsilon_j)$, we see that (b), (c), and (d) are trivial. Clearly (e) follows from (a), (b), (c), and (d). So it remains to prove (a).

Let $f_i(y) = x \cdot (\varepsilon_i \cdot y) - \varepsilon_i \cdot (x \cdot y)$. Then we must check that:

(1) $f_1(y) + f_2(y) = 0$ for all $y \in V$;

(2) $f_j(y) = 0$ for all $y \in V$, $j \ge 3$.

Since each f_i is linear, it suffices to do this for $y \in V_{rs}$ for various $1 \le r, s \le n$. We have several cases.

(i) $y \in V_{11} \oplus V_{12} \oplus V_{22}$. This is true if and only if $y \in \bigcap_{j \ge 3} V_0(\varepsilon_j)$. Since $V_0(\varepsilon) \cdot V_0(\varepsilon) \subset V_0(\varepsilon)$ for any idempotent ε, we see that $x \cdot y \in V_{11} \oplus V_{12} \oplus V_{22}$. Thus (1) follows since $\varepsilon_1 + \varepsilon_2$ is the identity on $V_{11} \oplus V_{12} \oplus V_{22}$, and (2) follows since ε_j is zero on $V_{11} \oplus V_{12} \oplus V_{22}$ if $j \ge 3$.

(ii) $y \in V_{13}$. Then $y \in V_0(\varepsilon_2) \cap V_{\frac{1}{2}}(\varepsilon_3)$. Since $V_0(\varepsilon) \cdot V_{\frac{1}{2}}(\varepsilon) \subset V_{\frac{1}{2}}(\varepsilon)$, we have $x \cdot y \in V_{23}$. Then (1) follows since $\varepsilon_1 + \varepsilon_2 = \frac{1}{2}$ on $V_{23} \oplus V_{13}$, and (2) follows since ε_3 is $\frac{1}{2}$ on $V_{23} \oplus V_{13}$ and $\varepsilon_j = 0$ on $V_{23} \oplus V_{13}$ for $j \ge 4$.

(iii) $y \in V_{1k}$ or V_{2k} for any $k \ge 3$ is similar to (ii).

(iv) $y \in V_{jk}$ with $j, k \ge 3$. Then $x \cdot y = 0$: if $j = k$, this follows from $y \in V_1(\varepsilon_j)$ and $x \in V_0(\varepsilon_j)$; if $j \ne k$, this follows from $x \cdot y \in V_{\frac{1}{2}}(\varepsilon_1) \cap V_{\frac{1}{2}}(\varepsilon_2) \cap V_{\frac{1}{2}}(\varepsilon_j) \cap V_{\frac{1}{2}}(\varepsilon_k)$. Hence also $x \cdot (\varepsilon_i \cdot y) = 0$, whence (1) and (2). $\square$

Summarizing this discussion, we find

$$\mathfrak{g} = Z(\mathfrak{A}) \oplus \bigoplus_{i \neq j} \mathfrak{g}_{ij} \, ,$$

where

$$\mathfrak{g}_{ij} = \{(D, x) \mid x \in V_{ij}, D = -2[L_x, L_{\varepsilon_i}]\} \, .$$

Applying the above formulae, one readily checks:

$$Z(\mathfrak{A}) = (Z(\mathfrak{A}) \cap \mathrm{Der}\, V) \oplus (Z(\mathfrak{A}) \cap V)$$

$$= \{D \in \mathrm{Der}\, V \mid D\varepsilon_i = 0 \text{ for all } i\} \oplus \bigoplus_{i=1}^{n} V_{ii} \, ,$$

and that, if we define for $i < j$

$$\mathrm{Der}\,(V)_{ij} = \{D \in \mathrm{Der}\, V \mid D = [L_x, L_{\varepsilon_i}], \text{ for some } x \in V_{ij}\} \, ,$$

then

$$\mathfrak{g}_{ij} \oplus \mathfrak{g}_{ji} \cong V_{ij} \oplus \mathrm{Der}\,(V)_{ij} \, ,$$

$$\dim \mathfrak{g}_{ij} = \dim \mathfrak{g}_{ji} = \dim V_{ij} = \dim \mathrm{Der}\,(V)_{ij} \, .$$

Finally, we study the set of (i, j) such that $V_{ij} \neq (0)$.

Proposition 3.13 *Let* $\gamma_i \in \mathrm{Hom}\,(\sum_{i=1}^{n} \mathbb{R}\varepsilon_i, \mathbb{R})$ *be the dual basis to* ε_i. *Let* $V = \prod_\alpha V_\alpha$ *be the decomposition into k-simple Jordan algebras and let* $\{\varepsilon_i\}_{i \in I_\alpha}$ *be the idempotents in* V_α. *Then:*

(i) *$V_{ij} \neq (0)$ if and only if $i \neq j$ and i, j are in the same I_α;*
(ii) *the k-roots are $\frac{1}{2}(\gamma_i - \gamma_j)$, where $i \neq j$ and i, j are in the same I_α;*
(iii) *the Weyl group is the group of all permutations of the $\{\varepsilon_i\}$ preserving the partition $\{I_\alpha\}$.*

Proof Since every element of the Weyl group is represented by an element of K, it acts on $\mathfrak{A}$ by a Jordan isomorphism, hence it acts by a permutation of the ε_i. Now if $V_{ij} \neq (0)$, then $\frac{1}{2}(\gamma_i - \gamma_j)$ is a root and the reflection w_{ij} in the hyperplane $\gamma_i - \gamma_j = 0$ is in the Weyl group. Since $(\gamma_i - \gamma_j)(\varepsilon_k) = 0$, for $k \neq i, j$, this reflection must be the permutation fixing ε_k for $k \neq i, j$ and interchanging ε_i, ε_j. Now any subgroup of the permutation group generated by transpositions is the group of all permutations preserving some partition $\{J_\alpha\}$. Thus if $V_{ij} \neq (0)$, then w_{ij} lies in the Weyl group and hence i, j are in the same J_α. Conversely, since the w_{ij} for i, j such that $V_{ij} \neq (0)$ generate the Weyl group, if J_α has at least two elements, then $V_{i_0, j_0} \neq (0)$ for some $i_0, j_0 \in J_\alpha$. Then for any $i \neq j$ with $i, j \in J_\alpha$, there exists $\sigma \in \mathrm{Norm}_K(\mathfrak{A})$ such that $\sigma \varepsilon_{i_0} = \varepsilon_i$, $\sigma \varepsilon_{j_0} = \varepsilon_j$. Then $V_{i,j} = \sigma(V_{i_0,j_0}) \neq (0)$. Thus $V_{ij} \neq (0)$ if and only if $i \neq j$ and i, j are in the same J_α. Finally,

$$V = \prod_{\alpha} \prod_{i,j \in J_\alpha} V_{ij},$$

and $\prod_{i,j \in J_\alpha} V_{ij}$ is k-simple. Hence $V_\alpha = \prod_{i,j \in J_\alpha} V_{ij}$ and $I_\alpha = J_\alpha$.   □

3.9

Let ε be an idempotent in V, and let $C_i = C_i(\varepsilon)$, and $V_i = V_i(\varepsilon)$ for $i = 0, 1$. Let $a(s)$ be the one-parameter subgroup corresponding to ε. Denoting by $Z(a)$ the centralizer of a in G, we know that $Z(a)$ is the Levi subgroup of $\text{Norm}\,(C_1)$, and also the Levi subgroup of $\text{Norm}\,(C_0)$. In fact, $Z(a) = \text{Norm}\,(C_1) \cap \text{Norm}\,(C_0)$. Setting $G_i = \text{Aut}\,(C_i, V_i)^o$ for $i = 0, 1$, we get a map

$$\varphi = \text{res}_{V_0} \times \text{res}_{V_1} : Z(a)^o \longrightarrow G_0 \times G_1.$$

Proposition 3.14 *We have a decomposition:*

$$Z(a)^o = G_0' \cdot G_1' \cdot K_0$$

(a direct product, modulo a finite abelian subgroup), where $K_0 = \ker \varphi$ is compact, $\text{res}\,\varphi : G_i' \longrightarrow G_i$ is surjective with finite kernel, and

$$\text{Lie}\,G_i' = [V_i, V_i] + V_i \subset \text{Lie}\,G.$$

If ε is k-rational, then so is this decomposition.

Proof By Section 2, we know that $\text{Lie}\,G = [V, V] \oplus V$. For $i = 0, 1$, define the subalgebra $\mathfrak{g}_i' = [V_i, V_i] \oplus V_i \subset \text{Lie}\,G$. Now any element of V_i strictly commutes with ε, so that actually $\mathfrak{g}_i' \subset \text{Lie}\,Z(a)$. There are restriction maps $\Psi_i : \text{Lie}\,Z(a) \longrightarrow \text{Lie}\,G_i$.

We claim that $\Psi_i|_{\mathfrak{g}_i'}$ is an isomorphism. We know by Section 2 that $\text{Lie}\,G_i = [V_i, V_i] \oplus V_i$. By definition, Ψ_i is just the identity on V_i and, since it preserves the Lie bracket, it is also the identity on $[V_i, V_i]$, hence the claim is proven.

Let $K_0 = \text{Ker}\,\varphi$, $\mathfrak{k}_0 = \text{Lie}\,K_0$. Note that if $g \in K_0$, then $g\varepsilon = \varepsilon$ and $g(p - \varepsilon) = p - \varepsilon$, hence $gp = p$; thus $K_0 \subset K$ and is compact. Now because $\text{Lie}\,Z(a)/\mathfrak{k}_0 \cong \text{Lie}\,G_0 \times \text{Lie}\,G_1 \cong \mathfrak{g}_0' \times \mathfrak{g}_1'$, it follows that we get a vector space decomposition:

$$\text{Lie}\,Z(a) = \mathfrak{k}_0 \oplus \mathfrak{g}_0' \oplus \mathfrak{g}_1'.$$

We claim these factors commute.

(i) To show $[\mathfrak{g}_0', \mathfrak{g}_1'] = (0)$, it suffices to show $[V_0, V_1] = (0)$, but this just says that V_0, V_1 strongly commute, which was shown in Section 2.

(ii) To show $[\mathfrak{k}_0, \mathfrak{g}'_i] = (0)$, it suffices to show $[\mathfrak{k}_0, V_i] = (0)$. But $\mathfrak{k}_0 = \mathrm{Ker}\,(d\varphi)$ is an ideal so $[\mathfrak{k}_0, V_i] \subset \mathfrak{k}_0$; while $[\mathfrak{k}, \mathfrak{p}] \subset \mathfrak{p}$ implies $[\mathfrak{k}_0, V_i] \subset \mathfrak{p}$.

$\square$

Corollary 3.15 $\mathrm{res}_{V_0} : Z(a)^o \longrightarrow G_0$ *is surjective.* $\square$

Corollary 3.16 $\mathrm{ker}(\mathrm{res}_{V_0}|_{Z(a)^o})^o = G'_1 \cdot K_0.$ $\square$

3.10

In this subsection, we assume G is k-simple. Let n be the k-rank of G and let $\varepsilon_1, \ldots, \varepsilon_n$ be a maximal set of mutually orthogonal k-idempotents. Call $a_j(s)$ the one-parameter subgroup corresponding to the idempotent $f_j = \varepsilon_1 + \cdots + \varepsilon_j$. Note that $a_j(0)$, defined as $\lim_{s \to 0} a_j(s)$, exists in $\mathrm{End}\,(V)$.

Lemma 3.17 $V_0(f_{j+1}) \subsetneq V_0(f_j).$

Proof Pick any $x \in V_0(f_{j+1})$. Then $0 = f_{j+1} \cdot x = f_j \cdot x + \varepsilon_{j+1} \cdot x$. But $\varepsilon_{j+1} \in V_1(f_{j+1})$ so $\varepsilon_{j+1} \cdot x = 0$. Thus $x \in V_0(f_j)$. Meanwhile $\varepsilon_{j+1} \in V_0(f_j) \setminus V_0(f_{j+1})$.

$\square$

Thus if we write $C_j = C_0(f_j) = a_j(0)C$, we have

$$0 = \overline{C}_n \subsetneq \overline{C}_{n-1} \subsetneq \overline{C}_{n-2} \subsetneq \cdots \subsetneq \overline{C}_1 \subsetneq \overline{C}_0 = \overline{C}.$$

We call this a *flag of k-boundary components*. If we fix $\varepsilon_1, \ldots, \varepsilon_n$ once and for all, we call it the *standard flag*, and its members the *standard rational boundary components*.

Let A be the maximal k-split torus with $\mathrm{Lie}\,A = \sum_{i=1}^n \mathbb{R}\varepsilon_i$.

Proposition 3.18 *If $b(s)$ is any k-split one-parameter subgroup in G such that $b(0) = \lim_{s \to 0} b(s)$ exists in $\mathrm{End}\,(V)$ and is non-zero, then $b(0)C$ is the image by some $g \in G(k)$ of a standard rational boundary component. In particular, it is a boundary component.*

Proof By conjugating by some $g \in G(k)$ we may assume that $b(s)$ is a one-parameter subgroup of A. We may always replace b by b^n for some positive integer n, since $b(0)$ does not change. From the explicit description of the root structure in Section 3.8, it is easy to check that $\{a_1^{m_1} a_2^{m_2} \cdots a_n^{m_n} \mid m_i \geq 0 \text{ for } i = 1, \ldots, n-1\}$ is a k-Weyl chamber in A. Thus conjugating with some $n \in N(k)$, where N is the normalizer of A, we may assume $b(s) = \prod_{i=1}^n a_i^{m_i}(s)$ with $m_i \geq 0$ for $i = 1, \ldots, n-1$. Note that $b(s)\varepsilon_n = s^{2m_n}\varepsilon_n$. Since $b(0)$ exists, it follows that

$m_n \geq 0$ as well. Recall that $a_i(0)$ is the orthogonal projection onto $V_0(f_i)$ and that

$$0 = V_0(f_n) \subsetneq V_0(f_{n-1}) \subsetneq V_0(f_{n-2}) \subsetneq \cdots \subsetneq V_0(f_1) \subsetneq V .$$

Then $a_1^{m_1} \cdots a_n^{m_n}(0)C = C_j$, where $m_{j+1} = \cdots = m_n = 0$ and $m_j > 0$. ☐

Note that $b(0)C$ is always a boundary component, even if G is not k-simple.

Every k-boundary component is the translate of a standard one. If the standard flag were a subflag of some larger flag of k-rational boundary components, any non-standard boundary component would have the same dimension as a standard one and thus be equal to it. So, the standard flag is a *maximal* flag of k-rational boundary components.

Proposition 3.19 *Any flag $\mathfrak{F}$ of k-rational boundary components is the image by some $g \in G(k)$ of some subflag of the standard flag.*

Proof The proof is similar to that of the previous proposition. Let

$$V_s \subsetneq V_{s-1} \subsetneq \cdots \subsetneq V_1 \subsetneq V$$

be the flag of k-boundary components in question. Let d_i be the Jordan identity in V_i. Then $E = \{p - d_1, d_1 - d_2, \ldots, d_{s-1} - d_s, d_s\}$ is a set of mutually orthogonal k-idempotents, and they generate an associative sub Jordan algebra. The corresponding torus B is k-split. So, after conjugating by an element $g \in G(k)$, we may assume that $B \subset A$.

Now, $\operatorname{Lie} A = \sum_{i=1}^n \mathbb{R}\varepsilon_i$ and $\operatorname{Lie} B \subset \operatorname{Lie} A$. Therefore,

$$E = \{\varepsilon_{11} + \cdots + \varepsilon_{1j_1}, \ldots, \varepsilon_{s1} + \cdots + \varepsilon_{sj_s}\} .$$

Since the Weyl group acting on $\{\varepsilon_1, \ldots, \varepsilon_n\}$ is the full group of permutations, after conjugating by some $n \in N(k)$ we may assume that $d_i = f_{\varphi(i)}$ for some increasing map $\varphi : \{1, \ldots, s\} \longrightarrow \{1, \ldots, n-1\}$. (Here, again, N is the normalizer of A.)

However, $f_{\varphi(i)}$ is the projection onto the $\varphi(i)$'th member of the standard flag. Since the idempotent determines the boundary component, we now have our flag as a subflag of the standard flag. ☐

Proposition 3.20 *There is a bijection between the set of flags $\mathfrak{F}$ of k-boundary components and k-parabolics $P \subset G$ given by $\Phi : \mathfrak{F} \longrightarrow \bigcap_{C_a \in \mathfrak{F}} \operatorname{Norm}(C_a)$.*

Proof From the root structure, we see easily that the minimal k-parabolic corresponding to the set of simple roots $\{\frac{1}{2}(\gamma_i - \gamma_{i+1}) \mid i = 1, \ldots, n-1\}$ normalizes the standard flag. Thus $\bigcap_{C_a \in \mathfrak{F}} \operatorname{Norm}(C_a)$ actually *is* a k-parabolic for any flag $\mathfrak{F}$, by use of the preceding corollary.

Define a standard k-parabolic to be one of the form $\Phi(\mathfrak{F})$, where $\mathfrak{F}$ is a sub-flag of the standard flag. Then $\text{Norm}(C_i)$ for standard C_i, $i = 1,\ldots,n-1$, are the maximal standard k-parabolics. The fact that *any* k-parabolic is conjugate via some $g \in G(k)$ to an intersection of maximal standard ones implies that Φ is surjective.

We show now that Φ is injective. Since any k-parabolic is uniquely the intersection of maximal k-parabolics, it suffices to show that, for any two k-boundary components, C_a and $C_{a'}$,

$$\text{Norm}(C_a) = \text{Norm}(C_{a'}) \Longrightarrow C_a = C_{a'} .$$

By Proposition 3.18, there exist $g, g' \in G(k)$ and standard k-boundary components C_i, C_j such that $gC_i = C_a$ and $g'C_j = C_{a'}$. Therefore,

$$g\text{Norm}(C_i)g^{-1} = \text{Norm}(C_a) = \text{Norm}(C_{a'}) = g'\text{Norm}(C_j)g'^{-1} .$$

So we get

$$g'^{-1}g\text{Norm}(C_i)(g'^{-1}g)^{-1} = \text{Norm}(C_j) .$$

However, the standard k-parabolics are not conjugate to each other, so $i = j$. Then, since the normalizer of any parabolic is equal to itself, we get that $g'^{-1}g \in \text{Norm}(C_i)$. Thus

$$C_{a'} = g'C_i = g'(g'^{-1}g)C_i = gC_i = C_a .$$

$\square$

Corollary 3.21 *A set of k-parabolics intersect in a k-parabolic if and only if their corresponding boundary components can be arranged into a flag.*

Corollary 3.22 *A real boundary component C' of C is k-rational if and only if $\text{Norm}(C')$ is k-rational.*

Proof Combine the theorem over k and the theorem over $\mathbb{R}$. $\square$

Lemma 3.23 *A is generated by $a_1(s),\ldots,a_n(s)$.*

Proof $a_i(s) = \exp t(\varepsilon_1 + \cdots + \varepsilon_i)$ where $s = e^{\frac{1}{2}t}$, so $a_1,\ldots,a_n$ generate a torus with Lie algebra $\sum_{i=1}^n \mathbb{R}\varepsilon_i$. $\square$

What is the orbit $Ap \subset C$? We compute: for any $q \in V$, which strongly commutes with the ε_i,

$$\exp\left(\sum t_j \varepsilon_j\right)q = \left(\prod \exp t_j \varepsilon_j\right)q = \sum(\exp t_j)\varepsilon_j q - \sum \varepsilon_j q + q$$
$$= \sum(\exp t_j)\varepsilon_j q .$$

Here we used $\varepsilon_i \cdot \varepsilon_j = \delta_{ij}\varepsilon_i$, $\sum \varepsilon_j = p$ and

$$\exp t\varepsilon = 1 + t\varepsilon + \frac{t^2}{2}\varepsilon + \cdots = e^t\varepsilon - \varepsilon + 1.$$

We are thinking of elements π of $V \cong \mathfrak{p}$ as being in $\mathrm{End}\,(V)$. Thus

$$\exp\left(\sum t_i \varepsilon_i\right)p = \sum(\exp t_i)\varepsilon_i p = \sum(\exp t_i)\varepsilon_i .$$

Hence:

Proposition 3.24 *The orbit of Λ is "linear," namely, $\Lambda p = \sum_{i=1}^n \mathbb{R}_{>0}\varepsilon_i$.* $\qquad\square$

We may use $\varepsilon_1, \ldots, \varepsilon_n$ as orthogonal coordinates in Ap, since $\langle \varepsilon_i, \varepsilon_j \rangle = \mathrm{Tr}\, L_{\varepsilon_i \cdot \varepsilon_j} = 0$ if $i \neq j$.

Now $\frac{1}{2}a'_j(1) = \varepsilon_1 + \cdots + \varepsilon_j$ and

$$a_j(s) = \begin{pmatrix} s^2 & \vdots & \vdots \\ -- -+ -- -+ -- \\ & \vdots & s & \vdots \\ -- -+ -- -+ -- \\ & \vdots & \vdots & 1 \end{pmatrix} \quad\begin{matrix} \} & V_1(f_j) \\ \\ \} & V_{\frac{1}{2}}(f_j) \\ \\ \} & V_0(f_j) \end{matrix} .$$

Clearly, $\varepsilon_k \in V_0(f_j)$ if $k > j$ and $\varepsilon_k \in V_1(f_j)$ if $k \leq j$. Thus

$$a_j(s)\varepsilon_k = \begin{cases} \varepsilon_k & \text{if } k > j \\ s^2\varepsilon_k & \text{if } k \leq j . \end{cases}$$

4 Siegel sets in self-adjoint cones

4.1

First we recall the general theory of Siegel sets. Let $\mathscr{G}$ be a reductive algebraic group defined over $\mathbb{Q}$ and $G = \mathscr{G}(\mathbb{R})^o$ and let X be the associated non-compact symmetric space (i.e., the homogeneous space G/K, for a maximal compact subgroup $K \subset G$). For every parabolic subgroup $\mathscr{P} \subset \mathscr{G}$ defined over $\mathbb{Q}$ and minimal among such subgroups ("minimal $\mathbb{Q}$-parabolic" for short), we wish to define a class of subsets $\mathfrak{S} \subset X$ called the Siegel sets associated to $P = \mathscr{P}(\mathbb{R})^o$.

Definition 4.1 Choose a basepoint $p \in X$. Let $A \subset P$ be the unique torus that is a conjugate of the maximal $\mathbb{Q}$-split torus of P and such that $\mathrm{Lie}\,A \perp \mathrm{Lie}\,(\mathrm{Stab}\,p)$ (cf. Lemma 3.12). Let $\Delta \subset \mathrm{Hom}\,(A, \mathbb{G}_m)$ be the simple positive roots (i.e., the minimal roots in the adjoint action of A on $\mathrm{Lie}\,\mathscr{P}$); let $A^+ = \{g \in A \mid \beta(g) \geq 1 \text{ for all } \beta \in \Delta\}$. Then the *Siegel sets* are the subsets of X of the form

$$\mathfrak{S}_\omega = \omega A^+ p ,$$

where $\omega \subset P = \mathscr{P}(\mathbb{R})^o$ is a compact subset.

We make three remarks. First of all, this definition does not depend on the choice of p. In fact, let p, p' be two basepoints, let $K = \text{Stab}(p)$. Since $G = P \cdot K$, P acts transitively on X and so we may write $p' = gp$ for some $g \in P$. Let $A, A' \subset P$ be the tori corresponding to p, p'. Then $A' = gAg^{-1}$; hence, for all $\omega \subset P$ compact,

$$\mathfrak{S}'_\omega = \omega(A')^+ p' = \omega(gA^+ g^{-1})(gp) = \mathfrak{S}_{\omega g} .$$

Secondly, this definition is slightly different from the usual one (see [3], pp. 85ff.). However, we claim that all "usual" Siegel sets are Siegel sets as above, and each of the above is contained in a usual Siegel set. In all applications, it is not the exact shape of the Siegel sets that is important, but rather the way they grow. To be precise, any class of subsets $\{X_\alpha\}$ *cofinal* with Siegel sets in the sense:

(i) for all α, there exists ω such that $X_\alpha \subset \mathfrak{S}_\omega$;

(ii) for all ω, there exists α such that $\mathfrak{S}_\omega \subset X_\alpha$,

would do just as well.

In our case, the "usual" Siegel sets are defined by decomposing P into

$$P = MAN ,$$

where N is the unipotent radical of P, A is some maximal $\mathbb{Q}$-split torus in P, and M is the anisotropic part of $Z(A)$, and by choosing p such that $\text{Lie}(\text{Stab } p) \perp \text{Lie } A$. Then one takes the set $\mathfrak{S}^*_{\omega,t} = \omega A_t p$, where $\omega \subset MN$ is compact and $A_t = \{g \in A \mid \beta(g) \geq t \text{ for all } \beta \in \Delta\}$. But if $a_t \in A$ satisfies $\beta(a_t) = t$ for all $\beta \in \Delta$, then

$$\mathfrak{S}^*_{\omega,t} = \mathfrak{S}_{\omega a_t} .$$

In the other direction, if $\omega \subset P$ is compact, then $\omega \subset \omega_1 \cdot \omega_2$, where $\omega_1 \subset MN$ and $\omega_2 \subset A$ are compact. Let

$$t = \inf_{\substack{\beta \in \Delta \\ g \in \omega_2}} \beta(g) ;$$

then

$$\mathfrak{S}_\omega \subset \mathfrak{S}^*_{\omega_1, t} .$$

The third remark is that if G is not $\mathbb{Q}$-simple, but rather $G = G_1 \times \cdots \times G_k$ over $\mathbb{Q}$, then all our sets decompose:

$$X = X_1 \times \cdots \times X_k ,$$
$$P = P_1 \times \cdots \times P_k ,$$
$$A = A_1 \times \cdots \times A_k ,$$
$$A^+ = A_1^+ \times \cdots \times A_k^+ .$$

So, among all Siegel sets, those with $\omega = \omega_1 \times \cdots \times \omega_k$ are cofinal and these decompose as

$$\mathfrak{S}_\omega = \mathfrak{S}_{\omega_1} \times \cdots \times \mathfrak{S}_{\omega_k} .$$

The virtue of Siegel sets lies in the following two fundamental results of reduction theory. Let $\Gamma \subset G(\mathbb{Q})$ be an arithmetic subgroup. Then:

(i) *there exist $\omega \subset P$ compact and a finite set $F \subset G(\mathbb{Q})$ such that*

$$X = \Gamma \cdot F \cdot \mathfrak{S}_\omega ;$$

(ii) *for all ω compact and all $g_1, g_2 \in G(\mathbb{Q})$, the set*

$$\{\gamma \in \Gamma \mid g_1 \mathfrak{S}_\omega \cap \gamma g_2 \mathfrak{S}_\omega \neq \emptyset\}$$

is finite.

4.2

Now return to the case of cones: let $G = \mathrm{Aut}\,(C,V)^o$, where C is a homogeneous self-adjoint cone, and suppose V and G are defined over $\mathbb{Q}$. Choose a rational basepoint p, let $K = \mathrm{Stab}\,(p)$, so that V is a Jordan algebra, and choose a maximal $\mathbb{Q}$-split torus A perpendicular to K. We also assume that G is $\mathbb{Q}$-simple. Let $\mathfrak{A} = \mathrm{Lie}\,A$, $\mathfrak{A} = \sum \mathbb{R}\varepsilon_i$, where $\varepsilon_1 + \cdots + \varepsilon_n = p$ as in Section 3. Let C_i be the boundary component containing $p_i = \varepsilon_{i+1} + \cdots + \varepsilon_n$ so that

$$\mathfrak{F} = \{\overline{C} \supset \overline{C}_1 \supset \cdots \supset \overline{C}_{n-1}\}$$

is a maximal flag of $\mathbb{Q}$-rational boundary components corresponding to a minimal $\mathbb{Q}$-parabolic P containing A. Let

$$a_i(e^{\frac{1}{2}t}) = \exp(t(\varepsilon_1 + \cdots + \varepsilon_i))$$

be the one-parameter subgroup such that $a_i(0)C = C_i$ and

$$a_i(s)\varepsilon_j = \begin{cases} \varepsilon_j & \text{if } j > i \\ s\varepsilon_j & \text{if } j \leq i . \end{cases}$$

Let

$$\tilde{C} = C \cup C_1 \cup C_2 \cup \cdots \cup C_{n-1} \cup \{0\} .$$

From our explicit description of the root structure, we see that the set Δ of simple roots consists in our case of $\beta_1, \ldots, \beta_{n-1}$, where on the Lie algebra level

$$d\beta_j(\varepsilon_i) = \begin{cases} -\frac{1}{2} & \text{if } i-j \\ \frac{1}{2} & \text{if } i = j+1 \\ 0 & \text{otherwise} . \end{cases}$$

Since $\frac{1}{2}a_i'(1) = \varepsilon_1 + \cdots + \varepsilon_i$, it is easy to see that

$$\beta_j(a_i(s)) = s^{-\delta_{ij}} .$$

Lemma 4.2 $\overline{A^+ p}$ *is the closed polyhedral cone* D^+ *generated by the points* p, *and* $p_1, \ldots, p_{n-1}$.

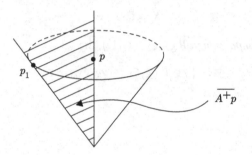

Proof Since $\beta_i(a_j(s)) = s^{-\delta_{ij}}$, we have

$$A^+ = \{a_1(s_1) \cdots a_n(s_n) \mid s_i \leq 1 \text{ for } i = 1, \ldots, n-1\} .$$

Therefore $A^+ p$ is the set of points

$$(s_1 \cdots s_n)^2 \varepsilon_1 + (s_2 \cdots s_n)^2 \varepsilon_2 + \cdots + s_n^2 \varepsilon_n ,$$

with $0 < s_i \leq 1$ for $i = 1, \ldots, n-1$ and $0 < s_n$. Moreover, $\overline{A^+ p}$ is the set of such points where now $0 \leq s_i \leq 1$ for $i = 1, \ldots, n-1$ and $0 \leq s_n$. The lemma follows easily. $\qquad\qquad\square$

Corollary 4.3 $\overline{\mathfrak{S}}_\omega \subset \tilde{C}$.

Proof Note that $\omega \subset P$, hence ω fixes $\tilde{C}$. $\qquad\qquad\square$

We want to know how the Siegel sets look in the boundary components. By induction, it will be apparent that it suffices to do this for the largest boundary component, C_1. Let $p_1 = p - \varepsilon_1 = \varepsilon_2 + \cdots + \varepsilon_n$ be the basepoint in C_1 and write $G_1 = \mathrm{Aut}\,(C_1, V_1)^o$. Let $\psi : P(a_1) \longrightarrow G_1$ be the restriction map. We know $\psi(A)p_1 = \sum_{k=2}^{n} \mathbb{R}_{>0}\varepsilon_k$.

We know that P is the normalizer of the standard flag, and P is a minimal $\mathbb{Q}$-parabolic in G. Let P_1 be the normalizer of the inherited standard flag in C_1.

By Proposition 3.14, we also have a subgroup $G_1' \subset Z(a_1) \subset P(a_1)$, where $\psi' = \psi|_{G_1'}$ is an isogeny onto G_1. Let

$$P_1' = \psi'^{-1}(P_1) = \{g \in G_1' \mid \psi(g) \text{ normalizes } C_1, \ldots, C_{n-1}\} = P \cap G_1' \,.$$

Proposition 4.4 *For any $g \in G_1$, there exists $g' \in Z(a_1)$ such that $\psi(g') = g$ and $g'\varepsilon_1 = \varepsilon_1$. If $g \in P_1$, we may take g' to be in P.*

Proof This follows from the above discussion and from the fact that G_1' is in the kernel of the restriction of $Z(a_1)$ to $V_1(\varepsilon_1)$. $\qquad\qquad\square$

Proposition 4.5 *For any Siegel set $\mathfrak{S}_\omega$, we have*

$$\overline{\mathfrak{S}}_\omega \cap \overline{C_1} = \overline{\mathfrak{S}}_{\psi(\omega)} \,.$$

Proof We simply compute:

$$
\begin{aligned}
\overline{\mathfrak{S}}_\omega \cap \overline{C_1} &= \overline{\omega A^+ p} \cap \overline{C_1} \\
&= \overline{\omega A^+ p \cap \overline{C_1}} \ (\text{since } \omega \text{ is compact}) \\
&= \omega(\overline{A^+ p \cap \overline{C_1}}) \ (\text{since } \omega \text{ normalizes } \overline{C_1}) \\
&= \psi(\omega)\overline{A_1^+ p_1} \ (\text{by Lemma 4.2}) \\
&= \overline{\psi(\omega)A_1^+ p_1} \ (\text{since } \psi(\omega) \text{ is compact}) \\
&= \overline{\mathfrak{S}}_{\psi(\omega)} \,.
\end{aligned}
$$

$\qquad\qquad\square$

4.3

We now compare Siegel sets with polyhedral cones. First of all, by a *polyhedral cone* $\pi \subset V$ we mean a closed set defined equivalently as

$$\{x \in V \mid \ell_i(x) \geq 0 \text{ for all } i = 1, \ldots, k\} \,,$$

for some finite set of linear functions ℓ_i, or as

$$\left\{ \sum_{i=1}^{k} \lambda_i y_i \in V \mid \lambda_i \geq 0 \right\} \,,$$

for some finite set of vectors $y_i \in V$. A word of caution: by the conventions we have been forced into

a polyhedral cone is, by definition, closed;
a homogeneous self-adjoint cone is, by definition, open.

Hopefully there are only a few places where this might cause confusion. A polyhedral cone is called $\mathbb{Q}$-*rational* if one can choose the ℓ_i (or the y_i) to be $\mathbb{Q}$-rational.

We assume, as above, that V is defined over $\mathbb{Q}$, and that C is a homogeneous self-adjoint cone in V, and that $G = \mathrm{Aut}(C,V)^o$ is defined over $\mathbb{Q}$ and is $\mathbb{Q}$-simple. Let $P \subset G$ be a minimal $\mathbb{Q}$-parabolic, let $\mathfrak{F} = \{\overline{C} \supset \overline{C}_1 \supset \cdots \supset \overline{C}_{n-1}\}$ be the associated flag, and let $\widetilde{C} = C \cup C_1 \cup \cdots \cup C_{n-1} \cup \{0\}$. Finally, let $D^+ = A^+ p$.

The main result will be as follows.

Theorem 4.6 *The closures of the Siegel sets $\overline{\mathfrak{S}}_\omega$ in $\overline{C}$ and the polyhedral cones $\pi \subset \widetilde{C}$ are cofinal, i.e., every $\overline{\mathfrak{S}}_\omega$ is contained in some π and every π is contained in some $\overline{\mathfrak{S}}_\omega$.*

Proof To prove that every Siegel set is in a polyhedral cone is easy. Start with ωD^+ with $\omega \subset P$ compact. Then, for $i = 1, \ldots, n-1$, let π_i be a polyhedral cone generated by a finite set of elements of C_i such that $\omega p_i \subset \pi_i$, and let π_0 be a polyhedral cone generated by elements of C such that $\omega p \subset \pi_0$. (This exists because ωp_i is a compact subset of C_i.)

Then (writing p_0 for p for simplicity of notation)

$$\omega D^+ = \left\{ b \left(\sum_{i=0}^{n-1} \lambda_i p_i \right) \mid \lambda_i \geq 0, b \in \omega \right\}$$

$$\subset \left\{ \sum_{i=0}^{n-1} \lambda_i (b_i p_i) \mid \lambda_i \geq 0, b_i \in \omega \right\}$$

$$\subset \left\{ \sum_{i=0}^{n-1} q_i \mid q_i \in \pi_i \right\} = \pi_0 + \pi_1 + \cdots + \pi_{n-1},$$

as asserted.

The other inclusion is harder: as it stands, it is not well adapted to proof by induction and, instead, we prove by induction the following stronger fact.

Proposition 4.7 *For every polyhedral cone $\pi \subset \widetilde{C}$, there is a compact set $\omega \subset P$ such that*

$$p + \pi \subset \omega(p + D^+).$$

Multiplying both sides by homotheties $\mathbb{R}_{>0}$, it follows that $\pi \subset \omega D^+$ as asserted. □

Proof (of Proposition 4.7) We proceed by induction on $\dim C$. Separating the generators of the π into those in C and those in $\tilde{C}_1 = C_1 \cup C_2 \cup \cdots \cup C_{n-1} \cup \{0\}$, we write $\pi = \pi_0 + \pi_1$, where $\pi_0 \subset C$, $\pi_1 \subset \tilde{C}_1$. Let $\mathbb{R}^{\geq 1}$ be the reals $[1, \infty)$. We claim that, for suitably large ω,

$$p + \pi_0 \subset \mathbb{R}^{\geq 1} \omega p .$$

Let H be the hyperplane of points of the form $p + q$, where $\langle p, q \rangle = 0$. Then $\pi_0 \cap H$ is compact and, if $\pi_0 \cap H \subset \omega p$, it follows that $p + \pi_0 \subset \mathbb{R}^{\geq 1} (\pi_0 \cap H) \subset \mathbb{R}^{\geq 1} \omega p$:

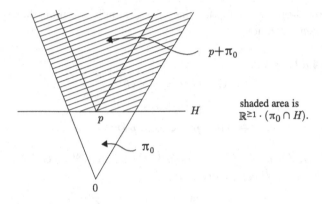

shaded area is $\mathbb{R}^{\geq 1} \cdot (\pi_0 \cap H)$.

Then

$$
\begin{aligned}
p + \pi &= (p + \pi_0) + \pi_1 \\
&\subset \mathbb{R}^{\geq 1} \omega p + \pi_1 \\
&\subset \mathbb{R}^{\geq 1} \omega (p + \omega^{-1} \pi_1) \\
&= \mathbb{R}^{\geq 1} \omega (\varepsilon_1 + (p_1 + \pi_1')) ,
\end{aligned}
$$

where, in the last step, π_1' is a larger polyhedral cone in $\tilde{C}_1$ containing $\omega^{-1} \pi_1$. Now use the induction hypothesis. Hence, there exists a compact subset

$$\omega_1 \subset P_1 = \{g \in \operatorname{Aut}(C_1)^o \mid g \text{ fixes the flag } \overline{C}_1 \supset \overline{C}_2 \supset \cdots \supset \overline{C}_{n-1}\} ,$$

such that

$$p_1 + \pi_1' \subset \omega_1 (p_1 + D_1^+) ,$$

where $D_1^+ = \overline{A_1^+ p_1}$ is the polyhedral cone generated by $p_1, p_2, \ldots, p_{n-1}$. Let $P_1' \subset P$ be the subgroup defined in Section 4.2 such that $P_1' \longrightarrow P_1$ is surjective with finite kernel and P_1' acts identically on the boundary component containing ε_1. Let ω_1' be the inverse image of ω_1 in P_1'. Then

$$
\begin{aligned}
p + \pi \subset \mathbb{R}^{\geq 1} \omega(\varepsilon_1 + \omega_1 (p_1 + D_1^+)) \\
= \mathbb{R}^{\geq 1} \omega \omega_1' (\varepsilon_1 + p_1 + D_1^+) \\
= \omega \omega_1' [\mathbb{R}^{\geq 1} (p + D_1^+)] \, .
\end{aligned}
$$

But

$$
\begin{aligned}
\mathbb{R}^{\geq 1}(p + D_1^+) &= \{ (1 + \lambda)(p + a) \mid a \in D_1^+, \lambda \geq 0 \} \\
&= \{ p + (\lambda p + a) \mid a \in D_1^+, \lambda \geq 0 \} \\
&= p + D^+ \, ,
\end{aligned}
$$

so $p + \pi \subset \omega \omega_1' (p + D^+)$, as asserted. $\square$

If we now drop the assumption that G is $\mathbb{Q}$-simple, we immediately get the following generalization.

Theorem 4.8 *Let*

$$
G = G^{(1)} \times \cdots \times G^{(k)} \, ,
$$

$$
C = C^{(1)} \times \cdots \times C^{(k)} \, ,
$$

$$
P = P^{(1)} \times \cdots \times P^{(k)} \, ,
$$

where $G^{(i)}$ are $\mathbb{Q}$-simple, $G^{(i)} = \mathrm{Aut}(C^{(i)}, V^{(i)})^o$, and $P^{(i)} \subset G^{(i)}$ is a minimal $\mathbb{Q}$-parabolic. Let $P^{(i)}$ correspond to a flag

$$
\mathfrak{F}^{(i)} = \{ \overline{C}^{(i)} = \overline{C}_0^{(i)} \supset \overline{C}_1^{(i)} \supset \cdots \supset \overline{C}_{n_i - 1}^{(i)} \} \, ,
$$

and let

$$
\tilde{C} = \prod_{i=1}^{k} \bigcup_{j=0}^{n_i - 1} \overline{C}_j^{(i)} \, .
$$

Then the closures of the Siegel sets $\overline{\mathfrak{S}}_\omega$ in $\tilde{C}$ and the polyhedral cones $\pi \subset \tilde{C}$ are cofinal. $\square$

Now let C^* be the union of C and all its $\mathbb{Q}$-rational boundary components. Then, combining Theorem 4.8 with the main results of reduction theory (via Siegel sets), we find:

Corollary 4.9

(i) *For every arithmetic subgroup $\Gamma \subset G$ and every pair of closed polyhedral cones $\pi_1, \pi_2 \subset C^*$, the set*

$$
\{ \gamma \in \Gamma \mid \gamma \pi_1 \cap \pi_2 \cap C \neq \emptyset \}
$$

is finite.

(ii) *For every arithmetic subgroup* $\Gamma \subset G$, *there exists a closed polyhedral cone* $\pi \subset C^*$ *such that* $(\Gamma\pi) \cap C = C$.

$\square$

Moreover it is now an easy matter to construct a polyhedral fundamental domain for the action of an arithmetic group Γ on C in the following sense.

Definition 4.10 A decomposition of C into rational polyhedral cones $\{\sigma_\alpha\}$ is called a Γ-*admissible polyhedral decomposition* of C if the following properties are satisfied:

(1) a face of a σ_α is a σ_β;
(2) $\sigma_\alpha \cap \sigma_\beta$ is a common face of σ_α and σ_β;
(3) $\gamma\sigma_\alpha$ is a σ_β, for all $\gamma \in \Gamma$;
(4) $\mathrm{mod}\,\Gamma$, there are only finitely many σ_α;
(5) $C = \bigcup_\alpha (\sigma_\alpha \cap C)$.

In fact, just choose a Siegel set $\mathfrak{S}$ and a finite set $F \subset G(\mathbb{Q})$ such that $C = \Gamma F \mathfrak{S}$. Let π be a polyhedral cone with rational vertices in $\widetilde{C}$ such that $\mathfrak{S} \subset \pi$. If $H_1, \ldots, H_m$ are the hyperplanes defining π, consider

$$\mathcal{H} = \{\gamma f H_i \mid \gamma f \pi \cap \pi \neq \emptyset, \text{ where } \gamma \in \Gamma, f \in F, i = 1, \ldots, m\}.$$

Since π is contained in some Siegel set, the property of Siegel sets guarantees that the images by Γ of the connected components of $\pi \setminus \bigcup_{H \in \mathcal{H}} H$ and their faces solve the problem.

5 Cores and co-cores

5.1

We have $C \subset V$, with a compatible lattice $L \subset V$ as usual. We also have the characteristic function $\varphi : C \longrightarrow \mathbb{R}_{>0}$, which is defined up to a constant (see Section 1). The following proposition will enable us to make a sensible normalization of φ.

Proposition 5.1 *The function* φ *is bounded on* $C \cap L$.

Proof For any open cone A in V, let A^* denote its dual in V^*, as in Section 1. Let π be an open simplicial cone in $\overline{C}$ with vertices $x_1, \ldots, x_r \in \overline{C} \cap L$. Then $\pi^* = \{\ell \in V^* \mid \ell(x_i) > 0 \text{ for all } i = 1, \ldots, n\}$. If we choose $\ell_1, \ldots, \ell_r \in V^*$ to be the dual basis to $x_1, \ldots, x_r$, then π^* is simplicial with vertices $\ell_1, \ldots, \ell_r$ and $\ell_1, \ldots, \ell_r \in L^* \otimes \mathbb{Q}$.

Now, for any cone A, let

$$\varphi_A(x) = \int_{A^*} e^{-\langle x, \ell \rangle} \, d\ell \, .$$

Thus our φ is just φ_C.

Since $\pi \subset C$ implies that $C^* \subset \pi^*$, we know

$$\varphi_C(x) \leq \varphi_\pi(x) = m \int_{\mathbb{R}^r_{>0}} e^{-\langle x, \Sigma a_i \ell_i \rangle} \, da \, ,$$

where m is some constant.

Let d be the volume of the parallelopiped spanned by $x_1, \ldots, x_r$. If now $x = \Sigma b_i x_i \in \pi \cap L$, then $b_i \geq \frac{1}{d}$ for all i, and

$$\varphi_C(x) \leq \varphi_\pi(x) = m \int_{\mathbb{R}^r_{>0}} e^{-\Sigma b_i a_i} \, da \leq m \int_{\mathbb{R}^r_{>0}} e^{-\frac{1}{d} \Sigma a_i} \, da < M \, ,$$

where M is some constant depending only on π.

By the reduction theory, we have $C = \bigcup \sigma_\alpha$, where the σ_α are polyhedral rational cones open in their linear span, and there are only a finite number of them modulo Γ. Since any polyhedron can be written as the union of simplices, we may assume each σ_α is simplicial. Since $\varphi(\gamma x) = \varphi(x)$ for all $\gamma \in \Gamma$ and $x \in C$, we have shown that

> φ is bounded on $L \cap$ (the union of the top-dimensional simplices).

We continue by induction. Assume that

> φ is bounded on $L \cap$ (the union of the σ_α with $\dim \sigma_\alpha \geq s$).

Suppose σ_β has dimension $s - 1$. Let $\text{Star}\,\sigma_\beta$ be the union of all simplices σ_α having σ_β as a face. Then $\text{Star}\,\sigma_\beta$ is a neighborhood of σ_β. By induction, φ is bounded on $L \cap (\text{Star}\,\sigma_\beta \setminus \sigma_\beta)$, say by the constant B.

There exists a finite set $\{\lambda_i\} \subset L^*$ such that, for any $y \in \partial \text{Star}\,\sigma_\beta$, we have $\lambda_i(y) = 0$ for some i, and $\lambda_i(z) > 0$ for every $z \in \sigma_\beta$ and all i. Because $\lambda_i(x) \geq 1$ for each i and each $x \in \sigma_\beta \cap L$, we see that

$$\text{dist}\,(L \cap \sigma_\beta, \partial \text{Star}\,\sigma_\beta) > \delta > 0 \, ,$$

where dist denotes the distance between the two sets.

Choose $v \in (\text{Star}\,\sigma_\beta \setminus \sigma_\beta) \cap L$, and pick $m > \frac{|v|}{\delta}$. We claim that, for any $x \in L \cap \sigma_\beta$, both $mx + v$ and $mx - v$ are in $\text{Star}\,\sigma_\beta \setminus \sigma_\beta$. For $mx + v$, the claim is trivial, but, since the $m\delta$-ball about mx is contained in $\text{Star}\,\sigma_\beta$ and $|v| < m\delta$, the claim is true for $mx - v$ as well.

Thus, $\varphi(mx) = \varphi(\frac{1}{2}(mx - v) + \frac{1}{2}(mx + v)) \leq B$ since φ is convex. So $\varphi(x) \leq$

$m^r B$ for any $x \in L \cap \sigma_\beta$, where $\dim V = r$. As there are only finitely many σ_β modulo Γ, we see that φ is bounded on

$$L \cap (\text{union of all } \sigma_\beta \text{ with } \dim \sigma_\beta \geq s - 1) .$$

$\square$

We normalize φ so that $\max\{\varphi(x) \mid x \in C \cap L\} = 1$. More generally, for any rational boundary component C_1, we have an orthogonal projection π_{C_1} onto C_1 and we normalize φ_{C_1} such that $\max\{\varphi(\pi_{C_1}x) \mid x \in C \cap L\} = 1$.

Now consider subsets (called *kernels*) $K \subset \overline{C}$ such that $\mathbb{R}^{\geq 1} K \subset K$, and $C \subset \mathbb{R}_{>0} K$, and $0 \notin \overline{K}$. We say that two kernels K_1, K_2 are *comparable* if there exist $\lambda_1, \lambda_2 \in \mathbb{R}_{>0}$ with $\lambda_1 K_1 \subset K_2 \subset \lambda_2 K_1$. Let L' denote $L \setminus \{0\}$.

Theorem 5.2 *The following kernels are comparable:*

(a) $\Gamma(p+C)$;

(b) *the convex hull of* $C \cap L$;

(c) $\{x \in C \mid \varphi_{C_1}(\pi_{C_1}x) \leq 1 \text{ for any rational boundary component } C_1\}$;

(d) $\{x \in C \mid \langle x, y \rangle \geq 1 \text{ for all } y \in \overline{C} \cap L'\}$;

(e) *the set*

$$\Gamma \cdot \left(\bigcup_i \omega_i(A_i^+ p + p) \right)$$

for any finite collection $\{\mathfrak{S}_i = \omega_i A_i^+ p\}$ *of Siegel sets with* $\Gamma \cdot (\bigcup_i \mathfrak{S}_i) = C$.

Note that in (c) we count C as a rational boundary component.

Definition 5.3 This class of comparable kernels are the *cores* of C.

In order not to interrupt the proof of the theorem, we first prove two lemmas. If C_1 is a rational boundary component corresponding to an idempotent ε, we denote by $C_1^\perp$ the boundary component corresponding to $p - \varepsilon$.

Lemma 5.4 *Let C_1 be a rational boundary component and set $C_0 = C_1^\perp$. Take $\gamma \in \Gamma$ arbitrary and let $C_1' = (\gamma^{-1}C_0)^\perp$. Then*

$$\varphi_{C_1} \pi_{C_1} \gamma = \varphi_{C_1'} \pi_{C_1'} .$$

Proof First we show that, for all $g \in \text{Norm}(C_0)$, there exists $\overline{g} \in \text{Norm}(C_1) \cap \text{Norm}(C_0)$ such that $\pi_{C_1} g = \overline{g} \pi_{C_1}$.

Let $a(s)$ be the one-parameter subgroup $\exp(t\varepsilon)$, where ε is the identity in the Jordan algebra of C_0 and $s = e^{\frac{1}{2}t}$. Then $\pi_{C_1} = \lim_{s \to 0} a(s)$ and hence $\mathrm{Norm}\,(C_0) = P(a^{-1})$. Then

$$\pi_{C_1}(gx) = \lim_{s \to 0} a(s)gx$$
$$= \lim_{s \to 0} a(s)ga^{-1}(s)a(s)x$$
$$= \bar{g}(\pi_{C_1}x)\,,$$

where $x \in C_1$ is arbitrary and $\bar{g} = \lim_{s \to 0} a(s)ga^{-1}(s)$ exists by the definition of $P(a^{-1})$ and is in $Z(a)$.

Further, if $g \in \mathrm{Stab}\,(p)$, then $\pi_{C_1}g = g\pi_{g^{-1}C_1}$. This is simply because g is an orthogonal transformation.

Now, $\mathrm{Norm}\,(C_0)$ and $\mathrm{Stab}\,(p)$ generate G, so write $\gamma = gh$ with $g \in \mathrm{Norm}\,(C_0)$ and $h \in \mathrm{Stab}\,(p)$. Then

$$\pi_{C_1}\gamma = \pi_{C_1}gh$$
$$= \bar{g}\pi_{C_1}h$$
$$= \bar{g}h\pi_{h^{-1}C_1}\,.$$

Meanwhile, $\bar{g}h : h^{-1}C_1 \longrightarrow C_1$ is an isomorphism of homogeneous cones, and, because the characteristic function is unique up to a constant,

$$\varphi_{h^{-1}C_1} = \mu\varphi_{C_1}\bar{g}h\,,$$

for some $\mu \in \mathbb{R}_{>0}$.

Thus $\mu\varphi_{C_1}\pi_{C_1}\gamma = \mu\varphi_{C_1}\bar{g}h\pi_{h^{-1}C_1} = \varphi_{h^{-1}C_1}\pi_{h^{-1}C_1}$. So, by the way we normalized φ_{C_1} and $\varphi_{h^{-1}C_1}$, and since γ fixes the lattice, it is clear that $\mu = 1$.

It remains to check that $h^{-1}C_1 = (\gamma^{-1}C_0)^\perp$. But $\gamma^{-1}C_0 = h^{-1}g^{-1}C_0 = h^{-1}C_0$, so if ε is the idempotent of C_0, it follows that $h^{-1}C_0$ has $h^{-1}\varepsilon_0$ as its idempotent, and thus $(\gamma^{-1}C_0)^\perp$ has $p - h^{-1}\varepsilon_0 = h^{-1}(p - \varepsilon_0)$, which is the idempotent of $h^{-1}C_1$.   □

Corollary 5.5 *The set in part* (c) *in the theorem is* Γ*-invariant.*

Proof This follows from the lemma and the fact that application of any $\gamma \in \Gamma$ sets up a bijection of rational boundary components.   □

Lemma 5.6 *Let* $\omega A^+ p$ *be a Siegel set. For a maximal set* $\varepsilon_1, \ldots, \varepsilon_n$ *of orthogonal idempotents of* $\mathrm{Lie}\,A$, *let*

$$\ell(y) = \frac{\langle \varepsilon_n, y \rangle}{\langle \varepsilon_n, \varepsilon_n \rangle}$$

for $y \in C$, *where the inner product is chosen as usual so that* A *consists of self-adjoint transformations.*

Then there exists a constant $m > 0$ such that

$$\ell(ghp) \geq R \text{ for some } g \in \omega, h \in A^+ \implies \ell(hp) \geq \frac{R}{m}.$$

Proof We know that $\bigcup_{h \in A^+} \{h^{-1}\omega h\}$ is relatively compact, so it makes sense to define

$$m = \sup\{\ell(h^{-1}ghp) \mid g \in \omega, h \in A^+\}.$$

Now, if $q \in C$ is arbitrary, we show next that $\ell(hq) = \ell(hp)\ell(q)$. In fact, write $q = \lambda \varepsilon_n + f$ with $\lambda \in \mathbb{R}_{>0}$ and f orthogonal to ε_n. Since $h \in A$ is self-adjoint, and ε_n is an eigenvector for anything in A, it follows that hf is orthogonal to ε_n. Thus,

$$\ell(hq) = \ell(h\lambda\varepsilon_n + hf) = \lambda\ell(h\varepsilon_n).$$

Apply this to $q = p$, remembering that the ε_i are mutually orthogonal: $\ell(hp) = \ell(h\varepsilon_n)$. By definition, $\ell(q) = \lambda$. Thus $\ell(hq) = \ell(hp)\ell(q)$. We conclude that if $\ell(hp) < \frac{R}{m}$, then $\ell(ghp) = \ell(h(h^{-1}gh)p) < \frac{R}{m}\ell(h^{-1}ghp) < R$, this being the contrapositive of what the lemma states. $\qquad\square$

Proof (of Theorem 5.2) We will prove the theorem in a chain, showing each set in parts (a) through (e) is contained in some dilation of the subsequent set.

(1) Let H be the convex hull of $C \cap L$. Choose $M \in \mathbb{Z}_{>0}$ such that $Mp = w$ for some lattice point w. Since H is Γ-invariant, it suffices to show that $p + C \subset \frac{1}{M}H$. By the reduction theory, C is covered by the Γ-translates of a finite union of rational polyhedra σ_α. So it is enough to show, given $\gamma \in \Gamma$ and σ_α, that $p + \gamma^{-1}\sigma_\alpha \subset \frac{1}{M}H$, or, equivalently, that $\gamma p + \sigma_\alpha \subset \frac{1}{M}H$.

If x is any point of σ_α, we can write $x = \sum_{i=1}^{k} t_i v_i$, where $v_i \in L \cap \overline{C}$ and $\sum t_i \leq 1$. Then

$$M\left(\gamma p + \sum t_i v_i\right) = \sum t_i(Mv_i + \gamma w) + \left(1 - \sum t_i\right)(\gamma w) \in H \cdot$$

because $Mv_i + \gamma w$ and γw are in $L \cap C$. We conclude that $\gamma p + \sigma_\alpha \subset \frac{1}{M}H$.

(2) Let J be the set described in part (c) of the theorem. Then $H \subset J$ simply by the way we normalized the φ_{C_1}s, and by their convexity.

(3) We must show that there exists $\lambda \in \mathbb{R}_{>0}$ such that, if $x \in C$ and $\varphi_{C_1}(\pi_{C_1}x) \leq 1$ for every rational boundary component C_1, then $\langle x, y \rangle \geq \lambda$ for all $y \in \overline{C} \cap L$.

We will prove this by induction on $\dim C$. If $\dim C = 0$, there is nothing to prove.

Now suppose $\dim C > 0$. If C_1 is a rational boundary component, the hypothesis for x is inherited by $\pi_{C_1} x$ and so, by induction, there exists λ_1 (depending on C_1) such that $\langle \pi_{C_1} x, y \rangle \geq \lambda_1$ for all $y \in \overline{C}_1 \cap L$. But $\langle x, y \rangle = \langle \pi_{C_1} x, y \rangle$, so $\langle x, y \rangle \geq \lambda_1$ in addition.

Now there are only a finite number of $^t\Gamma$-orbits of rational boundary components (see Proposition 15.6 in [3]). Let $C_1, \ldots, C_s$ be representatives for the $^t\Gamma$-orbits of rational boundary components and let $\lambda = \min(\lambda_1, \ldots, \lambda_s)$. If C_0 is another rational boundary component, pick $\gamma \in \Gamma$ such that $^t\gamma C_0 = C_i$ for some i between 1 and s. If $y \in \overline{C}_0 \cap L'$, then $^t\gamma y \in \overline{C}_i \cap L'$ and $\gamma^{-1} x \in K$ by Lemma 5.4 so that

$$\langle x, y \rangle = \langle \gamma^{-1} x, {}^t\gamma y \rangle \geq \lambda \ .$$

It remains to consider $y \in C \cap L$. We need only show that there exists some $\lambda' > 0$ such that $\varphi_C(x) \leq 1 \implies \langle x, y \rangle \geq \lambda' > 0$ for all $y \in \sigma \cap L \cap C$, where σ is spanned by $y_1, \ldots, y_r \in L \cap \overline{C}$. If $y = \sum a_i y_i$ and $d = \det(y_1, \ldots, y_r)$, then $y \in L \cap C \implies a_i \in \frac{1}{d}\mathbb{Z}$ and $a_i \geq 0$ for each i. Furthermore $y' = \sum_{i \mid a_i > 0} y_i \in L \cap C$, as otherwise $\{y_i \mid a_i > 0\}$ would be contained in a proper boundary component and hence so would y. But $\langle x, y \rangle \geq \frac{1}{d}\langle x, y' \rangle$ and $\langle x, y' \rangle$ is bounded away from zero because

(i) the set $\{x \mid \varphi_C(x) \leq 1\}$ is in the complement of some ball B around 0 by Proposition 1.6;

(ii) $\{\langle x, y' \rangle \mid x \in C \setminus B\}$ is bounded away from zero because $y' \in C$;

(iii) there are only finitely many possible y'.

This gives the desired $\lambda' > 0$ with the property that $\varphi_C(x) \leq 1 \implies \langle x, y \rangle \geq \lambda'$ for all $y \in \sigma \cap L \cap C$.

(4) We know that $C = \Gamma \cdot \bigcup_{i=1}^{n} \omega_i A_i^+ p$, so, for any $x \in C$, we have $x = \gamma g a p$ with $\gamma \in \Gamma$, $g \in \omega_i$ and $a \in A_i^+$ for some i. We must show that there exists $M > 0$ such that $\langle x, y \rangle \geq 1$ for all $y \in \overline{C} \cap L'$ implies that $M \cdot a p \in A_i^+ p + p$.

Let $\varepsilon_1, \ldots, \varepsilon_n$ be the maximal set of orthogonal idempotents of $\mathrm{Lie}\, A$, so that $A_i^+ p$ is the convex cone spanned by $\varepsilon_1, \varepsilon_1 + \varepsilon_2, \ldots, \varepsilon_1 + \cdots + \varepsilon_n = p$. Then it is easy to see that, if $x \in A_i^+ p$, then $x \in A_i^+ p + p$ if and only if $\langle \varepsilon_n, x \rangle \geq \langle \varepsilon_n, \varepsilon_n \rangle$.

Choose $M_1 \in \mathbb{Z}_{>0}$ so that $^t\gamma M_1 \varepsilon_n$ is a lattice point for all $\gamma \in \Gamma$. (This is possible because L and L^* are commensurate.) The hypothesis on x says that $\langle {}^t\gamma^{-1} M_1 \varepsilon_n, x \rangle \geq 1$. That is,

$$M_1 \langle \varepsilon_n, g a p \rangle = M_1 \langle {}^t\gamma^{-1} \varepsilon_n, \gamma g a p \rangle \geq 1 \ .$$

Letting m be the constant in Lemma 5.6, we get $M_1 \langle \varepsilon_n, a p \rangle \geq \frac{1}{m}$. If $M =$

$M_1 m \langle \varepsilon_n, \varepsilon_n \rangle$, then $\langle \varepsilon_n, Map \rangle \geq \langle \varepsilon_n, \varepsilon_n \rangle$, and we are done.

(5) For this last step, it clearly is enough to show $M \cdot \omega(A^+ p + p) \subset C + p$ for some $M > 0$.

Now, $A^+ p + p \subset C + p$, so that $\omega(A^+ p + p) \subset \omega C + \omega p = C + \omega p$. Since ω is compact, we may choose $M > 0$ so that $M \cdot \omega p \subset C + p$. Indeed, just pick M such that $q - \frac{1}{M} p \in C$ for all $q \in \omega p$. Then $M \cdot \omega(A^+ p + p) \subset M \cdot C + M \cdot \omega p \subset C + C + p \subset C + p$. $\qquad\square$

Corollary 5.7 *Any core is comparable to its closed convex hull.*

Proof Some of the cores in the list of the theorem are closed and convex. Let A be one of them and let B be any core. Then there are $\lambda_1, \lambda_2, \mu_1, \mu_2 \in \mathbb{R}_{>0}$ such that $\lambda_1 A \subset B \subset \lambda_2 A$ and $\mu_1 B \subset A \subset \mu_2 B$. The first implies that

$$\lambda_1 A \subset \text{closed convex hull of } B \subset \lambda_2 A ,$$

and so

$$\lambda_1 \mu_1 B \subset \text{closed convex hull of } B \subset \mu_2 \lambda_2 B .$$

Thus the closed convex hull of any core is again a core. $\qquad\square$

5.2

We need another class of comparable kernels that stand in duality to the cores of C. Recall that $\mathbb{R}^{\geq 1}$ denotes the set of real numbers $[1, \infty)$. We call $A \subset V$ a *semi-conical convex set* if A is convex and $\mathbb{R}^{\geq 1} A = A$. For any set $A \subset V$, define:

$$\text{semi-hull}(A) = \text{closed convex hull of } \mathbb{R}^{\geq 1} A ;$$
$$\check{A} = \{ x \in V \mid \langle x, a \rangle \geq 1 \text{ for all } a \in A \} .$$

Note that $\check{A} = \emptyset$ if $0 \in \overline{A}$.

Proposition 5.8 *If* semi-hull(A) *does not contain* 0, *then*

$$\check{\check{A}} = \text{semi-hull}(A) .$$

Proof Clearly $A \subset \check{\check{A}}$. Let $b \in \check{\check{A}}$ and suppose $b \notin$ semi-hull(A). Then by the separating hyperplane theorem, there is a $y \in V$ such that $\langle y, b \rangle < \lambda$ and $\langle y, a \rangle \geq \lambda$ for all $a \in$ semi-hull(A), for some $\lambda \in \mathbb{R}$.

If $\lambda > 0$, replace y by $\frac{1}{\lambda} y$. This gives $\langle y, b \rangle < 1$ and $\langle y, a \rangle \geq 1$ for all $a \in$ semi-hull(A), which implies that $y \in \check{A}$ and so $b \notin \check{\check{A}}$, a contradiction.

Meanwhile, if $\langle y,a \rangle < 0$ for some $a \in A$, then $\langle y,ta \rangle \longrightarrow -\infty$ as $t \longrightarrow +\infty$, which is impossible. Thus only the case $\lambda = 0$ remains. Now we have $\langle y,a \rangle \geq 0$ for all $a \in$ semi-hull(A) and $\langle y,b \rangle < 0$.

By the separating hyperplane theorem applied to 0 and semi-hull(A), there exists $z \in V$ with $\langle z,a \rangle \geq \mu$ for all $a \in$ semi-hull(A) and $0 = \langle z,0 \rangle < \mu$. That is, $\mu > 0$. Pick a small positive number δ so that $\langle y+\delta z,b \rangle < 0$ still holds. But $\langle y+\delta z,a \rangle \geq \delta\mu > 0$ for all $a \in$ semi-hull(A). So we are back to the previous case. $\qquad\square$

Proposition 5.9 *If K is a kernel of C, then so is $\check{K}$.*

Proof By definition, K is a kernel if and only if $K \subset \overline{C}$, and $\mathbb{R}^{\geq 1}K \subset K$, and $C \subset \mathbb{R}_{>0}K$, and $0 \notin \overline{K}$. Now, for $\check{K}$, it is automatic that $\mathbb{R}^{\geq 1}\check{K} \subset \check{K}$ and $0 \notin \check{K} = \check{\overline{K}}$. Also, $C \subset \mathbb{R}_{>0}K$ implies that, for any $x \in \check{K}$ and any $y \in C$, we have $\langle x,y \rangle > 0$, so that $\check{K} \subset \overline{C}$. Finally, for any $y \in C$, $\langle y,x \rangle$ is bounded away from 0 as x runs through K because $0 \notin \overline{K}$. Thus $C \subset \mathbb{R}_{>0}\check{K}$. $\qquad\square$

Proposition 5.10 *If two kernels K_1 and K_2 are comparable, then so are $\check{K}_1$ and $\check{K}_2$.*

Proof This follows because, if $\mu > 0$, then $(\mu K)^\vee = \mu^{-1}\check{K}$ for any set K, and, if $K_1 \subset K_2$, then $\check{K}_2 \subset \check{K}_1$ for any two sets K_1 and K_2. $\qquad\square$

So $K \mapsto \check{K}$ sets up a duality among closed convex kernels, and even among equivalence classes of comparable kernels which have the property of being comparable with their closed convex hulls (e.g., cores). Thus the duals of the cores (see Section 5.1) are a new class of comparable kernels, called *co-cores*.

In particular, denote two of the cores by

$$\begin{aligned} \Sigma_1 &= \text{closed convex hull of } C \cap L\,, \\ \Sigma_2 &= \{x \in C \mid \langle x,y \rangle \geq 1 \text{ for all } y \in \overline{C} \cap L'\}\,. \end{aligned} \tag{5.1}$$

Then the corresponding co-cores are

$$\begin{aligned} \check{\Sigma}_1 &= \{x \in \overline{C} \mid \langle x,y \rangle \geq 1 \text{ for all } y \in C \cap L\}\,, \\ \check{\Sigma}_2 &= \text{closed convex hull of } \overline{C} \cap L'\,. \end{aligned}$$

Using these co-cores we will construct some new Γ-admissible polyhedral decompositions of C.

First we must prove a few general propositions. Recall that e is an *extreme point* of a closed convex set A if and only if, for all $x,y \in A$, if $e = \frac{1}{2}(x+y)$, then $x = y = e$. We denote by $E(A)$ the set of all extreme points of A. The Krein–Milman theorem says that any compact convex set Σ is the closed convex hull of its extreme points. This fails if Σ is not compact, but we can say the following:

Proposition 5.11 *If Σ is a closed convex kernel, then Σ is the closed convex hull of $\bigcup_{e \in E(\Sigma)} (e + C)$.*

Proof For any $e \in E(\Sigma)$, $r \in C$, there is a $\lambda \in \mathbb{R}_{>0}$ such that $\lambda r \in \Sigma$, since Σ is a kernel. Then we have $\frac{\lambda}{\lambda+1} e + \frac{1}{\lambda+1} (\lambda r) \in \Sigma$; that is, $\frac{1}{\lambda+1} (e + r) \in \Sigma$. Since Σ is a semi-cone, $e + r \in \Sigma$. Thus Σ contains the closed convex hull of $\bigcup (e + C)$.

Conversely, suppose $q \in \Sigma$, but q is not contained in the closed convex hull of $\bigcup (e + C)$. By the separating hyperplane theorem, there exists $\lambda \in \mathbb{R}$, $w \in V$ such that $\langle w, e + r \rangle \geq \lambda$ for all $e \in E(\Sigma)$, $r \in C$, but $\langle w, q \rangle < \lambda$.

If we had $\langle w, r \rangle < 0$ for some $r \in C$, then $\langle w, e + tr \rangle \longrightarrow -\infty$ as $t \longrightarrow \infty$, which is impossible. Therefore $w \in \overline{C}$, and then $\langle w, q \rangle < \lambda$ implies that $\lambda > 0$. In the case when $w \in \partial C$, choose $y \in C$ of small enough norm so that $\langle w + y, q \rangle < \lambda$. Of course, $\langle w + y, e + r \rangle \geq \lambda$ still, so, replacing w by $w + y$, we may always assume $w \in C$.

Let $H = \{ z \in V \mid \langle z, w \rangle \leq \lambda \}$. Then $H \cap \Sigma$ is closed and convex. It is also compact, since $w \in C$. By the Krein–Milman theorem, since $q \in H \cap \Sigma$ and $\langle q, w \rangle < \lambda$, there must exist an extreme point e_0 of $H \cap \Sigma$ not on the hyperplane $\{ z \mid \langle z, w \rangle = \lambda \}$. Then e_0 is an extreme point of Σ and $\langle w, e_0 \rangle < \lambda$, a contradiction. $\qquad\square$

Corollary 5.12 *If Σ is a closed convex kernel, $\check{\Sigma} = E(\Sigma)^{\vee} \cap \overline{C}$.*

Proof Clearly, $E(\Sigma) \subset \Sigma$, so $\check{\Sigma} \subset E(\Sigma)^{\vee}$. Conversely, suppose $\langle w, e \rangle \geq 1$ for all $e \in E(\Sigma)$, for some $w \in \overline{C}$. Then $\langle w, e + r \rangle \geq 1$ for all $e \in E(\Sigma)$, $r \in C$, so, by the previous proposition, $\langle w, s \rangle \geq 1$ for all $s \in \Sigma$. $\qquad\square$

Next we show that among Γ-invariant kernels, cores and co-cores represent the two extremes. Precisely, we have:

Proposition 5.13 *If Σ is a Γ-invariant closed convex kernel, then there is a core K_1 and a co-core K_2 such that $K_1 \subset \Sigma \subset K_2$.*

Proof Since Σ is a kernel, we have $\lambda p \in \Sigma$ for some $\lambda \in \mathbb{R}_{>0}$. Then $\lambda(\Gamma p) \subset \Sigma$. Dualizing gives

$$\check{\Sigma} \subset [\lambda(\Gamma p)]^{\vee}.$$

We know a priori that $\check{\Sigma} \subset \overline{C}$, so, just as in the proof of the corollary above, we have

$$\check{\Sigma} \subset [\lambda(\Gamma p)]^{\vee} \cap \overline{C} = [\lambda(\Gamma p) + C]^{\vee} = [\lambda \Gamma(p + C)]^{\vee}.$$

Thus $K_1 = [\lambda \Gamma(p + C)]^{\vee\vee}$ is a core and $K_1 \subset \Sigma$. Now $\check{\Sigma}$ is itself a ${}^t\Gamma$-invariant closed convex kernel, so, by what we have just proved, there is a core $K_2 \subset \check{\Sigma}$. Then $K_3 = \check{K}_2$ is a co-core and $\Sigma = \overset{\vee}{\check{\Sigma}} \subset K_3$.   □

5.3

The idea is to make a Γ-admissible polyhedral decomposition by taking the cones over the faces of some Γ-invariant kernel Σ. Roughly speaking, Σ must be "locally polyhedral", so that its faces will be polygons. We also want the number of faces to be finite modulo Γ.

For example, take $C = \mathbb{R}^2_{\geq 0}$, $L = \mathbb{Z}^2$. In this case the cores Σ_1, Σ_2 in (5.1) coincide.

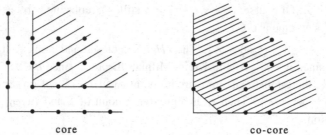

core   co-core

One sees that the problem with a core is that a face can be parallel to the boundary of the cone, and so contain an infinite number of vertices. We will deal with co-cores instead.

Let C^* be the union of C with all rational boundary components.

Definition 5.14 † A closed convex kernel Σ is called *rationally locally polyhedral* if, for any rational polyhedral cone Π with vertices in C^*, there exist $x_1, \ldots, x_s \in V(\mathbb{Q}) \cap \overline{C}$ and $\lambda_1, \ldots, \lambda_s \in \mathbb{Q}_{>0}$ such that

$$\Pi \cap \Sigma = \{y \in \Pi \mid \langle x_i, y \rangle \geq \lambda_i \text{ for } i = 1, \ldots, s\}.$$

Definition 5.15 We shall call a Γ-invariant rational locally polyhedral closed and convex kernel a *Γ-polyhedral kernel*.

Proposition 5.16 *If Σ is a Γ-polyhedral kernel, then there exists $M \in \mathbb{Z}$ such that $E(\Sigma) \subset \frac{1}{M}L$.*

---

† This definition differs from the corresponding definition in the first edition.

Proof Using our reduction theory, choose $\Pi_1, \ldots, \Pi_n$, a fundamental set of rational polyhedra for Γ. It is easy to see that they may be chosen so that their projections into any standard boundary component form a fundamental set in the boundary component. So,

$$\bigcup_{\substack{\gamma \in \Gamma \\ i=1,\ldots,n}} \gamma \Pi_i = C^* .$$

Clearly, $E(\Sigma) \cap \Pi_i \subset E(\Sigma \cap \Pi_i)$ for $i = 1, \ldots, n$.

By the definition of Γ-polyhedral, $\Sigma \cap \Pi_i$ is cut out of V by a finite number of rational hyperplanes. Therefore, $E(\Sigma \cap \Pi_i)$ is a finite set of rational points, and there exists $M \in \mathbb{Z}$ such that $E(\Sigma \cap \Pi_i) \subset \frac{1}{M}L$, for $i = 1, \ldots, n$.

For any $e \in E(\Sigma) \cap C^*$ there is a $\gamma \in \Gamma$ and $i \in \{1, \ldots, n\}$ such that $\gamma e \in \Pi_i$. Since Σ is Γ-invariant, we have $\gamma e \in E(\Sigma)$. Thus $E(\Sigma) \cap C^* = \bigcup \Gamma[E(\Sigma) \cap \Pi_i] \subset \frac{1}{M}L$.

Furthermore, $\Sigma \cap C$ is contained in the closed convex hull Σ' of the set $E(\Sigma) \cap C^* + C$, since, clearly, $\Sigma \cap \Pi_i$ is contained in Σ' and Σ' is Γ-invariant. Because $\Sigma \cap C$ is dense in Σ, also $\Sigma \subset \Sigma'$. But Σ is convex and closed, so that $\Sigma = \Sigma'$. Now we need the following lemma†:

Lemma 5.17 *Let D be a discrete subset of $\overline{C}$ and let X be the closed convex hull of $D + C$. Then every extreme point of X belongs to D.*

Proof Recall the definition of an exposed point: a point p in a closed convex set Y is said to be *exposed* if there exists a linear function ϕ such that $\phi(p) = 0$ but $\phi|_{Y \setminus \{p\}} > 0$. Then Straszewicz's theorem (see [12], Theorem 18.6) states that the exposed points are dense in the extreme points. Since D is discrete, it is enough to see that every exposed point p of X belongs to D. Since $p + \overline{C} \subset X$, we have $\phi|_{\overline{C}} \geq 0$ and $\phi_{\overline{C} \setminus \{0\}} > 0$. But then

$$\phi(p) = \inf(\phi|_X) = \inf(\phi|_{D+C}) = \inf(\phi|_D)$$

is assumed in some point $p' \in D$ and hence $p = p' \in D$. □

Taking $D = E(\Sigma) \cap C^*$ in the lemma, we have $X = \Sigma' = \Sigma$. We conclude that $E(\Sigma) = E(X) \subset D = E(\Sigma) \cap C^*$, and we have already seen that $E(\Sigma) \cap C^* \subset \frac{1}{M}L$.

□

For any $y \in C$, write

$$H_y = \{z \in V \mid \langle z, y \rangle = 1\} .$$

† This lemma is taken from [9] and replaces an incorrect argument in the first edition.

The following proposition will give us some explicit Γ-admissible decompositions of C. Recall that if X is a closed convex set and H is a hyperplane, then H is said to be a *supporting hyperplane* of X if X lies entirely in one of the two closed half-planes defined by H and $X \cap H \neq \emptyset$. In particular, $X \cap H$ is a face of X.

Proposition 5.18 *Let Σ be a Γ-polyhedral co-core, and let Y be the set of $y \in \overline{C}$ such that H_y is a supporting hyperplane of Σ which meets $E(\Sigma)$ in a set of points spanning V. Let σ_y be the cone over $H_y \cap \Sigma$. Then the set $\mathscr{S}$ of all faces of σ_y (including σ_y), as y ranges through Y, is a Γ-admissible polyhedral decomposition of C, cf. Definition 4.10.*

Proof First, we claim that in fact $Y \subset C$. Indeed, assume $y \in Y \cap \partial C$. By Proposition 5.16, y must be rational. Hence there exists some $z \in \overline{C} \setminus \{0\}$ rational with $\langle y, z \rangle = 0$. But then some multiple of z lies in $\overline{C} \cap L \setminus \{0\}$, and hence some multiple of z lies in the co-core Σ. But this contradicts the assumption that H_y is a supporting hyperplane of Σ.

We have several things to check.

(1) Since $y \in C$, it follows that $H_y \cap \Sigma$ is compact and so is supported by its extreme points. Since H_y is a supporting hyperplane, $E(H_y \cap \Sigma) \subset E(\Sigma)$. By Proposition 5.16, $E(H_y \cap \Sigma) \subset \frac{1}{M} L$ for some $M \in \mathbb{Z}$, and so $H_y \cap \Sigma$ is the closed convex hull of a finite set of points. Therefore σ_y is truly a polyhedral cone, and so is each of its faces.

(2) By definition, $\mathscr{S}$ is Γ-invariant and, if $\sigma \in \mathscr{S}$, then any face of σ is in $\mathscr{S}$.

(3) We show that if $\sigma, \tau \in \mathscr{S}$, then $\sigma \cap \tau$ is a face of both σ and τ. First, suppose both σ and τ are top-dimensional, so $\sigma = \sigma_y$ and $\tau = \sigma_z$ for some $y, z \in Y$. Then $\sigma_y = \mathbb{R}_{\geq 0}(H_y \cap \Sigma)$ and $\sigma_z = \mathbb{R}_{\geq 0}(H_z \cap \Sigma)$. Clearly, $\mathbb{R}_{\geq 0}(H_z \cap H_y \cap \Sigma) \subset \sigma_y \cap \sigma_z$. If $0 \neq w \in \sigma_y \cap \sigma_z$, there exist $\lambda, \mu \in \mathbb{R}_{>0}$ with $\lambda w \in H_y \cap \Sigma$ and $\mu w \in H_z \cap \Sigma$. We must have $\lambda = \mu$, for, if $\lambda > \mu$, say, then $\langle \lambda w, y \rangle = 1 \implies \langle \mu w, y \rangle < 1 \implies \mu w \notin \Sigma$. Thus $\sigma_y \cap \sigma_z = \mathbb{R}_{\geq 0}(H_z \cap H_y \cap \Sigma)$. Since $H_z \cap H_y \cap \Sigma$ is a face of $H_y \cap \Sigma$, we conclude that $\sigma_y \cap \sigma_z$ is a face of σ_y, and similarly of σ_z.

Now suppose σ and τ are arbitrary, so that $\sigma = \sigma_y \cap K_1 \cap K_2 \cap \cdots \cap K_n$, where $K_1, \ldots, K_n$ are supporting hyperplanes of σ_y, and $\tau = \sigma_z \cap K_{n+1} \cap \cdots \cap K_m$, where $K_{n+1}, \ldots, K_m$ are supporting hyperplanes of σ_z. Then $\sigma \cap \tau = \sigma_y \cap \sigma_z \cap K_1 \cap \cdots \cap K_m$. We have just seen that there is a hyperplane K supporting σ_y, so that $\sigma_y \cap \sigma_z = \sigma_y \cap K$. So

$$\sigma \cap \tau = \sigma_y \cap K \cap K_1 \cap \cdots \cap K_m = \sigma \cap K \cap K_{n+1} \cap \cdots \cap K_m .$$

Since $\sigma \cap K \subset \sigma_y \cap K = \sigma_y \cap \sigma_z$, we know that $K_{n+1}, \ldots, K_m$ all support $\sigma \cap K$, and K supports $\sigma \subset \sigma_y$. Therefore $\sigma \cap \tau$ is a face of σ. Similarly, it is a face of τ.

(4) Finally we show that $C \subset \bigcup_{\sigma \in \mathscr{S}} \sigma$ and that $\mathscr{S}$ is a finite set modulo Γ.

Let $\Pi_1, \ldots, \Pi_n$ be a fundamental set of rational polyhedra, as in the proof of the previous proposition.

For any $i \in \{1, \ldots, n\}$, we know that $\Sigma \cap \Pi_i$ is cut out by a finite number of half-spaces. In fact, there are $w_1, \ldots, w_m \in V$ such that

$$\Pi_i = \{z \in V \mid \langle w_k, z \rangle \geq 0 \text{ for } k = 1, \ldots, m\},$$

and $x_1, \ldots, x_t \in V$, $\lambda_1, \ldots, \lambda_t \in \mathbb{R}$ such that

$$\Sigma \cap \Pi_i = \{z \in \Pi_i \mid \langle x_\ell, z \rangle \geq \lambda_\ell \text{ for } \ell = 1, \ldots, t\}.$$

Clearly, $\Sigma \cap \Pi_i$ is the union of the semi-hulls over its faces of top dimension. A face giving such a top-dimensional semi-hull will be cut out by a hyperplane K which supports $\Sigma \cap \Pi_i$ such that there are $v_1, \ldots, v_N \in K \cap \Sigma \cap \Pi_i$ spanning V and either

(i) $\langle w_{k_0}, v_a \rangle = 0$ for some $k_0 \in \{1, \ldots, m\}$ and all $a = 1, \ldots, N$,

or

(ii) $\langle x_{\ell_0}, v_a \rangle = \lambda_{\ell_0}$ for some $\ell_0 \in \{1, \ldots, t\}$ and all $a = 1, \ldots, N$.

Obviously (i) is impossible because the v_a span V, and (ii) is possible only if $\lambda_{\ell_0} \neq 0$ and $K = \{z \in V \mid \langle x_{\ell_0}, z \rangle = \lambda_{\ell_0}\}$. Letting $y = \frac{1}{\lambda_{\ell_0}} x_{\ell_0}$, we have $K = H_y$. By definition, $y \in Y$. Let $Y_0 \subset Y$ be the set of all y obtained in this way. Clearly, Y_0 is finite.

For any $x \in C$, we have $\lambda x \in \Sigma$ for some $\lambda \in \mathbb{R}_{>0}$. Thus $\gamma \lambda x \in \Sigma \cap \Pi_i$ for some $\gamma \in \Gamma$ and $i \in \{1, \ldots, n\}$, or, in other words, $\gamma x \in \sigma_y$ for some $y \in Y_0$. Since Σ is Γ-invariant, this says $x \in \tau$ for some $\tau \in \mathscr{S}$.

Meanwhile, for any $\tau \in \mathscr{S}$, let $x \in \tau$ be in the interior. Then again $\gamma x \in \sigma_y$ for some $y \in Y_0$ and some $\gamma \in \Gamma$. By (3) above, $\gamma \tau \cap \sigma_y$ is a face of both $\gamma \tau$ and σ_y. Since γx lies in the interior of $\gamma \tau$ and in σ_y, this implies that $\gamma \tau$ is a face of σ_y. Since Y_0 is a finite set, we are done. $\square$

5.4

Let Σ be a Γ-polyhedral kernel. It is easy to see from the definition and the proof of Proposition 5.18 that there are a finite number of points $x_1, \ldots, x_n \in V(\mathbb{Q}) \cap \overline{C}$ such that

$$\Sigma = \{y \in \overline{C} \mid \langle y, {}^t\gamma x_i \rangle \geq 1 \text{ for all } \gamma \in \Gamma, i = 1, \ldots, n\}.$$

In general, let N be a positive integer and $T \subset \overline{C} \cap \frac{1}{N}L'$ (recall $L' = L \setminus \{0\}$). Denote $\Sigma_T = \{x \in \overline{C} \mid \langle x, y \rangle \geq 1 \text{ for all } y \in T\}$. Sometimes we will simply write Σ if the T is understood.

Proposition 5.19 *For any* $^t\Gamma$-*invariant subset* $T \subset \overline{C} \cap \frac{1}{N}L'$, Σ_T *is a* Γ-*polyhedral kernel.*

Proof First, Σ_T is clearly closed, convex, and Γ-invariant. Also $\mathbb{R}^{\geq 1}\Sigma_T \subset \Sigma_T$, $0 \notin \Sigma_T$, and $C \subset \mathbb{R}_{>0}\Sigma_T$. In other words, Σ_T is a kernel.

Let Π be a polyhedral cone with rational vertices contained in C^*. Since Π can be written as the union of simplicial cones with rational vertices, we may assume that Π is spanned by $z_1, \ldots, z_r \in NL^* \cap \overline{C}$, where $r = \dim V$. Then

$$\Pi = \left\{ \sum a_i z_i \mid a_i \geq 0 \text{ for } i = 1, \ldots, r \right\} .$$

We want to show that there exists a finite set $t_1, \ldots, t_k \in T$ such that, if $z \in \Pi$ and $\langle t_j, z \rangle \geq 1$ for all $j = 1, \ldots, k$, then $\langle t, z \rangle \geq 1$ for all $t \in T$. For then we would have,

$$\Pi \cap \Sigma_T = \{z \in \Pi \mid \langle z, t_j \rangle \geq 1 \text{ for all } j = 1, \ldots, k\} .$$

To do this we use the following. Let $\mathbb{Z}_{\geq 0}$ denote the non-negative integers. For $a = (a_1, \ldots, a_m) \in (\mathbb{Z}_{\geq 0})^m$ and $b = (b_1, \ldots, b_m) \in (\mathbb{Z}_{\geq 0})^m$, we write $b \geq a$ if and only if $b_i \geq a_i$ for all $i = 1, \ldots, m$.

Lemma 5.20 *For any set* $S \subset (\mathbb{Z}_{\geq 0})^m$, *there exists a finite subset* $S_0 \subset S$ *such that, for all* $b \in S$, *there exists some* $a \in S_0$ *with* $b \geq a$.

Proof For $m = 1$, it is clear. We proceed by induction on m, assuming the lemma for $m - 1$. Let $\pi_j : (\mathbb{Z}_{\geq 0})^m \longrightarrow (\mathbb{Z}_{\geq 0})^{m-1}$ be the projection omitting the j'th coordinate. Fix some $c = (c_1, \ldots, c_m) \in S$. For each $j = 1, \ldots, m$ and each $q = 0, \ldots, c_j - 1$, write $S_{j,q} = \{s \in S \mid s_j = q\}$. By induction applied to $\pi_j(S_{j,q})$, we get a finite set $S^0_{j,q} \subset S_{j,q}$ such that, for all $b \in S_{j,q}$, there exists some $a \in S^0_{j,q}$ with $b \geq a$. Now $S_0 = \bigcup_{j,q} S^0_{j,q} \cup \{c\}$ satisfies the lemma. $\square$

We continue with the proof of the proposition. Define $\Phi : L \cap \overline{C} \longrightarrow (\mathbb{Z}_{\geq 0})^r$ by $\Phi(y) = (\langle y, z \rangle, \ldots, \langle y, z_r \rangle)$. Let $S_0 \subset \Phi(T)$ be as in the lemma, and choose a section $\{t_1, \ldots, t_k\}$ of Φ over S_0.

For any $z = \sum a_i z_i \in \Pi$, suppose $\langle t_j, z \rangle \geq 1$ for all $j = 1, \ldots, k$. Then, for any $t \in T$, there is a $j \in \{1, \ldots, k\}$ such that $\langle t, z_i \rangle \geq \langle t_j, z_i \rangle$ for all $i = 1, \ldots, r$. Thus

$$\langle t, z \rangle = \sum a_i \langle t, z_i \rangle \geq \sum a_i \langle t_j, z_i \rangle = \langle t_j, z \rangle \geq 1$$

because $a_i \geq 0$ for all i. So $t_1, \ldots, t_k$ do the trick. $\square$

Proposition 5.21 *If Σ is a Γ-polyhedral kernel, then $\check{\Sigma}$ is a ${}^t\Gamma$-polyhedral kernel.*

Proof By Corollary 5.12, we have that $\check{\Sigma} = \{y \in \overline{C} \mid \langle y, \xi \rangle \geq 1 \text{ for all } \xi \in E(\Sigma)\}$. So what we want to show is that, given Π, there exists a finite number of extreme points $\xi_1, \ldots, \xi_t \in E(\Sigma)$ such that if $z \in \Pi$ and $\langle \xi_i, z \rangle \geq 1$ for $i = 1, \ldots, t$, then $\langle \xi, z \rangle \geq 1$ for every extreme point ξ. This will follow precisely as in the proof of Proposition 5.19, if we use Proposition 5.16. $\square$

We can summarize all this as follows:

Proposition 5.22 *The following three statements are equivalent:*

(a) Σ *is a Γ-polyhedral kernel;*
(b) $\Sigma = \{y \in \overline{C} \mid \langle y, {}^t\gamma x_i \rangle \geq 1 \text{ for all } \gamma \in \Gamma, i = 1, \ldots, n\}$ *for some* $x_1, \ldots, x_n \in V(\mathbb{Q}) \cap \overline{C}$;
(c) Σ *is the closed convex hull of*

$$\bigcup_{\substack{i=1,\ldots,n \\ \gamma \in \Gamma}} (\gamma x_i + C)$$

for some $x_1, \ldots, x_n \in V(\mathbb{Q}) \cap \overline{C}$. $\square$

Apply this to $\Sigma_1, \Sigma_2, \check{\Sigma}_1, \check{\Sigma}_2$ of (5.1):

$$\Sigma_2 = \{x \in \overline{C} \mid \langle x, y \rangle \geq 1 \text{ for all } y \in \overline{C} \cap L'\}$$

and

$$\check{\Sigma}_1 = \{x \in \overline{C} \mid \langle x, y \rangle \geq 1 \text{ for all } y \in C \cap L\}$$

are rationally locally polyhedral by Proposition 5.19. Then, by Proposition 5.21, the same is true of

$$\Sigma_1 = \check{\Sigma}_1 = \text{closed convex hull of } C \cap L$$

and

$$\check{\Sigma}_2 = \text{closed convex hull of } \overline{C} \cap L'.$$

To obtain Γ-admissible decompositions of the cone, we need only take into account Proposition 5.18.

Now $\check{\Sigma}_2$ is Γ-invariant, but $\check{\Sigma}_1$ is only ${}^t\Gamma$-invariant. We may use instead $\check{\Sigma}_1^* = \{x \in \overline{C} \mid \langle x, y \rangle \geq 1 \text{ for all } y \in C \cap L^*\}$. Then we may summarize:

Corollary 5.23 *Taking cones over the faces of the closed convex hull of $\overline{C} \cap L'$ (resp. $\{x \in \overline{C} \mid \langle x, y \rangle \geq 1 \text{ for all } y \in C \cap L^*\}$) yields a Γ-admissible polyhedral decomposition of C.* $\square$

The first kind is called a *Voronoi decomposition (of the first type)* and the second is called a decomposition into *central cones*, since in each case they generalize known constructions for the cone of positive-definite quadratic forms in a given number of variables.

6 Positive-definite forms in low dimensions

We consider the most classical example of the above theory:

V_n = vector space of symmetric $n \times n$ real matrices ;

C_n = cone of positive definite elements in V_n .

Then V_n has a natural $\mathbb{Q}$-structure given by the rational $n \times n$ matrices and is a Jordan algebra via

$$X \cdot Y = \tfrac{1}{2}(XY + YX).$$

So $\mathrm{Aut}(C_n)$ is just $\mathrm{GL}(n, \mathbb{R})/(\pm I_n)$ acting via

$$X \longmapsto {}^t A X A, \quad A \in \mathrm{GL}(n, \mathbb{R}) ,$$

and the characteristic function of the cone is given by

$$\varphi(X) = \frac{1}{\det(X)^{(n+1)/2}} .$$

There are two natural lattices in V_n:

L_n = integral $n \times n$-matrices X ,

L_n^* = semi-integral $n \times n$-matrices X ,

i.e.,

$$L_n^* = \left\{ X \in V_n \mid X_{ii} \in \mathbb{Z}, X_{ij} \in \tfrac{1}{2}\mathbb{Z} \text{ if } i \neq j \right\} .$$

If $I_n \in C_n$ is taken as the basepoint, we get the inner product

$$\langle X, Y \rangle = \mathrm{Tr}(XY)$$

on V_n for which:

(i) C_n is self-adjoint;
(ii) $\mathrm{Stab}(I_n) = O(n, \mathbb{R})/(\pm I_n)$ acts orthogonally;
(iii) its polar complement P, the set of symmetric matrices in $\mathrm{GL}(n, \mathbb{R})$, acts by self-adjoint maps; and
(iv) the notation L_n^* is justified:

$$L_n^* = \{ x \in V_n \mid \langle x, y \rangle \in \mathbb{Z} \text{ for all } y \in L_n \} .$$

The arithmetic group

$$\Gamma_n = \mathrm{GL}(n, \mathbb{Z})/(\pm I_n)$$

preserves both lattices. The rational boundary components of C_n correspond one-to-one with the subspaces $W \subset \mathbb{R}^n$ defined over $\mathbb{Q}$, the correspondence being given by

$$W \longleftrightarrow C(W) = \left\{ X \in V_n \,\middle|\, \begin{array}{l} X \text{ positive semi-definite} \\ \text{with null space } W \end{array} \right\}.$$

Thus

$$C_n^* = C_n \cup \bigcup_W C(W)$$

$$= \left\{ X \in V_n \,\middle|\, \begin{array}{l} X \text{ positive semi-definite} \\ \text{with rational null space } W \end{array} \right\}.$$

The problem of finding an *explicit* decomposition of C_n^* into rational poly-hedral cones $\{\sigma_\alpha\}$ invariant under Γ_n, and hence of finding a fundamental domain for Γ_n, is a very old one. For all n, an important role is played by the fundamental cone $\phi_n \subset C_n^*$:

$$\phi_n = \left\{ X \in V_n \,\middle|\, \begin{array}{l} \text{off-diagonal entries } X_{ij} \text{ non-positive} \\ \text{row sums } \sum_{i=1}^n X_{ij} \text{ non-negative} \end{array} \right\}.$$

In fact, ϕ_n is a simplicial cone, being expressible also as

$$\phi_n = \left\{ \sum_{i=1}^n \lambda_i x_i^2 + \sum_{1 \le i < j \le n} \lambda_{ij}(x_i - x_j)^2 \,\middle|\, \lambda_i \ge 0, \lambda_{ij} \ge 0 \right\}.$$

In some areas of applied mathematics, the cone ϕ_n plays a role – basically because the inequalities $x_{ij} \le 0$ for $i \ne j$ and $\sum_{i=1}^n x_{ij} \ge 0$ are the simplest *linear* way to force a matrix to be positive semi-definite. A basic result is:

$$C_n^* = \bigcup_{\gamma \in \mathrm{GL}(n,\mathbb{Z})} \gamma \phi_n \iff n \le 3$$

This illustrates the interesting fact that only in 4-space or higher do lattice packing problems and related geometry of numbers problems get interesting. For $n = 2$ or 3, however, all "reasonable" admissible polyhedral decomposi-tions of C_n^* are based on $\{\gamma \phi_n\}$ – either $\{\sigma_\alpha\}$ equals $\{\gamma \phi_n\}$, or it equals $\{\gamma \psi_i\}$, where $\{\psi_i\}$ is a decomposition of ϕ_n. Thus the standard fundamental domain for $n = 2$,

$$\tilde{\phi}_2 = \left\{ \begin{pmatrix} a & -b \\ -b & c \end{pmatrix} \,\middle|\, 0 \le 2b \le a \le c \right\},$$

comes about by barycentric subdivision of ϕ_2:

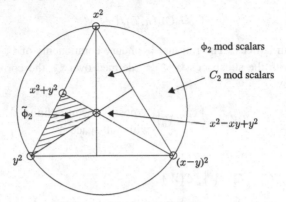

Since ϕ_2 is mapped into itself by the group of order six,

$$\left\{ I_2, \begin{pmatrix} 0 & 1 \\ 1 & 0 \end{pmatrix}, \begin{pmatrix} 0 & 1 \\ -1 & 1 \end{pmatrix}, \begin{pmatrix} 1 & -1 \\ 1 & 0 \end{pmatrix}, \begin{pmatrix} 1 & -1 \\ 0 & -1 \end{pmatrix}, \begin{pmatrix} 1 & 0 \\ 1 & -1 \end{pmatrix} \right\},$$

we see that $\tilde{\phi}_2 \cap \gamma\tilde{\phi}_2$ is at most a face of $\tilde{\phi}_2$, if $\gamma \neq$ id.

For $n \geq 4$, we need more cones. To find them systematically, we use a co-core. In fact, there are apparently two very natural co-cores in C_n, which we call the perfect and central co-cores:

$$K_{\text{perf}} = \text{the closed convex hull of } L_n^* \cap \overline{C}_n \setminus \{0\} \ ;$$
$$K_{\text{cent}} = \{X \in \overline{C}_n \mid \text{Tr}(XY) \geq 1 \text{ for all } Y \in L_n^* \cap C_n\} \ .$$

The dual cores are:

$$\check{K}_{\text{perf}} = \{X \in C_n \mid \text{Tr}(XY) \geq 1 \text{ for all } Y \in L_n^* \cap \overline{C}_n \setminus \{0\}\} \ ;$$
$$\check{K}_{\text{cent}} = \text{the closed convex hull of } L_n^* \cap C_n \ .$$

According to a result of Barnes and Cohn [2],

$$\min_{Y \in L_n^* \cap \overline{C}_n \setminus \{0\}} \text{Tr}(XY) = \min_{m \in \mathbb{Z}^n \setminus \{0\}} {}^t m X m \ ,$$

and hence we get a second definition of the perfect co-core and core:

$$\check{K}_{\text{perf}} = \{X \in C_n \mid {}^t m X m \geq 1 \text{ for all } m \in \mathbb{Z}^n \setminus \{0\}\} \ ;$$
$$K_{\text{perf}} = \text{closed convex hull of the rank-1 matrices } X_{ij} = m_i m_j$$
$$\text{with } m \in \mathbb{Z}^n \setminus \{0\} \ .$$

Furthermore,

$$\mu(X) = \min_{m \in \mathbb{Z}^n \setminus \{0\}} {}^t m X m$$

is the piecewise-linear function which is 1 on $\partial \check{K}_{\text{perf}}$.

Finally, we say that a form $X \in C_n$ is *perfect*, (resp. *central*), if it is a vertex of $\check{K}_{\text{perf}}$, (resp. $\check{K}_{\text{cent}}$). By the general theory, perfect (resp. central) forms correspond one-to-one with the top-dimensional faces of K_{perf}, (resp. K_{cent}), hence to the $\frac{n(n+1)}{2}$-dimensional cones in the decomposition of C_n defined by K_{perf}, (resp. K_{cent}), which we call the *perfect cones* (resp. *central cones*). In the case of perfect forms, we see easily that X is perfect if and only if $\mu(X) = 1$ and X is the only solution to the equations

$$^t m Y m = 1 \text{ for all } m \in \mathbb{Z}^n \text{ such that } ^t m X m = 1 ,$$

and that the corresponding perfect cone is the convex hull of the rank-1 forms $X_{ij} = m_i m_j$ for all $m \in \mathbb{Z}^n$ such that $^t m X m = 1$. Central cones, on the other hand, are given by

$$\sigma_Y = \{X \in C_n^* \mid \text{Tr}(XY) \leq \text{Tr}(XZ) \text{ for all } Z \in L_n^* \cap C_n\} ,$$

where Y is a central form.

For $n \leq 6$, all perfect forms have been listed; see [1] and [6]. For example, consider the form

$$\alpha_n = \frac{1}{2} \left(x_1^2 + \cdots + x_n^2 + (x_1 + \cdots + x_n)^2 \right)$$

corresponding to the matrix

$$\begin{pmatrix} 1 & \frac{1}{2} & \cdots & \frac{1}{2} \\ \frac{1}{2} & 1 & \cdots & \frac{1}{2} \\ \vdots & \vdots & \ddots & \vdots \\ \frac{1}{2} & \frac{1}{2} & \cdots & 1 \end{pmatrix}.$$

Note that α_n, up to unimodular equivalence, may be described as the Killing form for the root system A_n, written with respect to a basis of the lattice generated by the roots. This may be shown to be both perfect and central and it corresponds to a common face of K_{cent} and K_{perf}, namely the simplex in C_n^* with vertices x_i^2, $(x_i - x_j)^2$, where $1 \leq i, j \leq n$. This shows that ϕ_n is both a central and a perfect cone.

Another important form is

$$\delta_n = \frac{1}{2} \left((x_1 - x_2)^2 + x_3^2 + \cdots + x_n^2 + (x_1 + \cdots + x_n)^2 \right)$$

corresponding to the matrix

$$
\begin{pmatrix}
1 & 0 & \frac{1}{2} & \cdots & \frac{1}{2} \\
0 & 1 & \frac{1}{2} & \cdots & \frac{1}{2} \\
\frac{1}{2} & \frac{1}{2} & 1 & \cdots & \frac{1}{2} \\
\vdots & \vdots & \vdots & \ddots & \vdots \\
\frac{1}{2} & \frac{1}{2} & \frac{1}{2} & \cdots & 1
\end{pmatrix}.
$$

Note that δ_n, up to unimodular equivalence, may be described as the Killing form for the root system D_n, written with respect to a basis of the lattice generated by the roots. This form is also both perfect and central, but it does not seem to be known whether it defines identical faces of K_{cent} and K_{perf}.

Note that

$$
K_{\text{cent}} \supseteq K_{\text{perf}} \; ; \; \check{K}_{\text{cent}} \subseteq \check{K}_{\text{perf}} \, .
$$

It may be shown that equality holds if and only if $n \leq 5$. In particular, for $n = 4$, $K_{\text{cent}} = K_{\text{perf}}$, and forms are perfect if and only if they are central. It turns out that, in this case, α_4, δ_4, and their images under Γ_4 are the only perfect forms; hence, if ψ_4 is the perfect cone corresponding to δ_4, we conclude that, for $n = 4$, the cones $\gamma\phi_4$ and $\gamma\psi_4$ for $\gamma \in \Gamma_4$ form an admissible polyhedral decomposition of C_4^*.

References

[1] E. S. Barnes, The complete enumeration of extreme senary forms, *Phil. Trans. Royal Soc. London* **249** (1957), 461–506.

[2] E. S. Barnes and M. J. Cohn, On the inner product of positive quadratic forms, *J. London Math. Soc.* **12** (2) (1975/76), 32–36.

[3] A. Borel, *Introduction aux Groupes Arithmetiques*. Paris: Hermann, 1969.

[4] A. Borel and J.-P. Serre, Corners and arithmetic groups, *Comm. Math. Helv.* **48** (1973), 436–491.

[5] H. Braun and M. Koecher, *Jordan-Algebren*. Berlin: Springer-Verlag, 1966.

[6] H. S. M. Coxeter, Extreme forms, *Canad. J. Math.* **3** (1951), 391–441.

[7] N. Jacobson, *Structure and Representations of Jordan Algebras*. Providence, RI: American Mathematical Society, 1968.

[8] M. Koecher, *Jordan Algebras and their Applications*. Lecture Notes. University of Minnesota, 1962.

[9] E. Looijenga, *Semi-Toric Partial Compactifications I.* Preprint series of Department of Mathematics, Catholic University, Nijmegen, The Netherlands, June 1985.

[10] O. Loos, *Symmetric Spaces I: General Theory.* New York: Benjamin, 1969.

[11] D. Mumford, *Geometric Invariant Theory.* Berlin: Springer-Verlag, 1965.

[12] R. T. Rockafellar, *Convex Analysis.* Princeton, NJ: Princeton University Press, 1970.

[13] T. A. Springer, *Jordan Algebras and Algebraic Groups.* Berlin: Springer-Verlag, 1973.

[14] G. Voronoi, Nouvelles applications des paramètres continus à la théorie des formes quadratiques I, *J. Reine Angew. Math.* **133** (1907), 97–178.

III

Compactifications of locally symmetric varieties

This chapter presents the main results of this book. In order to make the book reasonably self-contained, and because of the difficulty in assembling the known results from a very diffuse and often hard-to-read literature, we have tried to include, with at least sketches of proof, a large proportion of the needed background on hermitian symmetric spaces. We have profited greatly from lectures on hermitian symmetric spaces given by P. Deligne in Paris in 1973. Large parts of Sections 2 and 3 are no more than an elaboration of his work. One of us (Rapoport) profited also from conversations with R. P. Langlands.

Section 1 explains the method of compactification in the case of a tube domain; some of the facts established there are used later. This is also done in Satake [10].

Sections 2, 3, and 4 review facts about hermitian symmetric domains and their boundary theory. In addition to help from Deligne and Langlands, we have used principally Helgason [6] and Wolf [12], but also Harish-Chandra [5], Langlands [8], Koranyi and Wolf [7], and Satake [11].

Section 5 states the Main Theorem; Section 6 contains its proof. Here basic facts from the fundamental paper by Baily and Borel [1] are used. For the reduction theory, we have found the book by Borel [2] very useful.

Finally Section 7 contains a more intrinsic formulation of the Main Theorem.

1 Tube domains and compactification of their cusps

Let $N_\mathbb{R}$ be a finite-dimensional $\mathbb{R}$-vector space and let $C \subset N_\mathbb{R}$ be a homogeneous self-adjoint cone. Set $\overline{G} = \text{Aut}(C, N_\mathbb{R})^o$ (this notation thus differs from that in Chapter II).

Definition 1.1 The open subset

$$U = N_{\mathbb{R}} + iC = \{x + iy \mid x \in N_{\mathbb{R}}, y \in C\}$$

of $N_{\mathbb{C}}$ is called a *tube domain*.

Let G be the group of complex-analytic automorphisms of U.

Proposition 1.2

(1) *U is a bounded symmetric domain.*
(2) *G, equipped with the compact open topology, is a Lie group.*

Proof (2) follows from (1) by general theory: U, as a bounded domain, is equipped with the Bergman metric. Moreover, G is a closed subgroup of the *Lie group* of isometries with respect to this metric. Thus G itself is a Lie group.

To see that U is a bounded domain, choose a basis of $N_{\mathbb{R}}$ such that $C \subset \{y_1 > 0, \ldots, y_n > 0\}$ (this is obviously possible). This embeds U into $\mathfrak{H}^n$, where as usual $\mathfrak{H}$ denotes the upper half-plane. Since $\mathfrak{H}$ is a bounded domain (isomorphic to the unit disc), so is $\mathfrak{H}^n$ and thus also U.

To see that U is a homogeneous domain, note that the real translations by $N_{\mathbb{R}}$ and the complex-linear extensions of the linear maps in $\overline{G}$ take U to U and act transitively on U.

Finally, to see that U is symmetric, it suffices to construct an involutive symmetry around the point $ie \in U$ (here $e = a$, the basepoint in C). As in Chapter II, Section 2, we describe C as the set of exponentials for a real Jordan algebra structure on $N_{\mathbb{R}}$ with identity e; this algebra structure extends to a complex Jordan algebra structure on $N_{\mathbb{C}}$. We claim that, in this algebra structure, every $x \in U$ is invertible, that $-x^{-1} \in U$, and that the map $x \longmapsto -x^{-1}$ is the required symmetry. But for any algebra structure with identity e, we have that $-(ie)^{-1} = ie$, and $x \longmapsto -x^{-1}$ is an involution near ie with no other fixed point. It remains to show that every $x \in U$ is Jordan-invertible and that $-x^{-1} \in U$. First of all, for every $g \in \overline{G}$, we claim that

$$x \text{ invertible implies } gx \text{ invertible, and } (gx)^{-1} = \sigma(g) \cdot x^{-1}, \qquad (1.1)$$

where $\sigma : \overline{G} \longrightarrow \overline{G}$ is the Cartan involution fixing $\overline{G} \cap \mathrm{Stab}\,(e)$. In fact, we saw in Chapter II, Section 2, that $(g \cdot e)^{-1} = \sigma(g) \cdot e$, hence if $g, g_1 \in \overline{G}$, then

$$(g \cdot g_1 \cdot (ie))^{-1} = -i\sigma(g \cdot g_1) \cdot e = -i\sigma(g) \cdot \sigma(g_1) \cdot e = \sigma(g) \cdot (g_1 \cdot (ie))^{-1},$$

i.e., (1.1) holds for $x \in iC$. By analytic continuation it holds for all invertible $x \in N_{\mathbb{C}}$. This reduces our problem to showing that, for all $x \in N_{\mathbb{R}}, x + ie$ is invertible and $-(x + ie)^{-1} \in U$. To see this, let $\mathbb{R}[x]$ be the real Jordan subalgebra of $N_{\mathbb{R}}$ generated by x and e. As in Chapter II, $\mathbb{R}[x] \cong \sum_{i=1}^n \mathbb{R}\varepsilon_i$, where ε_i are idempotents, and $\mathbb{R}[x] \cap C \cong \sum_{i=1}^{n} \mathbb{R}_{>0} \varepsilon_i$. Indeed, C is self-adjoint w.r.t. the pairing $\langle y, z \rangle = \mathrm{Tr}\,(L_{y \cdot z})$ by Chapter II, Theorem 2.13. Then, for $y = \sum_{i=1}^{n} y_i \varepsilon_i \in \mathbb{R}[x] \cap C$, we have

$$y_i \langle \varepsilon_i, \varepsilon_i \rangle = \langle y \varepsilon_i, \varepsilon_i \rangle = \langle y, \varepsilon_i^2 \rangle > 0 \,,$$

and hence $y \in \sum_{i=1}^{n} \mathbb{R}_{>0} \varepsilon_i$. Conversely, any such element is a square of an invertible element and thus lies in C. Since $x + ie$ lies in the subalgebra $\mathbb{C}[x]$, we are reduced to checking the one-dimensional case $N_{\mathbb{R}} = \mathbb{R}$. But it is obvious that, for all $a \in \mathbb{R}$, $a + i \in \mathbb{C}$ is invertible and $-(a+i)^{-1} \in \mathfrak{H}$, the upper half-plane. $\qquad \square$

Now start with

(1) an algebraic group $\mathscr{G}$ over $\mathbb{Q}$ such that
(2) its associated Lie group $G = \mathscr{G}(\mathbb{R})^o$ is the connected component of the group of complex-analytic automorphisms of U.

We also assume that

(3) the subgroup $P \subset G$, which is the semi-direct product of $\overline{G} = \mathrm{Aut}\,(C, N_{\mathbb{R}})^o$ by the group of real translations $N_{\mathbb{R}}$, is defined over $\mathbb{Q}$, i.e., $P = \mathscr{P}(\mathbb{R})^o$, where $\mathscr{P} \subset \mathscr{G}$ is an algebraic subgroup defined over $\mathbb{Q}$. This defines a rational structure on $N_{\mathbb{R}}$ and $\overline{G}$ such that $\overline{G}$ acts rationally on $N_{\mathbb{R}}$.

Let $\Gamma \subset \mathscr{G}(\mathbb{Q}) \cap G$ be an arithmetic subgroup. Let $N_{\mathbb{Z}}$ be the group of translations contained in Γ: this is a lattice in $N_{\mathbb{R}}$. We set $\overline{\Gamma} = (\Gamma \cap P)/N_{\mathbb{Z}}$, which is the image of Γ via $P \longrightarrow \overline{G}$.

Here is an example of the situation we consider.

Example 1.3 (Siegel case)

$N_{\mathbb{R}} = \mathrm{Sym}^2(\mathbb{R}^n)$;

C is the cone of positive-definite quadratic forms on $\mathbb{R}^n$;

U is the Siegel upper half space in $\mathbb{C}^{n(n+1)/2}$;

$G = \mathrm{Sp}\,(2n, \mathbb{R})/\{\pm 1\}$, $\Gamma = \mathrm{Sp}\,(2n, \mathbb{Z})/\{\pm 1\}$;

$\overline{G} = \mathrm{GL}\,(n, \mathbb{R})/\{\pm 1\}$, $\overline{\Gamma} = \mathrm{GL}\,(n, \mathbb{Z})/\{\pm 1\}$;

$\begin{pmatrix} a & b \\ a & d \end{pmatrix} \in G$ acts on U by $z \longmapsto (az+b) \cdot (cz+d)^{-1}$;

$\begin{pmatrix} a & 0 \\ 0 & {}^t a^{-1} \end{pmatrix} \in \overline{G}$ acts on $N_{\mathbb{R}}$ by $x \longmapsto a \cdot x \cdot {}^t a$;

$P = \begin{pmatrix} * & * \\ 0 & * \end{pmatrix} \cap \mathrm{Sp}\,(2n, \mathbb{R})/\{\pm 1\}$.

Let $\{\sigma_\alpha\}$ be a $\overline{\Gamma}$-admissible polyhedral decomposition of C. We now come to the main idea of this whole volume.

Let $T = N_{\mathbb{C}}/N_{\mathbb{Z}}$: this is a torus in the algebraic group sense with

$$N_{\mathbb{Z}} \cong \mathrm{Hom}\,(\mathbb{G}_m, T),$$

the group of one-parameter subgroups of T.

As in Chapter I, we have an exact sequence:

$$0 \longrightarrow T_c \longrightarrow T \xrightarrow{\;\mathrm{ord}\;} N_{\mathbb{R}} \longrightarrow 0\,.$$

Since U is $N_{\mathbb{Z}}$-invariant, it is the inverse image of an open subset $U' \subset T$; alternatively, $U' = \mathrm{ord}^{-1}(C)$. For every $\varepsilon \in C$, set

$$C_\varepsilon = C + \varepsilon\,,\ U_\varepsilon = N_{\mathbb{R}} + iC_\varepsilon\,,\ U'_\varepsilon = \mathrm{ord}^{-1}(C_\varepsilon) = U_\varepsilon/N_{\mathbb{Z}}\,.$$

(The U_ε are nothing but Piatetskii-Shapiro's *cylindrical sets*.) Next, by the theory of TE I†, $\{\sigma_\alpha\}$ defines a T-equivariant embedding:

$$T \subset X_{\{\sigma_\alpha\}}\,.$$

Let

$$U'' = \text{interior of the closure of } U' \text{ in } X_{\{\sigma_\alpha\}}\,;$$
$$U''_\varepsilon = \text{interior of the closure of } U'_\varepsilon \text{ in } X_{\{\sigma_\alpha\}}\,.$$

Now, $\overline{\Gamma}$ acts on $\{\sigma_\alpha\}$, as well as on T. So $\overline{\Gamma}$ acts on $X_{\{\sigma_\alpha\}}$, prolonging its action on T; this action preserves U''.

Theorem 1.4

(i) $\overline{\Gamma}$ *acts properly discontinuously on U''.*

(ii) $(\overline{\Gamma} \cdot U''_\varepsilon)/\overline{\Gamma}$ *is open and relatively compact in $U''/\overline{\Gamma}$.*

We use the commutative diagram introduced in Chapter I, Section 1:

$$
\begin{array}{ccc}
X_{\{\sigma_\alpha\}} & \xrightarrow{\;\mathrm{ord}\;} & N_{\{\sigma_\alpha\}} \\
\cup & & \cup \\
T & \xrightarrow{\;\mathrm{ord}\;} & N_{\mathbb{R}} \\
\cup & & \cup \\
U' & \longrightarrow & C
\end{array}
$$

Proof of (i) We know that ord is a continuous $\overline{\Gamma}$-equivariant map. It thus suffices to show that $\overline{\Gamma}$ acts properly discontinuously on $\mathrm{ord}(U'')$. Because ord is a quotient map by the compact group T_c by Chapter I, and, in particular, an *open* map, $U'' = \mathrm{ord}^{-1}(C'')$, where C'' is the interior of the closure of C inside

† Recall this reference from p. x.

$N_{\{\sigma_\alpha\}}$. As in Chapter I, Section 1, we can describe the points of $N_{\{\sigma_\alpha\}}$ by symbols $x + \infty \cdot \sigma_\alpha$, for $x \in N_{\mathbb{R}}$, and a fundamental system of neighborhoods of $x + \infty \cdot \sigma_\alpha$ meets C in the sets $x + y + B_\varepsilon + \sigma_\alpha$, where B_ε is the ε-ball around 0, and $y \in \sigma_\alpha$. It follows that if $x + \infty \cdot \sigma_\alpha \in C''$, then $x + y + B_\varepsilon + \sigma_\alpha \subset C$ for suitable y, ε. But $x + y + \infty \cdot \sigma_\alpha = x + \infty \cdot \sigma_\alpha$, so all points of C'' are represented by symbols $x + \infty \cdot \sigma_\alpha$ with $x \in C$. Conversely, if $x \in C$, then, for small enough ε, we have $x + B_\varepsilon + \sigma_\alpha \subset C$; hence,

$$x + y + z + \infty \cdot \sigma_\beta \in \text{closure of } C \text{ in } N_{\{\sigma_\alpha\}} \,,$$

for all $y \in \sigma_\alpha$ and $z \in B_\varepsilon$, and for all faces σ_β of σ_α. This means that $x + \infty \cdot \sigma_\alpha \in C''$. Thus

$$C'' = \bigcup_\alpha \bigcup_{x \in C} \{x + \infty \cdot \sigma_\alpha\} \,.$$

Next, for all $x + \infty \cdot \sigma_\alpha$, we claim that there is a finite set $\sigma_{\alpha_1}, \ldots, \sigma_{\alpha_n}$ of polyhedral cones such that

$$\bigcup_{l=1}^n \sigma_{\alpha_i} \text{ is a neighborhood of } x + \infty \cdot \sigma_\alpha \,. \tag{1.2}$$

In fact, let $y_1, \ldots, y_m \in C$ have the property that x is in the interior of the polyhedral cone $\langle y_1, \ldots, y_m \rangle$ spanned by the y_i; more precisely, suppose $x + B_\varepsilon \subset \langle y_1, \ldots, y_m \rangle$. Let τ be the polyhedral cone spanned by σ_α and the y_i. We apply the main theorem of reduction theory (Chapter II, Corollary 4.9), plus the fact that mod $\overline{\Gamma}$ there are only finitely many σ_α, to conclude that

$$\{\alpha \mid \sigma_\alpha \cap \tau \cap C \neq \emptyset\}$$

is finite. Since the σ_α cover C, it follows that

$$\tau \cap C \subset (\sigma_{\alpha_1} \cup \cdots \cup \sigma_{\alpha_n})$$

for suitable σ_{α_i}. Therefore

$$x + B_\varepsilon + \sigma_\alpha \subset (\sigma_{\alpha_1} \cup \cdots \cup \sigma_{\alpha_n}) \,.$$

This proves (1.2).

Now let $x_i \in C''$ be a sequence of points converging to $x \in C''$ and let $\gamma_i \in \overline{\Gamma}$ be such that $y_i := \gamma_i \cdot x_i$ converges to a point $y \in C''$. We have to show that the set of γ_i for $i \gg 0$ consists of only finitely many elements $\{\gamma^1, \ldots, \gamma^n\}$ and that $\gamma^j \cdot x = y$, for $j = 1, \ldots, n$. Since C is a $\overline{\Gamma}$-invariant open dense subset of C'', we may suppose that $x_i \in C$.

Now, by the preceding discussion, there exist finitely many polyhedra σ_{α_i} and $\sigma_{\alpha'_j}$ such that

$$x_k \in \sigma_{\alpha_i} \,, \quad y_k \in \sigma_{\alpha'_j} \,.$$

By taking a subsequence if necessary, we may suppose that $x_i \in \sigma$, $y_i \in \sigma'$. It follows that $\gamma_i \sigma \cap \sigma' \cap C \neq \emptyset$. By Corollary 4.9 of Chapter II, this implies that the set of the γ_i, for $i \gg 0$, consists of only finitely many elements $\{\gamma^1, \ldots, \gamma^n\}$ which necessarily satisfy $\gamma^j \cdot x = y$, for $j = 1, \ldots, n$.

Proof of (ii) The openness is clear. Since C' is the quotient of U' by a compact group we have to show the following statement:

(ii') *Let* $\bar{z}_1, \bar{z}_2, \ldots$ *be a sequence of points in* C_ε; *then there exist elements* $\gamma_i \in \overline{\Gamma}$, *such that, after passing to a subsequence, the sequence* $\gamma_i \bar{z}_i$ *converges to a point in* C''.

Now, by conditions (4) and (5) of a $\overline{\Gamma}$-admissible decomposition, applied to $\{\sigma_\alpha\}$ (see Definition 4.10 in Chapter II), we can find $\bar{z}'_i = \gamma_i \cdot \bar{z}_i$ such that, after passing to a subsequence, all $\bar{z}'_i$ lie in one and the same σ_α. It follows from the description of the topology of N_{σ_α} given in Chapter I that a subsequence of the $\{\bar{z}'_i\}$ converges to a point $\bar{z}' \in N_{\sigma_\alpha}$. It remains to show that $\bar{z}' \in C''$. This is clear if $\sigma_\alpha \setminus \{0\} \subset C$.

In general, $\sigma_\alpha \cap \partial C$ is contained in a finite union of rational boundary components $\overline{C}_k = N_{k,\mathbb{R}} \cap \overline{C}$ of C for $k = 1, \ldots, n$. Here $N_{k,\mathbb{R}} \subset N_\mathbb{R}$ is a rational subspace and we may choose a rational linear functional ℓ_k on $N_\mathbb{R}$ such that $\ell_k \geq 0$ on $\overline{C}$ and $(\ell_k = 0) \cap \overline{C} = \overline{C}_k$. We may assume that $\varepsilon \in L$, where L is a lattice fixed by $\overline{\Gamma}$ and commensurable with $N_\mathbb{Z}$. It follows that $\gamma_i \varepsilon \in L$. Furthermore, ℓ_k on $L \cap C$ is bounded away from zero, say by c_k, as the image of L in $N_\mathbb{R}/N_{k,\mathbb{R}}$ is a lattice. Since

$$\bar{z}'_i \in \gamma_i \varepsilon + C,$$

it follows that $\ell_k(\bar{z}'_i) > \ell_k(\gamma_i \varepsilon) \geq c_k$. Since this is true for all k, the limit $\bar{z}'$ of the sequence $\{\bar{z}'_i\}$ lies in C''. $\square$

Now, the basic idea to compactify U/Γ is to glue U/Γ and $(\overline{\Gamma} \cdot U'_\varepsilon)/\overline{\Gamma}$ along the set $(\overline{\Gamma} \cdot U'_\varepsilon)/\overline{\Gamma} = ((\Gamma \cap P) \cdot U_\varepsilon)/(\Gamma \cap P)$. Roughly speaking, this will add to U/Γ points at infinity at the cusp $i\infty \cdot C$ of U. To glue, we must know that $(\overline{\Gamma} \cdot U'_\varepsilon)/\overline{\Gamma}$ is a subset of U/Γ, i.e., that if ε is large enough, two Γ-equivalent points of U_ε are in fact $\Gamma \cap P$-equivalent. This is a consequence of "Piatetskii-Shapiro's lemma."

We will not pursue the construction further at this stage because we will also have to treat the "higher-dimensional cusps," i.e., the cusps at ∞ in Siegel domains of the third kind. And we will then use Siegel sets instead of Piatetskii-Shapiro's cylindrical sets U_ε to carry out the details. However, before launching into the morass of the general cusps, it is nice to look at the case where the above, relatively simple, construction is sufficient – the case of $\mathbb{Q}$-rank 1. This is what we will do in the following appendix.

Appendix: Groups of $\mathbb{Q}$-rank 1 acting on tube domains

We keep the notations from Section 1.

Proposition 1.5 *The following conditions are equivalent:*

(i) $C/\overline{\Gamma} \cdot \mathbb{R}_{>0}$ *is compact;*

(ii) *the $\mathbb{Q}$-rank of $\overline{G}$ is 1;*

(iii) *the only rational point on ∂C is 0;*

(iv) *the $\mathbb{Q}$-rank of G is 1;*

(v) *the (proper) rational boundary components of the hermitian symmetric domain U are all zero- dimensional.*

Proof That (ii) $\Longleftrightarrow$ (iv) follows from the facts that, first, a $\mathbb{Q}$-parabolic contains a maximal $\mathbb{Q}$-split torus of G and, second, that $\overline{G}$ is the reductive part of the parabolic subgroup P of G; hence $\overline{G}$ contains a maximal $\mathbb{Q}$-split torus of G.

The equivalence of (ii) and (iii) comes from the correspondence between parabolic subgroups and rational boundary components, see Chapter II, Section 3.10. Similarly, the equivalence of (iv) and (v) comes from the general theory of bounded domains (see Section 3, Proposition 3.9, below). Finally, the equivalence (iii) $\Longleftrightarrow$ (i) is a consequence of the reduction theory for cones. $\qquad \square$

Example 1.6 Let k be a totally real extension of degree n of $\mathbb{Q}$. Let

$$C = \mathbb{R}_{>0}^n \subset \mathbb{R}^n \;;$$
$$\overline{\mathscr{G}} = R_{k/\mathbb{Q}}(\mathbb{G}_m) \quad \text{(Weil's restriction of scalars)} \;;$$
$$\mathscr{G} = R_{k/\mathbb{Q}}(\mathrm{SL}(2)) \;.$$

In this case U is the n-fold product of $\mathfrak{H}$ with itself. Also Γ is commensurable with $\mathrm{SL}(2, \mathcal{O})$, where $\mathcal{O}$ is the ring of integers in k, and $\overline{\Gamma}$ is commensurable with the group of units of k. This example is called the *Hilbert case*, and for $n = 2$ was treated in Chapter I, Section 5.

In this appendix, we want to look more closely at the $\mathbb{Q}$-rank 1 case. In this case, we see immediately that $U'' \setminus U' \subset U''_\varepsilon$ for every ε.

Therefore, $U''/\overline{\Gamma} = (U'/\overline{\Gamma}) \cup ((\overline{\Gamma} \cdot U''_\varepsilon)/\overline{\Gamma})$. Let

$$E = (U''/\overline{\Gamma}) \setminus (U'/\overline{\Gamma})$$

be the locus at infinity that is added on. Then $E \subset (\overline{\Gamma} \cdot U''_\varepsilon)/\overline{\Gamma}$ for every ε, so, by Theorem 1.4, E is relatively compact in $U''/\overline{\Gamma}$. But E is closed in $U''/\overline{\Gamma}$, so

E is itself compact. Put another way, we have an analytic space $U''/\overline{\Gamma}$ and a compact analytic subset E, and $U'/\overline{\Gamma}$ is just $(U''/\overline{\Gamma}) \setminus E$:

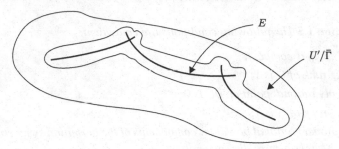

The main result of this appendix is as follows.

Theorem 1.7 *The set E is exceptional, i.e., can be blown down in $U''/\overline{\Gamma}$ to a point.*

Proof We may, if we like, pass to a subgroup of finite index $\overline{\Gamma}'$ of $\overline{\Gamma}$ and prove that $(U'' \setminus U')/\overline{\Gamma}'$ can be blown down.

So we can assume from the beginning that $\overline{\Gamma}$ acts freely on U''; in particular, $U'/\overline{\Gamma}$ is a manifold.

We are going to apply the following fact (cf. [4]).

Let $A \subset X$ be a compact analytic subset of an analytic space, where $X \setminus A$ is a manifold. If A possesses arbitrarily small strictly Levi-pseudoconvex neighborhoods, then A can be blown down to a point.

(Recall that an open subset $U \subset X$ is called *strictly Levi-pseudoconvex* if, for all $y \in \partial U$, there exists a real C^2-function φ defined in a neighborhood V of y such that

(i) $U \cap V = \{x \in V \mid \varphi(x) < 0\}$;

(ii) *(convexity condition)* for $0 \neq t \in T_y$ with $d\varphi(t) = 0$, one has $\partial \bar{\partial} \varphi(t) > 0$.)

As in the previous section, let C'' be the interior of the closure of C in $N_{\{\sigma_\alpha\}}$. We will use the characteristic function φ of the cone C introduced in Chapter II, Section 1. It is easy to see that φ extends to a continuous function φ on C'' with $\varphi \equiv 0$ on $C'' \setminus C$. Moreover, φ is $\overline{\Gamma}$-invariant, and hence comes from a function $\overline{\varphi}$ on $C''/\overline{\Gamma}$. Let $f = \varphi \circ \mathrm{ord}$ be the induced function on U''. Then f comes from a function $\overline{f}$ on $U''/\overline{\Gamma}$. Since $\varphi(x) = 0$ if and only if $x \in C'' \setminus C$, we get $\overline{f}(x) = 0$ if and only if $x \in E$. Therefore the open sets

$$V_c = \{X \in U''/\overline{\Gamma} \mid \overline{f}(x) < c\}$$

form a family of arbitrarily small neighborhoods of E. Hence it suffices to show that V_c is pseudoconvex, or to check the Levi condition at the points of the boundary of V_c.

We can choose coordinates $z_i = x_i + i \cdot y_i$ on U' such that f becomes

$$f(z_1, \ldots, z_n) = \varphi(y_1, \ldots, y_n) .$$

Since φ is strictly convex (see Chapter II, Proposition 1.4), f is strictly Levi-convex, and the theorem is proven. $\qquad\qquad\qquad\qquad\qquad\qquad\qquad$ $\square$

Remarks (i) In the Hilbert case, the function $\varphi : \mathbb{R}^n \longrightarrow \mathbb{R}$ considered in the above proof is simply $\varphi(x_1, \ldots, x_n) = 1/(x_1 \cdots x_n)$. For $n = 2$, the constructed neighborhoods coincide with the ones considered in Chapter I, Section 5.

(ii) By applying the above procedure to all cusps we obtain in the end a *compactification* of U/Γ, which is just U/Γ with a finite number of points added: we recover in the $\mathbb{Q}$-rank 1 case the so-called Baily–Borel compactification.

2 The structure of bounded symmetric domains

In this section we summarize the standard theory of symmetric spaces. One of its purposes is to introduce notation.

2.1

A *Riemannian symmetric space* is a connected Riemannian manifold D such that, for every point $x \in D$, there exists an involutive automorphism s_x which has x as an isolated fixed point.

If M is a complex hermitian manifold, then D is a *hermitian symmetric space* if, for every point $x \in D$, there exists an involutive automorphism s_x which has x as an isolated fixed point (here, of course, the condition that s_x is an automorphism means that s_x is holomorphic as well as isometric).

If D is a hermitian symmetric space, then the Riemannian manifold D decomposes† as

$$D = D_0 \times D_1 \times \cdots \times D_n ,$$

where:

- D_0 is the quotient of a complex vector space with a translation-invariant metric by a discrete group of translations (such a hermitian symmetric space is said to be of *euclidean type*);
- D_i, $i \neq 0$, is an irreducible and non-euclidean hermitian symmetric space.

† If D is simply connected, this is well known and can be found, for instance, in Helgason [6]. In the hermitian case, simply connectedness is not needed: see Wolf [13], p. 490, Lemma 1.

Those factors D_i, for $i \neq 0$, which are compact are said to be of *compact type* and are rational projective varieties; a non-compact factor D_i with $i \neq 0$ is said to be of *non-compact type* and is a bounded domain in $\mathbb{C}^n$ (we will prove this later).

The space D is called *non-euclidean*, resp. *symmetric domain*, resp. *of compact type*, if, respectively, the factor D_0 is absent in the above decomposition, or all the D_i are of non-compact type, or all D_i are of compact type. In the latter two cases D is simply connected. If, in addition, there is only one factor present, D is called *simple*.

Let D be a non-euclidean hermitian symmetric space and denote by G the identity component of its (Lie) group of automorphisms. Then the group G acts transitively on the space D and, choosing a basepoint $o \in D$, we may write

$$D = G/K \,,$$

where $K \subset G$ is a compact subgroup.

The symmetry s_o induces an automorphism σ of the group G such that $K = K^\sigma$. Letting

$$\mathfrak{g} = \mathrm{Lie}\,(G) \,,$$
$$\mathfrak{k} = \mathrm{Lie}\,(K) = \text{subspace of } \mathfrak{g}, \text{ where } \sigma = \mathrm{Id} \,,$$
$$\mathfrak{p} = \text{subspace of } \mathfrak{g}, \text{ where } \sigma = -\mathrm{Id} \,,$$

we get a decomposition

$$\mathfrak{g} = \mathfrak{k} \oplus \mathfrak{p} \,.$$

Note that there is a canonical isomorphism $\mathfrak{p} \xrightarrow{\sim} T_o$, the tangent space to D at o.

If D is now a hermitian symmetric *domain*, then the group G is semi-simple and adjoint, K is a maximal compact subgroup, and the above direct sum decomposition is a Cartan decomposition.

Note that G is the connected component $\mathscr{G}(\mathbb{R})^o$ of the set of real points of a unique algebraic group $\mathscr{G}$ defined over $\mathbb{R}$: via the adjoint representation

$$\mathrm{Ad} : G \longrightarrow \mathrm{GL}(\mathfrak{g}) \,,$$

$\mathscr{G}$ is the Zariski closure of $\mathrm{Im}(\mathrm{Ad})$.

Set

$$\mathfrak{k}_c = \mathfrak{k} \,,$$
$$\mathfrak{p}_c = i\mathfrak{p}(\subset \mathfrak{g}_{\mathbb{C}}) \,,$$
$$\mathfrak{g}_c = \mathfrak{k}_c \oplus \mathfrak{p}_c \,.$$

Then $\mathfrak{g}_c$ is a compact real form of $\mathfrak{g}_{\mathbb{C}}$. Passing to the group level, we can define the compact dual of D by

$$\check{D} = G_c/K_c \;.$$

Conversely, starting with a hermitian symmetric space of compact type $\check{D}$, one gets back to the space D by the same construction.

Let D be a non-euclidean hermitian symmetric space. Then there exists a morphism

$$u_o : U^1 \longrightarrow G$$

from the circle group U^1 into G such that $u_o(z) \in K$ for any $z \in U^1$ and $u_o(z)$ induces the multiplication by z on the tangent space T_o of D at o. The group $K \subset G$ is the centralizer of $u_o(U^1)$ in G, and hence is connected. Further, if D is simple, $u_o(U^1)$ is the center of $K \subset G$.

We set

$$h_o = u_o^2 \;.$$

Noting that, via the isomorphism $\mathfrak{p} \xrightarrow{\sim} T_o$ the adjoint action of K on $\mathfrak{p}$ corresponds to the action of K on T_o, we see that

$$J = \operatorname{Ad}\left(h_o\left(e^{2\pi i/8}\right)\right)\Big|_{\mathfrak{p}}$$

defines the given complex structure on T_o, whereas

$$\sigma = \operatorname{Ad}\left(h_o(i)\right) \;.$$

Let

$$\mathfrak{p}_{\mathbb{C}} = \mathfrak{p}_+ \oplus \mathfrak{p}_-$$

be the decomposition into $\pm i$-eigenspaces for J. Note that these are abelian subalgebras of $\mathfrak{g}_{\mathbb{C}}$ since, e.g.,

$$[\mathfrak{p}_+, \mathfrak{p}_+] \subset \{x \in \mathfrak{g} \mid Jx = -x\} = (0) \;.$$

Denote by $P_{\pm}$ the subgroup of $G_{\mathbb{C}}$ generated by $\exp(\mathfrak{p}_{\pm})$. Then $K_{\mathbb{C}}$ normalizes $P_{\pm}$ and $K_{\mathbb{C}} \cdot P_-$ is a parabolic subgroup of $G_{\mathbb{C}}$ with unipotent radical P_-; hence $G_{\mathbb{C}}/K_{\mathbb{C}} \cdot P_-$ is a projective algebraic variety, which we call for the moment X.

Theorem 2.1 (Borel and Harish-Chandra embedding theorem)

(a) *The map $P_+ \times K_{\mathbb{C}} \times P_- \longrightarrow G_{\mathbb{C}}$ given by multiplication is injective, G is contained in the image, and $(K_{\mathbb{C}} \cdot P_-) \cap G = K$.*

(b) *Hence we have the following maps:*

$$G/K \longrightarrow P_+ \times K_{\mathbb{C}} \times P_-/K_{\mathbb{C}} \cdot P_- \longrightarrow G_{\mathbb{C}}/K_{\mathbb{C}} \cdot P_-$$

$$\wr\| \qquad\qquad \wr\| \qquad\qquad \wr\|$$

$$D \qquad\qquad\qquad P_+ \qquad\qquad\qquad\qquad X$$

$$\wr\uparrow \exp$$

$$\mathfrak{p}_+$$

These maps are holomorphic open immersions, the image of D in $\mathfrak{p}_+$ is a bounded domain, and the image of $\mathfrak{p}_+$ in X is a dense Zariski open set.

(c) *Finally, the compact form $G_c \subset G_{\mathbb{C}}$ of G acts transitively on X; further-more, $G_c \cap (K_{\mathbb{C}} \cdot P_-) = K$ so that*

$$X \cong \check{D} := G_c/K ,$$

i.e., X is the compact dual of D.

For a full proof of this, we refer the reader to Helgason [6], Ch. 8, §7, and we give here only a brief indication; thus, on the Lie algebra level, we have

$$\mathfrak{g}_{\mathbb{C}} = \mathfrak{k}_{\mathbb{C}} \oplus \mathfrak{p}_+ \oplus \mathfrak{p}_- ,$$

$$\mathfrak{k} = \mathfrak{g} \cap (\mathfrak{k}_{\mathbb{C}} \oplus \mathfrak{p}_-) = \mathfrak{g}_c \cap (\mathfrak{k}_{\mathbb{C}} \oplus \mathfrak{p}_-) ,$$

and

$$\dim D = \dim \check{D} = \dim \mathfrak{p}_+ = \dim P_+ = \dim G_{\mathbb{C}}/K_{\mathbb{C}} \cdot P_- ,$$

from which we deduce that all the assertions are "true locally," e.g., the natural maps $D \longrightarrow X$, and $P_+ \longrightarrow X$, and $G_c/K \longrightarrow X$ are local immersions. The main step is to check

$$G \subset (\text{bounded subset of } P_+) \cdot K_{\mathbb{C}} \cdot P_- .$$

This can be done by using Theorem 2.4 below to reduce to the simplest case $G = \mathrm{SL}(2,\mathbb{R})/\{\pm 1\}$, where it follows by an explicit explicit calculation. The rest is straightforward.

2.2

We want next to look at holomorphic maps between bounded symmetric do-mains $f : D_1 \longrightarrow D_2$. The maps which have good Lie-theoretic meaning are the *symmetric maps*; this means that, for every $x \in D_1$,

$$f \circ s_x^{(1)} = s_{f(x)}^{(2)} \circ f , \tag{2.1}$$

where $s_x^{(1)}$, resp. $s_{f(x)}^{(2)}$, are the symmetries of D_1, resp. D_2, with respect to x, resp. $f(x)$. It is readily checked that this implies that, if $G_i = \mathrm{Aut}\,(D_i)^o$, then there is a *covering* G_1' of G_1 and a homomorphism

$$\varphi : G_1' \longrightarrow G_2$$

for which f is equivariant:

$$f(g \cdot x) = \varphi(g) \cdot f(x) .$$

To see this, let $G_1'' \subset G_1 \times G_2$ be the connected component of the set of pairs (g_1, g_2) such that $f(g_1 \cdot x) = g_2 \cdot f(x)$; via (2.1), check that $p_1 : G_1'' \longrightarrow G_1$ is surjective; now let G_1' be the product of those simple factors of G_1'' that map non-trivially to G_1. Note that we may assume $G_1' \subset G_1 \times G_2$, hence $G_1' = \mathscr{G}'(\mathbb{R})^o$, where $\mathscr{G}'$ is an algebraic group over $\mathbb{R}$, and $G_1' \longrightarrow G_1$ is a *finite* covering.

Proposition 2.2 † *Let D_1 and D_2 be bounded symmetric domains, let $o_i \in D_i$ be basepoints, and let $G_i = \mathrm{Aut}\,(D_i)^o$. There are natural bijections between the following sets:*

(a) *holomorphic symmetric maps $f : D_1 \longrightarrow D_2$ such that $f(o_1) = o_2$;*
(b) *maps $f : D_1 \longrightarrow D_2$ such that $f(o_1) = o_2$ and, for all $x \in D_1$, $\theta \in \mathbb{R}$,*

$$f \circ u_x(e^{i\theta}) = u_{f(x)}(e^{i\theta}) \circ f ,$$

 where $u_x(e^{i\theta})$, resp. $u_{f(x)}(e^{i\theta})$, are the automorphisms fixing x, resp. $f(x)$, and multiplying by $e^{i\theta}$ in the tangent spaces;
(c) *connected coverings $\pi : G_1' \longrightarrow G_1$ and homomorphisms*

$$\varphi = \varphi_1 \times \varphi_2 : \mathbb{R} \times G_1' \longrightarrow G_2 ,$$

 such that $\pi \times \varphi_2 : G_1' \longrightarrow G_1 \times G_2$ is injective and such that, if we lift the homomorphism $\theta \longmapsto u_{o_1}(e^{i\theta})$ as follows:

 then $u_{o_2}(e^{i\theta}) = \varphi(\theta, u'(\theta))$.

Proof First of all, note that any map $f : D_1 \longrightarrow D_2$ is φ_2-equivariant for at most one $\varphi_2 : G_1' \longrightarrow G_2$. Indeed, introduce $G_1'' \subset G_1 \times G_2$ as above. Then

† A slight modification of a result of Satake and Kuga.

Lie $G_1'' = $ Lie $G_1 \times$ Lie $\widetilde{K}$, where $\widetilde{K} = \{g_2 \in G_2 \mid g_2 \circ f = f\}$. Now $\widetilde{K} \subset \mathrm{Stab}_{o_2}$, so $\widetilde{K}$ is compact. Furthermore, since f is φ_2-equivariant, the map $\pi \times \varphi_2 : G_1' \to G_1 \times G_2$ factors through G_1''. We thus get a map

$$\mathrm{Lie}\, G_1 = \mathrm{Lie}\, G_1' \to \mathrm{Lie}\, G_1'' = \mathrm{Lie}\, G_1 \times \mathrm{Lie}\, \widetilde{K} \,.$$

As G_1 is semisimple without compact factors, there are no homomorphisms from G_1 to $\widetilde{K}$, so it follows that the image of $\mathrm{Lie}\, G_1'$ in $\mathrm{Lie}\, G_1 \times \mathrm{Lie}\, G_2$ is uniquely determined. Since G_1' is connected and $\pi \times \varphi_2$ is injective, G_1' and φ_2 are uniquely determined.

Now start with $f : D_1 \longrightarrow D_2$, holomorphic and symmetric. Construct $\varphi_2 : G_1' \to G_2$ as above. Let $\sigma_i : G_i \longrightarrow G_i$ be the Cartan involution with respect to $K_i = \mathrm{Stab}_{o_i}$. Then f is also $\sigma_2 \circ \varphi_2 \circ \sigma_1$-equivariant, so $\varphi_2 \circ \sigma_1 = \sigma_2 \circ \varphi_2$. Therefore $d\varphi_2$ preserves the Cartan decomposition, in particular $d\varphi_2(\mathfrak{p}_1) \subset \mathfrak{p}_2$. Also, since f is holomorphic, $d\varphi_2 : \mathfrak{p}_1 \longrightarrow \mathfrak{p}_2$ is complex-linear. Now, every element of D_1 equals $\exp(a) \cdot o_1$, with $a \in \mathfrak{p}_1$. So calculate:

$$f \circ u_{o_1}(e^{i\theta})(\exp a \cdot o_1) = f\left(\exp(e^{i\theta}a) \cdot o_1\right)$$
$$= \varphi\left(\exp(e^{i\theta}a)\right) \cdot o_2$$
$$= \exp\left(d\varphi_2(e^{i\theta} \cdot a)\right) \cdot o_2$$
$$= \exp\left(e^{i\theta} \cdot d\varphi_2(a)\right) \cdot o_2$$
$$= u_{o_2}(e^{i\theta}) \cdot \varphi_2(\exp a) \cdot o_2$$
$$= u_{o_2}(e^{i\theta}) \cdot f(\exp a \cdot o_1) \,.$$

This proves that f has the property in (b) for $x = o_1$; the general case follows by conjugating.

To go from (b) to (c), let $\alpha_\theta : G_1' \longrightarrow G_1'$ be conjugation by $u'(\theta)$, and let $\beta_\theta : G_2 \longrightarrow G_2$ be conjugation by $u_{o_2}(e^{i\theta})$. Then (b) shows that f is both φ_2-equivariant and $\beta_\theta^{-1} \circ \varphi_2 \circ \alpha_\theta$-equivariant. This means that $\varphi_2(G_1)$ commutes with $u_{o_2}(e^{i\theta}) \cdot \varphi_2(u'(\theta))^{-1}$ for all θ, which means precisely that $\varphi : \mathbb{R} \times G_1' \longrightarrow G_2$ as in (c) can be constructed.

Conversely, given $\varphi : \mathbb{R} \times G_1' \longrightarrow G_2$, let K_1' be the centralizer of u', and let K_2 be the centralizer of u_{o_2}. Then $\varphi(\mathbb{R} \times K_1') \subset K_2$, so φ defines $f : (\mathbb{R} \times G_1')/(\mathbb{R} \times K_1') \longrightarrow G_2/K_2$, i.e., $f : D_1 \longrightarrow D_2$. Using the fact that $\varphi_1(\theta) = u_{o_2}(e^{i\theta}) \cdot \varphi_2(u'(\theta))^{-1}$ centralizes $\mathrm{Im}\,\varphi$, we check easily the identity in (b). $\qquad \square$

Corollary 2.3 *There is a bijection between the following two sets:*

(a) *holomorphic symmetric maps* $f : \mathfrak{H} \longrightarrow D$ *such that* $f(\mathrm{i}) = o$; *and*

(b) *homomorphisms* $\varphi : U^1 \times SL(2,\mathbb{R}) \longrightarrow G$ *such that*

$$\varphi\left(e^{i\theta}, h^{SL}(e^{i\theta})\right) = h_o(e^{i\theta}),$$

where

$$h^{SL}(e^{i\theta}) = \begin{pmatrix} \cos\theta & \sin\theta \\ -\sin\theta & \cos\theta \end{pmatrix}.$$

Proof The only new point here is that $SL(2)$ is simply connected as an algebraic group, so, if $G_1 = \text{Aut}\,(\mathfrak{H})$, we can assume $G'_1 = SL(2,\mathbb{R})$. $\quad\square$

We remark that, if

$$f : D_1 \longrightarrow D_2,$$
$$\varphi_2 : G'_1 \longrightarrow G_2$$

have the properties of Proposition 2.2, then $d\varphi_2 : \mathfrak{g}_1 \longrightarrow \mathfrak{g}_2$ commutes with the Cartan involutions, and hence

$$d\varphi_2(\mathfrak{k}_1) \subset \mathfrak{k}_2,$$
$$d\varphi_2(\mathfrak{p}_1) \subset \mathfrak{p}_2,$$
$$d\varphi_2(\mathfrak{p}_{1,\pm}) \subset \mathfrak{p}_{2,\pm};$$

hence, extending φ_2 to $\varphi'_2 : G'_{1,\mathbb{C}} \longrightarrow G_{2,\mathbb{C}}$, we have

$$\varphi'_2(K'_{1,\mathbb{C}}) \subset K_{2,\mathbb{C}},$$
$$\varphi'_2(P'_{1,\pm}) \subset P_{2,\pm}.$$

Therefore we can extend f, getting the following commutative diagram:

$$
\begin{array}{ccc}
D_1 & \xrightarrow{\ f\ } & D_2 \\
\cap & & \cap \\
P_{1,+} & \xrightarrow{\ f\ } & P_{2,+} \\
\cap & & \cap \\
\check{D}_1 & \xrightarrow{\ \check{f}\ } & \check{D}_2
\end{array}
$$

2.3

We now take up the study of the roots of G and $G_{\mathbb{C}}$.

Let $\mathfrak{t} \subset \mathfrak{k}$ be a Cartan subalgebra of $\mathfrak{k}$. Then $\mathfrak{t}$ contains $\text{Im}\,(dh_o)$, so that $\mathfrak{t}$ is also a Cartan subalgebra of $\mathfrak{g}$. Let

$$\Psi = \mathfrak{t}_{\mathbb{C}}\text{-root system of } \mathfrak{g}_{\mathbb{C}},$$

so that

$$\mathfrak{g}_{\mathbb{C}} = \mathfrak{t}_{\mathbb{C}} + \sum_{\varphi \in \Psi} \mathfrak{g}^{\varphi} \ .$$

A root φ is called *compact* if $\mathfrak{g}^{\varphi} \subset \mathfrak{k}_{\mathbb{C}}$; it is called *non-compact* if $\mathfrak{g}^{\varphi} \subset \mathfrak{p}_{\mathbb{C}}$. We denote:

$$\Psi_K = \text{the compact roots} \ ,$$
$$\Psi_p^+ = \text{the non-compact roots with } \mathfrak{g}^{\varphi} \subset \mathfrak{p}_+ \ ,$$
$$\Psi_p^- = \text{non-compact roots with } \mathfrak{g}^{\varphi} \subset \mathfrak{p}_- \ ,$$

and

$$\Psi_p = \Psi_p^+ \cup \Psi_p^- \ .$$

One can choose a linear ordering on Ψ such that $\Psi_p^+ \subset \{\text{positive roots}\}$ and $\Psi_p^- \subset \{\text{negative roots}\}$. For example, we may choose as positive roots those corresponding to a Weyl chamber in it whose closure contains $(dh_o)(i)$.

For $\varphi \in \Psi$, define

$$h_\varphi \in \text{it}$$

by

$$2\frac{\langle \varphi, \psi \rangle}{\langle \varphi, \varphi \rangle} = \psi(h_\varphi) \text{ for all } \psi \in \mathfrak{t}^* \ ,$$

where $\langle \cdot, \cdot \rangle$ denotes the Killing form. Choose root vectors

$$e_\varphi \in \mathfrak{g}^{\varphi}$$

such that

$$[e_\varphi, e_{-\varphi}] = h_\varphi \ ,$$

and such that the complex conjugation of $\mathfrak{g}_{\mathbb{C}}$ with respect to $\mathfrak{g}$ permutes e_φ and $e_{-\varphi}$, whenever $\varphi \in \Psi_p$. Let

$$x_\varphi = e_\varphi + e_{-\varphi} \ ,$$
$$y_\varphi = i(e_\varphi - e_{-\varphi}) \text{ for } \varphi \in \Psi_p^+ \ .$$

These elements form a basis over $\mathbb{R}$ of $\mathfrak{p}$ such that $Jx_\varphi = y_\varphi, Jy_\varphi = -x_\varphi$.

Two roots φ and ψ are called *strongly orthogonal*, denoted by

$$\varphi \perp\!\!\!\perp \psi \ ,$$

if neither of $\varphi \pm \psi$ is a root; in this case φ and ψ are orthogonal. Harish-Chandra [5], p. 583, chooses in the following inductive way a maximal set of strongly orthogonal roots:

$$\gamma_1, \ldots, \gamma_r$$

is the maximal set of roots such that each γ_i' is the smallest element of Ψ_p^+ strongly orthogonal to $\gamma_1', \ldots, \gamma_{i-1}'$.† We write h_i, x_i, e_i, e_{-i}, y_i for $h_{\gamma_i'}$, $x_{\gamma_i'}$, $e_{\gamma_i'}$, $e_{-\gamma_i'}$, $y_{\gamma_i'}$ (where $i = 1, \ldots, r$).

The following theorem is basic to the theory of hermitian symmetric domains.

Theorem 2.4 (Harish-Chandra)

(i) *The subspace* $\mathfrak{a} = \sum_{i=1}^{r} \mathbb{R}x_i \subset \mathfrak{p}$ *is a maximal commutative subalgebra of* $\mathfrak{p}$; *hence* $A = \exp(\mathfrak{a})$ *is the connected component of the group of real points of a maximal split torus* $\mathscr{A}$ *in* $\mathscr{G}$, *i.e.,* $A = \mathscr{A}(\mathbb{R})^\circ$. *Moreover, the action of* $K \cdot A$ *is transitive, i.e.,* $K \cdot A \cdot o = D$.

(ii) *There exists a morphism*

$$\varphi : U^1 \times \mathrm{SL}(2, \mathbb{R})^r \longrightarrow G$$

such that:

(a) $\varphi\left(u, h^{SL}(u), \ldots, h^{SL}(u)\right) = h_o(u)$;

(b) $\mathrm{d}\varphi$ *on the ith factor* $\mathrm{SL}(2, \mathbb{R})$ *is given by*

$$\mathrm{d}\varphi \begin{pmatrix} a & b \\ c & -a \end{pmatrix} = a \cdot x_i - \frac{b+c}{2} \cdot y_i + \frac{b-c}{2} \cdot ih_i \ ,$$

and hence

$$\mathrm{d}\varphi\left(\mathfrak{sl}(2, \mathbb{R})^r\right) = \mathfrak{a} + J\mathfrak{a} + [\mathfrak{a}, J\mathfrak{a}] \ ;$$

in particular, $\mathrm{d}\varphi$ *induces an isomorphism between the subalgebra of "diagonal matrices" in* $\mathfrak{sl}(2, \mathbb{R})^r$ *and* $\mathfrak{a}$;

(c) φ *induces a symmetric holomorphic map*

$$\widetilde{\varphi} : \mathfrak{H}^r \longrightarrow D$$

equivariant with respect to φ, *where* $\mathrm{SL}(2, \mathbb{R})$ *acts on* $\mathfrak{H}$ *as usual and* U^1 *acts trivially, taking* $(\mathrm{i}, \mathrm{i}, \ldots, \mathrm{i}) \in \mathfrak{H}^r$ *to* $o \in D$.

Proof (i) We refer the reader to Helgason [6], p. 314, for the somewhat delicate, but elementary, verification that $\mathfrak{a}$ is maximal. Let $A = \exp(\mathfrak{a}) \subset G$. Then, by Cartan,

$$G = K \cdot A \cdot K \ ,$$

---

† These roots are called γ_i' to distinguish them from the closely related roots γ_i defined on a *real split torus* below. Incidentally, it should be pointed out that $\{\gamma_i'\}$ is a very special maximal set of strongly orthogonal roots: Theorem 2.4 below definitely does not hold for every such maximal set.

and hence

$$D = K \cdot A \cdot K \cdot o = K \cdot A \cdot o .$$

(ii)   Let $\mathfrak{g}_i = \mathbb{R}x_i + \mathbb{R}y_i + \mathbb{R}\mathrm{i}h_i$; then

$$\mathfrak{g}_i \xrightarrow{\ \sim\ } \mathfrak{sl}(2,\mathbb{R})$$

by the map indicated in (ii), (b). We thus obtain

$$\varphi_2 : \mathrm{SL}(2,\mathbb{R})^r \longrightarrow G$$

such that $d\varphi_2$ is given by (ii), (b). Moreover, noting that $\mathrm{Ad}\,h_o(u)$ fixes each h_i, and multiplies e_i by u^2 and e_{-i} by u^{-2}, one checks easily that

$$\mathrm{Ad}\,h_o(u)\Big|_{\mathfrak{g}_i} = \mathrm{Ad}\,\varphi_2 \left(I,\dots, \left(\begin{array}{cc} \cos\theta & \sin\theta \\ -\sin\theta & \cos\theta \end{array} \right),\dots,I \right)\Big|_{\mathfrak{g}_i} \ \ \text{if } u = e^{i\theta} ,$$

and hence that

$$\mathrm{Ad}\,h_o(u)\Big|_{\Sigma\,\mathfrak{g}_i} = \mathrm{Ad}\,\varphi_2 \left(\left(\begin{array}{cc} \cos\theta & \sin\theta \\ -\sin\theta & \cos\theta \end{array} \right),\dots, \left(\begin{array}{cc} \cos\theta & \sin\theta \\ -\sin\theta & \cos\theta \end{array} \right) \right)\Big|_{\Sigma\,\mathfrak{g}_i} .$$

Since $h_o(u)$ and $\varphi_2\left(\left(\begin{array}{cc} \cos\theta & \sin\theta \\ -\sin\theta & \cos\theta \end{array} \right),\dots \right)$ both lie in $T = \exp\mathfrak{t}$, they commute, and we may write:

$$h_o(u) = \varphi_1(u) \cdot \varphi_2 \left(\left(\begin{array}{cc} \cos\theta & \sin\theta \\ -\sin\theta & \cos\theta \end{array} \right),\dots \right) ,$$

where $\varphi_1(u)$ centralizes $\mathrm{Im}\,\varphi_2$. Defining φ by

$$\varphi = \varphi_1 \times \varphi_2 : U^1 \times \mathrm{SL}(2,\mathbb{R})^r \longrightarrow G ,$$

it satisfies both (ii), (a) and (b).

Finally, (ii), (c) follows from Proposition 2.2. $\square$

Definition 2.5 The number r appearing in the above theorem is called the *rank* of D or the $\mathbb{R}$-*rank* of G.

The map $\widetilde{\varphi}$ has the following strong universal property.

Proposition 2.6 (Satake) *Every symmetric holomorphic map*

$$\widetilde{\psi} : \mathfrak{H} \longrightarrow D$$

such that $\widetilde{\psi}(\mathrm{i}) = o$ *is of the form*

$$\widetilde{\psi}(z) = k \cdot \widetilde{\varphi}(\dots,z,\dots,\mathrm{i},\dots)$$

for some $k \in K$ *and some distribution of* z's *and* i's *among the* r *variables of* $\widetilde{\varphi}$.

Proof According to Corollary 2.3, $\widetilde{\psi}$ is ψ_2-equivariant for some homomorphism

$$\psi = \psi_1 \times \psi_2 : U^1 \times SL(2,\mathbb{R}) \longrightarrow G$$

such that $h_o(e^{i\theta}) = \psi\left(e^{i\theta}, h^{SL}(e^{i\theta})\right)$. It follows that:

$$d\psi\left(\mathbb{R} \cdot \begin{pmatrix} 1 & 0 \\ 0 & -1 \end{pmatrix}\right) \subset \mathfrak{p}.$$

We use the fact that every element of $\mathfrak{p}$ is in $\mathrm{Ad}\, K \cdot \mathfrak{a}$ (the polar decomposition in $\mathfrak{g}$). Therefore, replacing $\widetilde{\psi}$ by $k \cdot \widetilde{\psi}$ and ψ by $k\psi k^{-1}$, we can assume

$$d\psi\left(\mathbb{R} \cdot \begin{pmatrix} 1 & 0 \\ 0 & -1 \end{pmatrix}\right) \subset \mathfrak{a}.$$

Then

$$d\psi\left(\mathbb{R} \cdot \begin{pmatrix} 0 & 1 \\ 1 & 0 \end{pmatrix}\right) \subset J\mathfrak{a}$$

and

$$d\psi\left(\mathbb{R} \cdot \begin{pmatrix} 0 & 1 \\ -1 & 0 \end{pmatrix}\right) \subset [\mathfrak{a}, J\mathfrak{a}].$$

This shows that ψ factors through $\mathrm{Im}\, \varphi \subset G$, and hence that ψ_2 factors through

$$\psi_2^* : SL(2,\mathbb{R}) \longrightarrow SL(2,\mathbb{R})^r.$$

Now, $\mathrm{pr}_i \circ \psi_2^*$ is either trivial or conjugate to the identity by some element of $k_i \in SO(2,\mathbb{R})$, from which we see that $\varphi_2(k_1,\ldots,k_r) \cdot \widetilde{\psi}$ has the required form.
□

Next, we want to decompose $\mathfrak{g}$ into irreducible pieces under the restriction of the adjoint representation to $U^1 \times SL(2,\mathbb{R})^r$. The irreducible real representations of $U^1 \times SL(2,\mathbb{R})^r$ are easily enumerated as follows.

(a) The irreducible real representations of U^1 are:

- $U_0 = \mathbb{R}$, trivial representation;
- $U_k = \mathbb{R}^2$, with action

$$e^{i\theta} \longmapsto \begin{pmatrix} \cos k\theta & \sin k\theta \\ -\sin k\theta & \cos k\theta \end{pmatrix},$$

for $k = 1,2,\ldots$.

Note that these representations are only irreducible as *real* representations: they split over $\mathbb{C}$ into the direct sum of two one-dimensional representations with characters

$$e^{i\theta} \longmapsto (e^{i\theta})^{\pm k},$$

respectively.

(b) The irreducible representations of $\mathrm{SL}(2,\mathbb{R})$ are

- W_k, the k'th symmetric power of the standard representation, for $k = 0, 1, 2, \ldots$.

Note that these are complex-irreducible as well.

It follows easily that the irreducible representation of $U^1 \times \mathrm{SL}(2,\mathbb{R})^r$ are of the form

$$U_i \otimes W_{j_1} \otimes \ldots \otimes W_{j_r} \, .$$

The fact that U^1, acting via h_o on $\mathfrak{g}_\mathbb{C}$, has only the characters z^2, 1, z^{-2} implies easily that, if this representation appears in $\mathfrak{g}_\mathbb{C}$, then

$$i + \sum_{k=1}^r j_k = 0 \text{ or } 2 \, ,$$

and we get:

Proposition 2.7 *The irreducible representations of $U^1 \times \mathrm{SL}(2,\mathbb{R})^r$ which appear in $\mathfrak{g}$ are at most the following:*

(a$_i$) $U_0 \otimes (W_0 \otimes \cdots \otimes W_2 \otimes \cdots \otimes W_0)$ (one W_2) ;
(b$_{ij}$) $U_0 \otimes (W_0 \otimes \cdots \otimes W_1 \otimes \cdots \otimes W_1 \otimes \ldots \otimes W_0)$ (two W_1) ;
(c$_i$) $U_1 \otimes (W_0 \otimes \cdots \otimes W_1 \otimes \cdots \otimes W_0)$ (one W_1) ;
(d) $U_2 \otimes$ trivial ;
(e) $U_0 \otimes$ trivial .

Moreover, the image via $d\varphi$ of the ith copy of $\mathrm{SL}(2,\mathbb{R})^r$ defines a representation of type (a$_i$), *and this is the only one of this type. Furthermore,* (d) *does not occur at all.*

Proof Let V be a representation of type (a$_i$) or (d). Then, in both cases, V contains a 2-dimensional subrepresentation W, where h_0 acts by the representation U_2, containing some $0 \neq w \in W$ on which $\mathfrak{a}$ acts trivially. The first condition implies that $w \in \mathfrak{p}$, whence by the maximality of $\mathfrak{a}$ we get $w \in \mathfrak{a}$. This leads to the given factors of type (a$_i$). □

This proposition gives us a bird's-eye view of the complex root decomposition of $\mathfrak{g}_\mathbb{C}$ in which $\mathrm{ad}\,\mathfrak{t}_\mathbb{C}$ is diagonalized, and, simultaneously, of the real root decomposition of $\mathfrak{g}$ in which $\mathrm{ad}\,\mathfrak{a}$ is diagonalized. To be precise, let

$$\mathfrak{a}' = \sum_{i=1}^r \mathbb{R} \cdot h_i \subset i\mathfrak{t} \, ,$$

and define

$$\varphi \sim \psi \Longleftrightarrow \varphi - \psi|_{\mathfrak{a}'} \equiv 0 \text{ , for } \varphi, \psi \in \Psi \text{ .}$$

Proposition 2.7 tells us which representations of $\mathfrak{a}'$ may occur in $\mathfrak{g}$ and, in particular, tells us that, for all $\varphi \in \Psi$, one of the following occurs:

(I) $\varphi \sim \pm\gamma_i'$, for some i, in which case

$$\mathfrak{g}^\varphi \subset \text{factor of type } (a_i), \text{ hence in fact } \varphi = \pm\gamma_i' \text{ ;}$$

(II) $\varphi \sim \frac{1}{2}(\pm\gamma_i' \pm \gamma_j')$, for some $i \neq j$, in which case

$$\mathfrak{g}^\varphi \subset \text{factor of type } (b_{ij}) \text{ ;}$$

(III) $\varphi \sim \pm\frac{1}{2}\gamma_i'$, for some i, in which case

$$\mathfrak{g}^\varphi \subset \text{factor of type } (c_i) \text{ ;}$$

(IV) $\varphi \sim 0$, in which case

$$\mathfrak{g}^\varphi \subset \text{factor of type } (e) \text{ .}$$

But because we also know how h_o acts on each of these factors, we can say more:

(I) each (a_i)-factor is just $\mathbb{R}(ih_i) + \mathbb{R}x_i + \mathbb{R}y_i$, giving one positive non-compact root γ_i' and one negative non-compact root $-\gamma_i'$;

(II) each of the (b_{ij})-factors is 4-dimensional and each gives one positive non-compact root, $\sim \frac{\gamma_i' + \gamma_j'}{2}$; two compact roots, one positive and one negative, $\sim \pm\frac{\gamma_i' - \gamma_j'}{2}$, respectively, and one negative non-compact root, $\sim -\frac{\gamma_i' + \gamma_j'}{2}$;

(III) each of the (c_i)-factors is 4-dimensional and each gives one positive non-compact root, $\sim \gamma_i'/2$, two compact roots, one positive and one negative, $\sim \pm\gamma_i'/2$, respectively, and one negative non-compact root, $\sim -\gamma_i'/2$;

(IV) each of the (e)-factors is 1-dimensional and gives a compact root, ~ 0.

We can coarsen our complex root decomposition

$$\mathfrak{g} = \mathfrak{t}_{\mathbb{C}} \oplus \sum_{\varphi \in \Psi} \mathfrak{g}^\varphi$$

by lumping together root spaces with equivalent roots. Let $_{\mathbb{R}}\Psi'$ be the set of non-zero linear maps $\mathfrak{a}' \longrightarrow \mathbb{R}$ given by restricting roots $\varphi \in \Psi$ to $\mathfrak{a}'$, so that

$$\mathfrak{g}_{\mathbb{C}} = Z(\mathfrak{a}')_{\mathbb{C}} \oplus \sum_{\psi' \in _{\mathbb{R}}\Psi'} \mathfrak{g}^{\psi'} \text{ ,} \qquad (*')$$

where

$$Z(\mathfrak{a}')_{\mathbb{C}} = \mathfrak{a}_{\mathbb{C}} \oplus \sum_{\varphi \sim 0} \mathfrak{g}^{\varphi} ,$$

$$\mathfrak{g}^{\psi'} = \sum_{\varphi \in \Psi, \varphi \sim \psi'} \mathfrak{g}^{\varphi}$$

= eigenspace in $\mathfrak{g}_{\mathbb{C}}$, where ad $\mathfrak{a}'$ is given by the character ψ'.

Then this is essentially the *real root decomposition* of $\mathfrak{g}$ with respect to the split torus $A = \exp \mathfrak{a}$. In fact, if

$$c = \varphi \left(\dots, \frac{1}{\sqrt{2}} \begin{pmatrix} 1 & i \\ i & 1 \end{pmatrix}, \dots \right) ,$$

then $\mathrm{Ad}\, c(\mathfrak{a}) = \mathfrak{a}'$. Therefore, if $_{\mathbb{R}}\Psi = (\mathrm{Ad}\, c)^*(_{\mathbb{R}}\Psi')$ is the induced set of linear maps $\mathfrak{a} \longrightarrow \mathbb{R}$, we get an isomorphic decomposition of $\mathfrak{g}$ via ad $\mathfrak{a}$. But now $\mathfrak{a}$ is real, so it is a *real* decomposition:

$$\mathfrak{g} = Z(\mathfrak{a}) \oplus \sum_{\psi \in _{\mathbb{R}}\Psi} \mathfrak{g}^{\psi} , \qquad (*)$$

where

$$\mathfrak{g}^{\psi} = \mathrm{Ad}\, c(\mathfrak{g}^{(\mathrm{Ad}\, c)^{*-1}(\psi)})$$

= eigenspace in $\mathfrak{g}$, where ad $\mathfrak{a}$ is given by the character ψ.

What is $_{\mathbb{R}}\Psi$, i.e., which combinations of factors actually occur in $\mathfrak{g}$ in the above proposition? First of all, (a_i) occurs once. We will denote the corresponding roots of $\mathfrak{a}$ by $\pm \gamma_i$: these are simply the roots of $\mathfrak{a}$ occurring in the image under $d\varphi$ of the ith copy of $\mathfrak{sl}(2, \mathbb{R})$; in fact, in $\mathbb{R}x_i + \mathbb{R}y_i + \mathbb{R}ih_i$. As for the rest, the result is:

Proposition 2.8 *Let D be simple of rank r. Then either of the following two possibilities occurs:*

Case C_r: $_{\mathbb{R}}\Psi = \{\pm\frac{1}{2}(\gamma_i + \gamma_j)$ *for* $i \geq j; \pm\frac{1}{2}(\gamma_i - \gamma_j)$ *for* $i > j\}$,
all (b_{ij})-factors occur, but no (c_i)-factors occur;

Case BC_r: $_{\mathbb{R}}\Psi = \{\pm\frac{1}{2}(\gamma_i + \gamma_j)$ *for* $i \geq j; \pm\frac{1}{2}(\gamma_i - \gamma_j)$ *for* $i > j; \pm\frac{1}{2}\gamma_i\}$,
all (b_{ij})- and all (c_i)-factors occur.

In both of these cases, the Weyl group (the automorphisms of $\mathfrak{a}$ induced by $\mathrm{Ad}\,\mathrm{Norm}(A)$) is the group of all signed permutations $\gamma_i \longmapsto \pm\gamma_{\sigma(i)}$ with σ a permutation of $\{1, \dots, r\}$. If we order the real roots so that $\gamma_1 > \dots > \gamma_r$, the simple roots $_{\mathbb{R}}\Delta$ are:

$$\alpha_i = (\gamma_i - \gamma_{i+1})/2,\ 1 \leq i \leq r-1; \quad and \quad \alpha_r = \begin{cases} \gamma_r & \text{Case } C_r \\ \gamma_r/2 & \text{Case } BC_r \end{cases} .$$

Proof In the image under φ of the ith factor of $\mathrm{SL}(2,\mathbb{R})$ there exists an element w_i which normalizes $\mathfrak{a}$, sends γ_i into $-\gamma_i$, and fixes the γ_j for $j \neq i$. Let $s_i = \mathrm{Ad}(w_i)$: then s_i belongs to the Weyl group W. In particular, this implies that the γ_i are orthogonal to each other. By Proposition 2.7, the other roots are among the following:

$$(\mathrm{b}_{ij}) : \tfrac{1}{2}(\pm\gamma_i \pm \gamma_j) \,,$$
$$(\mathrm{c}_i) : \pm\tfrac{1}{2}\gamma_i \,.$$

The Weyl group W is generated by the reflections around the roots, i.e., by the s_i and, if $\tfrac{1}{2}(\pm\gamma_i\pm\gamma_j)$ occurs as a root, by the symmetry which interchanges the two roots γ_i and γ_j, leaving fixed the roots γ_k, for $k \neq i,j$. Since D is simple, W acts irreducibly on $\mathfrak{a}$; i.e., $W/$(sign changes) is a transitive group of permutations of $\{1,\ldots,r\}$ generated by transpositions. It follows that W contains all permutations of the γ_i among themselves. This implies that all $\tfrac{1}{2}(\pm\gamma_i\pm\gamma_j)$ are roots and that either *all* or *none* of the $\tfrac{1}{2}\gamma_i$ are roots. The first case leads to the type C_r, and the second case leads to the type BC_r. The simple roots are now easily written down. $\qquad\square$

2.4

We now return to the general case, i.e., drop the assumption that D is simple. Theorem 2.9 below determines the position of D inside $\mathfrak{p}_+$, with D sitting inside $\mathfrak{p}_+$ via the Harish-Chandra embedding.

For $X \in \mathfrak{p}_+$, define the linear operator

$$T(X) : \mathfrak{p}_- \longrightarrow \mathfrak{k}_{\mathbb{C}} \,,$$
$$Y \longmapsto [Y,X] \,.$$

If

$$\tau : \mathfrak{g}_{\mathbb{C}} \longrightarrow \mathfrak{g}_{\mathbb{C}}$$

denotes complex conjugation with respect to $\mathfrak{g}_c$, we put a positive-definite hermitian form on $\mathfrak{g}_{\mathbb{C}}$:

$$B_\tau(u,v) = -B(u,\tau v) \,, \text{ for } u,v \in \mathfrak{g}_{\mathbb{C}} \,.$$

Let

$$T^*(X) : \mathfrak{k}_{\mathbb{C}} \longrightarrow \mathfrak{p}_-$$

be the adjoint of $T(X)$ with respect to B_τ.

Theorem 2.9 (Harish-Chandra, Hermann) *Let*

$$D_0 = D \cap \sum_{i=1}^{r} \mathbb{C} \cdot e_i \, .$$

Then

(i) *with $\widetilde{\varphi}$ defined as in Theorem 2.4,*

$$D_0 = \left\{ \sum_{i=1}^{r} a_i e_i \mid |a_i| < 1 \right\} = \mathrm{Im}\left(\widetilde{\varphi} : \mathfrak{H}^r \longrightarrow D\right) ;$$

(ii) $D = \mathrm{Ad}\, K(D_0) = \{ X \in \mathfrak{p}_+ \mid T^*(X) \circ T(X) < 2\mathrm{Id}_{\mathfrak{p}_-} \} \, .$

In particular, if $a, b \in \mathbb{C}$, with $|a| + |b| \le 1$, then

$$X, Y \in D \Longrightarrow aX + bY \in D \, .$$

(Part (ii) is called the *Hermann convexity theorem*.)

Proof We use the fact that $D \longrightarrow \mathfrak{p}_+$ is K-equivariant, where K acts via Ad on $\mathfrak{p}_+$. Thus $D = K \cdot A \cdot o$, so D can be described inside $\mathfrak{p}_+$ as $D = \mathrm{Ad}\, K(A \cdot o)$. Now, a calculation with $\mathrm{SL}(2)$ shows that

$$A \cdot o = \left\{ \sum_{i=1}^{r} a_i e_i \mid a_i \in \mathbb{R}, \ -1 < a_i < 1 \right\}$$

and

$$\mathrm{Im}\, \widetilde{\varphi} = \left\{ \sum_{i=1}^{r} a_i e_i \mid a_i \in \mathbb{C}, \ |a_i| < 1 \right\} .$$

The isomorphism $\mathfrak{p}_+ \cong \mathfrak{p}$ is also K-equivariant and it is well-known that in $\mathfrak{p}$, for every $k \in K$, either

(a) $\mathrm{Ad}\, k(\mathfrak{a}) \cap \mathfrak{a} \subset$ singular subalgebra of $\mathfrak{a}$;
   or
(b) $\mathrm{Ad}\, k(\mathfrak{a}) = \mathfrak{a}$, and hence k lies in the Weyl group.

Thus, if $D \cap \sum_{i=1}^{r} \mathbb{R} e_i$ were bigger than $\prod_{i=1}^{r}(-1, +1)e_i$, then one could find $x \in D \cap \sum_{i=1}^{r} \mathbb{R} e_i$ such that x lies in no singular subalgebra and $x \notin \prod_{i=1}^{r}(-1, +1)e_i$ with

$$x = (\mathrm{Ad}\, k)(y) , \quad y \in \prod_{i=1}^{r}(-1, +1)e_i \, .$$

Then k would be in the Weyl group, and hence $\mathrm{Ad}\, k$ would be a permutation plus sign change, and hence $x \in \prod_{i=1}^{r}(-1, +1)e_i$, a contradiction. Therefore $D \cap \sum_{i=1}^{r} \mathbb{R} e_i = \prod_{i=1}^{r}(-1, +1)e_i$. Since $\mathrm{Ad}\, K$ contains rotations in the planes $\mathbb{C} \cdot e_i$ (via $\varphi(\mathrm{SO}(2, \mathbb{R})^r)$), it follows that D_0 is rotation-invariant and (i) is proven.

To prove (ii), note that the sets on both sides are K-invariant, so it suffices to consider elements

$$X = \sum_{i=1}^{r} a_i e_i , \quad a_i \in \mathbb{R} ,$$

and show that

$$T^*(X) \circ T(X) < 2\mathrm{Id} \iff |a_i| < 1 \text{ for } i = 1,\dots,r .$$

Expressed in terms of operator norms,

$$\| S \|^2 = \sup\{B_\tau(S(Y),S(Y)) \mid B_\tau(Y,Y) = 1\} ,$$

the condition on the LHS is just

$$\| T(X) \| < \sqrt{2} .$$

We thus have to show that

$$\| T(X) \| < \sqrt{2} \iff |a_i| < 1 \text{ for } i = 1,\dots,r .$$

This is done by an explicit computation: write $Y \in \mathfrak{p}_-$ as

$$Y = \sum_{i=1}^{r} b_i e_{-i} + \sum_{i=1}^{r} \sum_{\alpha \in P_i} b_\alpha e_{-\alpha} + \sum_{i<j} \sum_{\alpha \in P_{ij}} b_\alpha e_{-\alpha} , \tag{2.2}$$

where

$$P_i = \left\{ \text{positive non-compact roots } \varphi \text{ with } \varphi \sim \tfrac{1}{2}\gamma_i' \right\} ,$$
$$P_{ij} = \left\{ \text{positive non-compact roots } \varphi \text{ with } \varphi \sim \tfrac{1}{2}(\gamma_i' + \gamma_j') \right\} .$$

We compute $[X,Y]$, using the vanishing of various brackets:

$$[X,Y] = \sum_{i} a_i b_i [e_i, e_{-i}] + \sum_{i} \sum_{\alpha \in P_i} a_i b_\alpha [e_i, e_{-\alpha}] + \sum_{i \neq j} \sum_{\alpha \in P_{ij}} a_i b_\alpha [e_i, e_{-\alpha}] . \tag{2.3}$$

Furthermore, by our normalization of the e_α,

$$\tau(e_\alpha) = -e_{-\alpha} ;$$

hence $\alpha \neq \beta \implies B_\tau(e_\alpha, e_\beta) = 0$, and all terms in the sum (2.2) are orthogonal to each other. Using this, together with the Jacobi identity, one computes

$$B_\tau([e_i, e_{-\alpha}], [e_i, e_{-\beta}]) = \alpha(h_i) \cdot B(e_{-\alpha}, e_\beta) ,$$

and this expression vanishes unless $\alpha = \beta$. Thus the brackets in sum (2.3) are orthogonal to each other.

We may now compare

$$\| Y \|^2 = \sum_i |b_i|^2 \| e_{-i} \|^2 + \sum_i \sum_{\alpha \in P_i} |b_\alpha|^2 \| e_{-\alpha} \|^2$$
$$+ \sum_{i<j} \sum_{\alpha \in P_{ij}} |b_\alpha|^2 \| e_{-\alpha} \|^2$$

with

$$\| [X,Y] \|^2 = \sum_i |b_i|^2 |a_i|^2 |\gamma_i(h_i)| \| e_{-i} \|^2$$
$$+ \sum_i \sum_{\alpha \in P_i} |b_\alpha|^2 |a_i|^2 |\alpha(h_i)| \| e_{-\alpha} \|^2$$
$$+ \sum_{i<j} \sum_{\alpha \in P_{ij}} \left(|a_i|^2 |b_\alpha|^2 |\alpha(h_i)| + |a_j|^2 |b_\alpha|^2 |\alpha(h_j)| \right) \| e_{-\alpha} \|^2 \ .$$

Thus, if $|a_i| < 1$, for $i = 1, \ldots, r$, it is evident that $\| T(X) \|^2 < 2$. Conversely, if $\| [X,Y] \| < \sqrt{2} \| Y \|$, for all $Y \in \mathfrak{p}_-$, then, as one sees by plugging $Y = e_{-i}$ into the above expression, we must have $|a_i| < 1$ for $i = 1, \ldots, r$.   □

Corollary 2.10 (of proof) *Let $X \in D$ and write*

$$X = \sum a_i e_i + \sum b_\alpha e_\alpha \ ,$$

where α runs through all positive non-compact roots restricting to either $\frac{1}{2}\gamma_i$ or $\frac{1}{2}(\gamma_i + \gamma_j)$ (i.e., not restricting to γ_i).
   Then $|a_i| < 1$ for $i = 1, \ldots, r$.

This can be seen by calculating $[X, e_{-i}]$ in the same way as above; cf. Langlands [8], p. 110.

2.5

As a final topic, suppose $\mathscr{G}$ is a semi-simple algebraic group defined over $\mathbb{Q}$ such that $D = \mathscr{G}(\mathbb{R})^o/K$, for $K \subset \mathscr{G}(\mathbb{R})^o$ maximal compact, is a bounded symmetric domain. (This *usually* means that $\mathscr{G}(\mathbb{R})^o/(\text{center}) = \text{Aut}(D)^o$, but it also allows $\mathscr{G}(\mathbb{R})^o$ to have compact factors, i.e.,

$$\mathscr{G}(\mathbb{R})^o/(\text{compact normal subgroup}) = \text{Aut}(D)^o \ .)$$

In this case, an important role is played by the maximal $\mathbb{Q}$-split tori $\mathscr{A} \subset \mathscr{G}$. (If $\mathscr{A}$ is non-trivial and $\mathscr{G}$ is $\mathbb{Q}$-simple, then, in fact, $\mathscr{G}(\mathbb{R})^o$ has no compact factors.) Choose $\mathscr{B} \subset \mathscr{G}$, a maximal $\mathbb{R}$-split torus such that $\mathscr{A} \subset \mathscr{B}$. Let $s = \dim \mathscr{A} = \mathbb{Q}$-rank $\mathscr{G}$ and $r = \dim \mathscr{B} = \mathbb{R}$-rank $\mathscr{G}$. Note that compact factors do not change the root structure so that the preceding results are still applicable. The roots of $\mathscr{G}$ with respect to $\mathscr{B}$ are a basis

$$\gamma_i \in X_{\mathbb{Q}}(\mathscr{B}) = \mathrm{Hom}\,(\mathscr{B}, \mathbb{G}_m) \otimes \mathbb{Q} \text{ for } 1 \leq i \leq r$$

plus some subset of the further elements,

$$\tfrac{1}{2}(\pm\gamma_i \pm \gamma_j), \quad 1 \leq i < j \leq r; \quad \pm\tfrac{1}{2}\gamma_i, \quad 1 \leq i \leq r\,.$$

Baily and Borel [1], pp. 467–468, prove by a purely root-theoretic analysis the following basic fact.

Proposition 2.11 *There is a partition*

$$\{1,\dots,s\} = I_0 \cup I_1 \cup I_2 \cup \dots \cup I_r$$

such that the subtorus $\mathscr{A}$ is defined by

$$\mathscr{A} = \left\{\gamma_i = 1 \text{ for } i \in I_0\;; \quad \gamma_i = \gamma_j \text{ for } i, j \in I_k, \text{ where } k > 0\right\}\;;$$

or, in additive notation,

$$X_{\mathbb{Q}}(\mathscr{A}) \cong X_{\mathbb{Q}}(\mathscr{B})/ \left\{ \begin{array}{l} \text{subspace spanned by } \gamma_i \text{ for } i \in I_0 \\ \text{and } \gamma_i - \gamma_j \text{ for } i, j \in I_k, \text{ where } k > 0 \end{array} \right\}\;;$$

$$\cong \sum_{i=1}^{s} \mathbb{Q}\beta_i, \text{ where } \beta_i = \text{image of any } \gamma_j, j \in I_i\,.$$

(Their result is stated differently, but boils down to the above.)

Corollary 2.12 *The $\mathbb{Q}$-roots are $\pm\beta_1, \dots, \pm\beta_s$ plus a subset of the further elements $\pm\beta_i/2$, $(\pm\beta_i \pm \beta_j)/2$ for $i \neq j$. If $\mathscr{G}$ is $\mathbb{Q}$-simple, then all $(\pm\beta_i \pm \beta_j)/2$ occur and either all $\pm\beta_i/2$ occur or none do. The $\mathbb{Q}$-Weyl group is then the full group of permutations and sign changes of the β_i.* $\qquad\qquad\square$

3 Boundary components

In this section we decompose the closure $\overline{D}$ of D inside $\mathfrak{p}_+$ into the disjoint union of lower-dimensional bounded domains, the *boundary component*. We analyze the structure of boundary components and determine their normalizers and centralizers.

3.1

Start with

$$\mathfrak{a} = \sum \mathbb{R}x_i \subset \mathfrak{p}$$

and

$$\mathfrak{sl}(2,\mathbb{R})^r \cong \sum \mathbb{R}x_i + \sum \mathbb{R}y_i + \sum \mathbb{R}ih_i \subset \mathfrak{g}$$

as in Theorem 2.4. Let $S \subset \{1,\dots,r\}$ be any subset.† Define the subalgebra $\mathfrak{l}_S$ of $\mathfrak{g}$:

$$\mathfrak{l}_S = \sum_{\substack{\varphi \in_{\mathbb{R}} \Psi \\ \varphi = \sum_{j \notin S} a_j \gamma_j}} \left(\mathfrak{g}^\varphi + [\mathfrak{g}^\varphi, \mathfrak{g}^{-\varphi}]\right) .$$

In the decomposition of Proposition 2.7, $\mathfrak{l}_S$ may be described as follows:

$$\mathfrak{l}_S = \sum_{i \notin S} (a_i\text{-factor}) \oplus \sum_{\substack{i < j \\ i,j \notin S}} (b_{ij}\text{-factors}) \oplus \sum_{i \notin S} (c_i\text{-factor})$$

$$\oplus \left\{ \begin{array}{l} \text{the part of the e-factors spanned by } [x,y] \\ x,y \in \text{ some } b_{ij}\text{-factor or some } c_i\text{-factor} \end{array} \right\} .$$

Hence $\mathfrak{l}_{S,\mathbb{C}}$ can be written as a sum over the complex roots $\varphi \in \Psi$:

$$\mathfrak{l}_{S,\mathbb{C}} = \sum_{\substack{\varphi \sim \sum_{j \notin S} a_j \gamma_j \\ \varphi \not\approx 0}} \left(\mathfrak{g}^\varphi + [\mathfrak{g}^\varphi, \mathfrak{g}^{-\varphi}]\right) .$$

This shows that $\mathfrak{l}_S$ is stable under $\mathrm{Ad}\, h_o(e^{i\theta})$; hence

(a) $\mathfrak{l}_S = \mathfrak{k} \cap \mathfrak{l}_S \oplus \mathfrak{p} \cap \mathfrak{l}_S$

and

(b) $\mathfrak{p}_{\mathbb{C}} \cap \mathfrak{l}_{S,\mathbb{C}} = \mathfrak{p}_+ \cap \mathfrak{l}_{S,\mathbb{C}} \oplus \mathfrak{p}_- \cap \mathfrak{l}_{S,\mathbb{C}} ,$

which we abbreviate as follows:

$$\mathfrak{p}_{S,\mathbb{C}} = \mathfrak{p}_{+,S} \oplus \mathfrak{p}_{-,S} .$$

By (a), $\mathfrak{l}_S$ is a reductive subalgebra of $\mathfrak{g}$, and, since it is generated by nilpotent elements, $\mathfrak{l}_S$ is semi-simple without compact factors. By (b), $\mathfrak{l}_S$ is of hermitian type. Let $L_S \subset G$ be the corresponding subgroup. Then

$$D_S = L_S / L_S \cap K$$

is a bounded symmetric domain and we have a symmetric embedding of D_S in D. In fact, we have even more, because L_S commutes with the subgroup (modulo center)

$$\prod_{i \in S} \mathrm{SL}(2,\mathbb{R})_i$$

† In what follows, we consider S interchangeably as $S \subset \{\gamma_1,\dots,\gamma_r\}$ or $S \subset \{1,\dots,r\}$.

arising from the subalgebra

$$\sum_{i \in S} (\mathbb{R}x_i + \mathbb{R}y_i + \mathbb{R}ih_i) \,.$$

This shows that we get a whole equivariant diagram of symmetric holomorphic maps:

$$
\begin{array}{ccc}
\Delta^s \times D_S & \xrightarrow{\ f_1\ } & D \\
\cap & & \cap \\
\mathbb{C}^s \times \mathfrak{p}_{+,S} & \xrightarrow{\ f_2\ } & \mathfrak{p}_+ \\
\cap & & \cap \\
(\mathbb{P}^1)^s \times \check{D}_S & \xrightarrow{\ f_3\ } & \check{D}
\end{array}
$$

where $s = |S|$. (Here we identify the domain $\mathrm{SL}(2,\mathbb{R})/\mathrm{SO}(2)$ with the open unit disc Δ instead of $\mathfrak{H}$; we will only do this for the purpose of the discussion that follows in Subsection 3.1.)

Define

$$F_S = f_2((1,\dots,1) \times D_S) \subset \mathfrak{p}_+ \,,$$
$$\check{F}_S = f_3((1,\dots,1) \times \check{D}_S) \subset \check{D} \,.$$

Note that F_S lies in ∂D and is a complex submanifold of $\mathfrak{p}_+$. In fact, f_2 is linear and is just the inclusion map of the following subspace of $\mathfrak{p}_+$:

$$\sum_{i \in S} \mathbb{C}e_i + \underbrace{\left[\sum_{i \notin S} \mathbb{C}e_i + \sum_{\substack{i<j \\ i,j \notin S}} \sum_{\alpha \in P_{ij}} \mathbb{C}e_\alpha + \sum_{i \notin S} \sum_{\alpha \in P_i} \mathbb{C}e_\alpha \right]}_{\mathfrak{p}_{+,S}} \,.$$

Here, as in Subsection 2.4,

$$P_{ij} = \left\{ \varphi \in \Psi_p^+ \mid \varphi \sim \tfrac{1}{2}(\gamma_i' + \gamma_j') \right\} \,,$$
$$P_i = \left\{ \varphi \in \Psi_p^+ \mid \varphi \sim \tfrac{1}{2}\gamma_i' \right\} \,.$$

Thus by Hermann convexity applied to D_S.

$$F_S = \sum_{i \in S} e_i + \left\{ \overline{X} \in \mathfrak{p}_{+,S} \mid T_S(\overline{X})^* \circ T_S(\overline{X}) < 2\mathrm{Id}_{\mathfrak{p}_{-,S}} \right\} \,,$$

where

$$T_S(\overline{X}) : \mathfrak{p}_{-,S} \longrightarrow \mathfrak{l}_{S,\mathbb{C}} \cap \mathfrak{k}_\mathbb{C}$$

is given by

$$T_S(\overline{X})(\overline{Y}) = [\overline{Y},\overline{X}] \,.$$

We use this last description to prove:

Lemma 3.1 $\overline{F}_S = \check{F}_S \cap \overline{D}$ *and all holomorphic maps*

$$\lambda : \Delta \longrightarrow \mathfrak{p}_+$$

such that

$$\mathrm{Im}\,(\lambda) \subset \overline{D}\,, \qquad \mathrm{Im}\,\lambda \cap F_S \neq \emptyset\,,$$

map Δ *into* F_S.

Proof Let $X = \sum_i a_i e_i + \sum_i \sum_{\alpha \in P_i} a_\alpha e_\alpha + \sum_{ij} \sum_{\alpha \in P_{ij}} a_\alpha e_\alpha \in \mathfrak{p}_+$ and assume that $X \in \overline{D}$ and $a_i = 1$, for $i \in S$. Then we claim that $a_\alpha = 0$ for $\alpha \in P_i$, $i \in S$, and $a_\alpha = 0$ for $\alpha \in P_{ij}$, i or $j \in S$.

For this, calculate $[X, e_{-i}]$ as in the proof of the Hermann convexity theorem:

$$[X, e_{-i}] = a_i[e_i, e_{-i}] + \sum_{\alpha \in P_i} a_\alpha[e_\alpha, e_{-i}] + \sum_{\substack{\alpha \in P_{ij} \\ i \neq j}} a_\alpha[e_\alpha, e_{-i}]\,,$$

hence

$$\| [X, e_{-i}] \|^2 = 2|a_i|^2 \| e_{-i} \|^2$$
$$+ \left[\sum_{\alpha \in P_i} |a_\alpha|^2 \| e_\alpha \|^2 + \sum_{\substack{\alpha \in P_{ij} \\ i \neq j}} |a_\alpha|^2 \| e_\alpha \|^2 \right].$$

Since $X \in \overline{D}$, we must have $\| [X, e_{-i}] \|^2 \leq 2 \| e_{-i} \|^2$ by the Hermann convexity theorem. Hence, if $a_i = +1$, then

$$2 \| e_{-i} \|^2 + \left[\sum_{\alpha \in P_i} |a_\alpha|^2 \| e_\alpha \|^2 + \sum_{\substack{\alpha \in P_{ij} \\ i \neq j}} |a_\alpha|^2 \| e_\alpha \|^2 \right] \leq 2 \| e_{-i} \|^2\,,$$

which shows the claim.

Hence, for such X, we may write

$$X = \sum_{i \in S} e_i + \overline{X}\,,$$

where $\overline{X} \in \mathfrak{p}_{+,S}$. But, since, for $\overline{Y} \in \mathfrak{p}_{-,S}$, $[X, \overline{Y}] = [\overline{X}, \overline{Y}]$, we conclude that $T_S(\overline{X})^* \circ T_S(\overline{X}) \leq 2\mathrm{Id}_{\mathfrak{p}_{-,S}}$, i.e., $\overline{X} \in \overline{D}_S = $ closure of D_S inside $\mathfrak{p}_{+,S}$. This proves:

$$\overline{D} \cap \{X \mid \text{coeff. } a_i \text{ of } e_i \text{ is } 1, \text{ for } i \in S\} = \overline{F}_S\,.$$

We can now prove the lemma. The function f_i, which to

$$x = \sum_{i \in S} a_i e_i + \sum_i \sum_{\alpha \in P_i} b_\alpha e_\alpha + \sum_{i<j} \sum_{\alpha \in P_{ij}} b_\alpha e_\alpha \in \mathfrak{p}_+$$

associates $a_i \in \mathbb{C}$, is a linear function bounded above by 1 on $\overline{D}$; if $i \in S$, then f_i

takes on the value 1 on F_S. The maximum principle, applied to the holomorphic function $f_i \circ \lambda : \Delta \longrightarrow \mathbb{C}$, implies now that $f_i \circ \lambda \equiv 1$ on Δ.

The preceding considerations show that $\lambda(\Delta) \subset \overline{F}_S$. But, by the Hermann convexity theorem, D_S is a convex subset of $\mathfrak{p}_{+,S}$, thought of as a *real* vector space; hence, for every $x \in \partial F_S \subset \sum_{i \in S} e_i + \mathfrak{p}_{+,S}$, there exists a linear functional ℓ on $\mathfrak{p}_+$ and a real number a such that $\ell > a$ on F_S and $\ell(x) = a$. Now, again by the maximum principle, this time applied to $\ell \circ \lambda$, we conclude that

$$\operatorname{Im}(\lambda) \cap \partial F_S \neq \emptyset \Longrightarrow \operatorname{Im} \lambda \subset \partial F_S , \quad \text{i.e., } \operatorname{Im} \lambda \cap F_S = \emptyset ,$$

which contradicts our assumptions; hence $\operatorname{Im} \lambda \subset F_S$. $\qquad \square$

This proves that F_S is a boundary component of D in the following sense.

Definition 3.2 A *boundary component* of a bounded symmetric domain D is an equivalence class in $\overline{D}$ under the equivalence relation generated by $x \sim y$ if there exists a holomorphic map

$$\lambda : \Delta \longrightarrow \mathfrak{p}_+$$

such that $\operatorname{Im}(\lambda) \subset \overline{D}$, and $x, y \in \operatorname{Im} \lambda$.

We now have:

Theorem 3.3

(i) *$\overline{D}$ is the disjoint union of boundary components.*

(ii) *The boundary components of D are just the sets*

$$k \cdot F_S ; \qquad k \in K , \quad S \subset \{1, \ldots, r\} ,$$

with possible repetitions. They are hermitian symmetric domains of rank $r - |S|$.

(iii) *Decompose*

$$D = D_1 \times \cdots \times D_n$$

into simple factors. Then the boundary components of D are the products of boundary components of the simple factors D_i.

(iv) *A boundary component of a boundary component is a boundary component.*

(v) *For every boundary component F, there are holomorphic symmetric maps*

$$
\begin{array}{ccc}
\mathfrak{H} & \xrightarrow{\ f_F\ } & D \\
\cap & & \cap \\
\mathbb{P}^1 & \xrightarrow{\ f_F\ } & \check{D}
\end{array}
$$

such that $f_F(\mathrm{i}) = o$, $f_F(\infty) \in F$, *and equivariant with respect to a morphism*

$$\varphi_F : U^1 \times \mathrm{SL}(2, \mathbb{R}) \longrightarrow G \,,$$

such that $\varphi_F(e^{\mathrm{i}\theta}, h^{SL}(e^{\mathrm{i}\theta})) = h_o(e^{\mathrm{i}\theta})$.†

Proof Part (i) is trivial. For part (ii), let

$$X \in F = \text{boundary component of } D \,.$$

Since K is maximal compact, $\overline{D} = \mathrm{Ad}\,K \cdot (\prod_{i=1}^{r}[-1, +1] \cdot e_i)$; hence we can find $k \in K$ and a subset $S \subset \{\gamma_1, \ldots, \gamma_r\}$ such that

$$\mathrm{Ad}\,(k)(X) = \sum_{i=1}^{r} a_i e_i \,, \quad a_i \in [-1, +1] \,.$$

But K also contains elements normalizing $\mathfrak{a}$ and inducing arbitrary sign changes, so we may assume $a_i \geq 0$. Let $S = \{i \mid a_i = 1\}$. Then $\mathrm{Ad}\,(k)(X) \in F_S$.

Part (iii) is straightforward; check that $x = (x_1, \ldots, x_n)$ and $y = (y_1, \ldots, y_n)$ in $\overline{D}$ are equivalent if and only if $x_i \sim y_i$ in $\overline{D}_i$ for all i. Part (iv) is an immediate consequence of (ii): let D_1 be a boundary component of D, and let D_2 be a boundary component of D_1. Then

$$D_1 = k \cdot F_S = k \cdot \left(\sum_{i \in S} e_i + D_S \right) \,,$$

and hence

$$D_2 = k \cdot \left(\sum_{i \in S} e_i + D_2' \right) \,,$$

where D_2' is a boundary component of D_S. But then

$$D_2' = k' \cdot \left(\sum_{i \in S'} e_i + D_{S \cup S'} \right)$$

for some $k' \in L_S \cap K$, $S' \subset \{1, \ldots, r\} \setminus S$, by (ii) applied to $D = D_S$. Thus

$$D_2 = k \cdot k' \cdot \left(\sum_{i \in S \cup S'} e_i + D_{S \cup S'} \right)$$
$$= k \cdot k' \cdot F_{S \cup S'}$$

is a boundary component of D.

---

† The question of uniqueness of the pair (f_F, φ_F) is taken up in Theorem 3.7 below.

Part (v), when $F = F_S$, follows by taking $\varphi_F = \varphi_S$, where

$$\varphi_S(e^{i\theta}, x) = \varphi\left(e^{i\theta}; \ldots, \underbrace{x}_{\text{if } i \in S}, \ldots, \underbrace{e^{i\theta}}_{\text{if } i \notin S}, \ldots\right),$$

where $\varphi : U^1 \times \mathrm{SL}(2, \mathbb{R})^r \longrightarrow G$ is Harish-Chandra's map (see Section 2.3), and taking $f_F = f_S$ to be the corresponding map of symmetric spaces. For general $F = k \cdot F_S$, let $\varphi_F = k\varphi_S k^{-1}$ and $f_F = kf_S$. $\square$

This theorem shows that the F_S are good models for studying any one boundary component. They are equally good for studying arbitrary flags of boundary components because of the next result.

Proposition 3.4

(i) *If $S_1 \subset S_2$, then $\overline{F}_{S_1} \supset \overline{F}_{S_2}$.*

(ii) *If*

$$\overline{D} \supset \overline{F}_1 \supset \overline{F}_2 \supset \cdots \supset \overline{F}_t$$

is any flag of boundary components, then there are subsets

$$S_1 \subset S_2 \subset \cdots \subset S_t \subset \{1, \ldots, r\}$$

and an element $k \in K$ such that

$$k \cdot F_i = F_{S_i}, \quad 1 \leq i \leq t.$$

Proof If $S_1 \subset S_2$, then $L_{S_1} \supset L_{S_2}$, hence $D_{S_1} \supset D_{S_2}$ and

$$\overline{D} \supset \underbrace{\left(\sum_{i \in S_1} e_i + \overline{D}_{S_1}\right)}_{\overline{F}_{S_1}} \supset \underbrace{\left(\sum_{i \in S_2} e_i + \overline{D}_{S_2}\right)}_{\overline{F}_{S_2}},$$

since $\sum_{i \in S_2 \setminus S_1} e_i \in \overline{D}_{S_1}$. The second part is proved just as part (iv) above: first find $k_1 \in K$ such that $k_1 \cdot F_1 = F_{S_1}$; then find $k_2 \in K \cap L_{S_1}$ such that $k_2 k_1 \cdot F_2 = F_{S_2}$, and so on. $\square$

3.2

Our next purpose is to determine the *normalizer* $N(F)$ of a boundary component F:

$$N(F) = \{g \in G \mid gF = F\},$$

and to show that the pair (f_F, φ_F) in part (v) is unique. It is easiest to attack $N(F_S)$ first. As in the proof of the theorem, let

$$\varphi_S : U^1 \times \mathrm{SL}(2, \mathbb{R}) \longrightarrow G$$

be given by

$$\varphi_S(e^{i\theta}, x) = \varphi(e^{i\theta}; \ldots, \underbrace{e^{i\theta}}_{\text{if } i \notin S}, \ldots, \underbrace{x}_{\text{if } i \in S}, \ldots),$$

where again φ is Harish-Chandra's map. Let $w_S : \mathbb{G}_m \longrightarrow \mathcal{G}$ be

$$w_S(t) = \varphi_S\left(1, \begin{pmatrix} t & 0 \\ 0 & t^{-1} \end{pmatrix}\right).$$

Consider the associated parabolic subgroup

$$P_S = P(w_S^{-1}) = \left\{ g \in G \mid \lim_{t \to 0} w_S(t) g w_S(t)^{-1} \text{ exists} \right\},$$

which is the intersection of G with a real parabolic subgroup $\mathcal{P}_S \subset \mathcal{G}$.

It is easy to calculate $\mathrm{Lie}\, P_S$ using the real root decomposition:

$$\mathrm{Lie}\, P_S = Z(\mathfrak{a}) + \sum_{\substack{\varphi \in {}_\mathbb{R}\Psi \\ \langle dw_S, \varphi \rangle \geq 0}} \mathfrak{g}^\varphi,$$

where $\langle dw_S, \varphi \rangle$ is the inner product of $dw_S(1) \in \mathfrak{a}$ and $\varphi \in \mathrm{Hom}\,(\mathfrak{a}, \mathbb{R})$. Since

$$\langle dw_S, \gamma_i \rangle = \begin{cases} 1 & i \in S \\ 0 & i \notin S, \end{cases}$$

it follows that

$$\{\varphi \in {}_\mathbb{R}\Psi \mid \langle dw_S, \varphi \rangle \geq 0\} = \{\tfrac{1}{2}(\pm\gamma_i \pm \gamma_j), \pm\tfrac{1}{2}\gamma_i, \ i, j \notin S\}$$
$$\cup \{\tfrac{1}{2}(\gamma_i \pm \gamma_j), \tfrac{1}{2}\gamma_i, \ i \in S, \text{ any } j\}.$$

Lemma 3.5 $P_S \subset N(F_S)$.

Proof We use the Cayley transformation

$$c_S = \varphi_S\left(1, \frac{1}{1-\mathrm{i}} \begin{pmatrix} 1 & \mathrm{i} \\ 1 & -\mathrm{i} \end{pmatrix}\right).$$

Here $c_S \in G_\mathbb{C}$, and one checks, via the equivariant map $\mathfrak{H}^s \times D_S \xrightarrow{fs} D$ and its extension $(\mathbb{P}^1)^s \times \check{D}_S \xrightarrow{fs} \check{D}$, that

$$c_S(D_S) = F_S \quad \text{and} \quad c_S(\check{D}_S) = \check{F}_S,$$

where we identify D_S with $L_S \cdot o$ in D.

Now, every $g \in P_S$ either carries F_S to F_S or to a boundary component disjoint from F_S. Therefore it suffices to show that

$$g \cdot c_S(o) \in c_S(\check{D}_S)$$

for every $g \in P_{S,\mathbb{C}}$, because then, for $g \in P_S$, $g \cdot c_S(o) \in \overline{D} \cap c_S(\check{D}_S) = c_S(\overline{D}_S)$, hence $g(F_S) \subset \overline{F}_S$; since $\dim g(F_S) = \dim F_S$, it then follows that $gF_S = F_S$.

Therefore we only need to show that

$$c_S^{-1} g c_S(o) \in \check{D}_S \ ,$$

or

$$c_S^{-1} g c_S \in L_{S,\mathbb{C}} \cdot K_\mathbb{C} \cdot P_- \ .$$

But define

$$P'_{S,\mathbb{C}} = c_S^{-1} P_{S,\mathbb{C}} c_S = \left\{ g \in G_\mathbb{C} \mid \lim_{t \to 0} w'_S(t) g w'_S(t)^{-1} \ \text{exists} \right\} \ ,$$

where

$$w'_S = c_S^{-1} w_S c_S : \mathbb{G}_{m,\mathbb{C}} \longrightarrow G_\mathbb{C} \ .$$

Then

$$w'_S(e^{i\theta}) = \varphi_S \left(1, \begin{pmatrix} \cos\theta & -\sin\theta \\ \sin\theta & \cos\theta \end{pmatrix} \right) \ ,$$

and $\mathrm{Lie}\, P'_{S,\mathbb{C}}$ is easy to calculate using the complex root decomposition:

$$\mathrm{Lie}\, P'_{S,\mathbb{C}} = \mathfrak{t}_\mathbb{C} \oplus \sum_{\substack{\varphi \in \Psi \\ \langle dw'_S, \varphi \rangle \geq 0}} \mathfrak{g}^\varphi$$

$$= \mathfrak{t}_\mathbb{C} \oplus \sum_{\substack{\varphi \sim \frac{\pm\gamma'_i \pm \gamma'_j}{2} \ \text{or} \ \frac{\pm\gamma'_i}{2} \\ i,j \notin S}} \mathfrak{g}^\varphi \ \oplus \ \sum_{\substack{\varphi \sim \frac{-\gamma'_i - \gamma'_j}{2} \ \text{or} \ \frac{-\gamma'_i}{2} \\ i \in S, \ \text{any} \ j}} \mathfrak{g}^\varphi \ \oplus \ \sum_{\substack{\varphi \sim \frac{-\gamma'_i + \gamma'_j}{2} \\ i \in S, \ \text{any} \ j}} \mathfrak{g}^\varphi \ \oplus \sum_{\varphi \sim 0} \mathfrak{g}^\varphi .$$

Then $\mathfrak{t}_\mathbb{C} \subset \mathfrak{k}_\mathbb{C}$, the second term generates $\mathfrak{l}_{S,\mathbb{C}}$, the third term is in $\mathfrak{p}_-$, and the fourth and fifth terms are in $\mathfrak{k}_\mathbb{C}$. Let $\mathfrak{b}$ be the subalgebra generated by the first, third, and fourth term, and let $\mathfrak{c}$ be the subalgebra generated by the first four terms. Then $\mathfrak{b}$ and $\mathfrak{c}$ are ideals, and $\mathfrak{b}$ generates a normal subgroup contained in $K_\mathbb{C} \cdot P_-$. Therefore, $\mathfrak{c}$ generates a normal subgroup contained in $L_{S,\mathbb{C}} \cdot K_\mathbb{C} \cdot P_-$, and, since the fifth term is contained in $\mathfrak{k}_\mathbb{C}$, we finally get $P'_{S,\mathbb{C}} \subset L_{S,\mathbb{C}} \cdot K_\mathbb{C} \cdot P_- \cdot K_\mathbb{C} = L_{S,\mathbb{C}} \cdot K_\mathbb{C} \cdot P_-$, as required. $\qquad\square$

Proposition 3.6 $P_S = N(F_S)$.

Proof It suffices to show this when G is simple because, in the general case, everything decomposes into a product. But when G is simple, applying the Weyl group, we can assume that $S = \{1, \ldots, b\}$. Then note that all but one of the simple roots α_i is zero on $\operatorname{Im} dw_S$. Therefore the associated parabolic $\mathscr{P}_S$ is a *maximal* real parabolic subgroup of $\mathscr{G}$. Since

$$P_S \subset N(F_S) \subsetneq G \,,$$

the connected components of P_S and $N(F_S)$ coincide. But then $N(F_S)$ normalizes $N(F_S)^o$, and hence normalizes $\operatorname{Lie} N(F_S)$, which equals the Lie algebra $\operatorname{Lie} P_S$. But $P_{S,\mathbb{C}}$ is the full normalizer of $\operatorname{Lie} P_S$ inside $G_{\mathbb{C}}$, so

$$N(F_S) \subset G \cap P_{S,\mathbb{C}} = P_S \,,$$

hence the asserted equality. $\qquad\square$

We now prove:

Theorem 3.7 *For all boundary components $F \subset \overline{D}$, the equivariant pair (f_F, φ_F)*

$$f_F : \mathfrak{H} \longrightarrow D \,,$$
$$\varphi_F : U^1 \times \operatorname{SL}(2, \mathbb{R}) \longrightarrow G \,,$$

where f_F is symmetric with $f_F(i) = o$ and $f_F(\infty) \in F$, is unique, and, if w_F is defined by

$$w_F(t) = \varphi_F \left(1, \begin{pmatrix} t & 0 \\ 0 & t^{-1} \end{pmatrix} \right) \,,$$

then

$$N(F) = P(w_F^{-1}) = \left\{ g \in G \mid \lim_{t \to 0} w_F(t) g w_F(t)^{-1} \text{ exists} \right\} \,.$$

Finally, if $N(F_1) = N(F_2)$, then $F_1 = F_2$.

Proof Because of Proposition 3.6, the middle part follows from the first part. Moreover, because of Theorem 3.3 (iii), to prove the rest we may assume G simple. Now, by Proposition 2.6, all such φ_F arise as $k\varphi_S k^{-1}$, for some $k \in K$. The first part therefore amounts to the following statement:

$$k \cdot F_{S_1} = F_{S_2} \,, \ k \in K \Longrightarrow k\varphi_{S_1} k^{-1} = \varphi_{S_2} \,.$$

But $k \cdot F_{S_1} = F_{S_2}$ implies $\operatorname{rank}(F_{S_1}) = \operatorname{rank}(F_{S_2})$, hence $|S_1| = |S_2|$. Therefore, by Proposition 2.8, there is an element w of the Weyl group $\operatorname{Norm}(\mathfrak{a})/\operatorname{Cent}(\mathfrak{a})$ inducing a permutation of the x_i such that $wS_1 = S_2$: we may realize w as an element of $K \cap \operatorname{Norm}(\mathfrak{a})$. This reduces us to proving

$$k \in K \cap N(F_S) \Longrightarrow k\varphi_S = \varphi_S k \,.$$

Let σ be the Cartan involution. Then

$$k \in N(F_S) \implies \lim_{t \to 0} w_S(t) k w_S(t)^{-1} \text{ exists}$$

$$\implies \lim_{t \to 0} \sigma(w_S(t)) \sigma(k) \sigma(w_S(t))^{-1} \text{ exists}$$

$$\implies \lim_{t \to 0} w_S(t)^{-1} k w_S(t) \text{ exists}.$$

Thus $t \mapsto w_S(t) k w_S(t)^{-1}$ extends to a morphism $\mathbb{P}^1 \longrightarrow \mathscr{G}$. This must be constant, i.e., $k w_S = w_S k$. But $k \in K$ implies $k h_o = h_o k$. Thus

$$\varphi_S\left(1, \begin{pmatrix} t & 0 \\ 0 & t^{-1} \end{pmatrix}\right) \quad \text{and} \quad \varphi_S\left(e^{i\theta}, \begin{pmatrix} \cos\theta & \sin\theta \\ -\sin\theta & \cos\theta \end{pmatrix}\right)$$

both centralize k. Since these elements generate $U^1 \times \mathrm{SL}(2,\mathbb{R})$, we also get $k\varphi_S = \varphi_S k$. As for the final assertion, when G is simple, we saw in (the proof of) Proposition 3.6 that all the $N(F)$ are maximal parabolic. Let $g \in G$ with $gF_1 = F_2$. Then $gN(F_1)g^{-1} = N(F_2) = N(F_1)$, hence g normalizes $N(F_1)$ so that $g \in N(F_1)$, whence $F_1 = gF_1 = F_2$. □

Corollary 3.8 *If $\overline{F}_1 \supset F_2$ are two boundary components of D, then there is a unique symmetric holomorphic map*

$$f: \mathfrak{H}^2 \longrightarrow D$$

such that

$$f(i, i) = o,$$
$$f(i, \infty) \in F_1,$$
$$f(\infty, \infty) \in F_2.$$

In particular, w_{F_1} and w_{F_2} commute with each other.

Proof Combine Proposition 3.4 with Theorem 3.7. □

Let $o_F = f_F(\infty)$: this is the *natural basepoint* of F determined by the basepoint $o \in D$.

Proposition 3.9 *When G is simple and hence D is irreducible, the association $F \longmapsto N(F)$ defines a bijection between the set of boundary components of D and the set of maximal real parabolic subgroups of G.*

In general, if $G = G_1 \times \cdots \times G_k$, $D = D_1 \times \cdots \times D_k$, where $G_i = \mathrm{Aut}(D_i)^o$, with G_i simple, then $F \longmapsto N(F)$ defines a bijection between the set of boundary components $F = F_1 \times \cdots \times F_k$ of D, where here we allow $F_i = D_i$, and the set of real parabolics $P = P_1 \times \cdots \times P_k$ of G, with P_i either maximal real parabolic in G_i, or $P_i = G_i$.

Proof The general case reduces immediately to the case where G is simple. Theorem 3.7 shows injectivity. Proposition 3.6 shows that $N(F)$ is maximal. Moreover, by the proof of Proposition 3.6, the maximal real parabolic corresponding to any simple root do occur as $N(F_S)$ for suitable S, hence we get surjectivity. □

3.3

Pursuing the same ideas, we can determine the *centralizer* $Z(F)$ of a boundary component F:

$$Z(F) = \{g \in G \mid gx = x \text{ for all } x \in F\} \ .$$

The result is (for the connected component $Z(F)^o$):

Theorem 3.10 *Let* $F \subset \overline{D}$ *be a boundary component and let* $w_F : \mathbb{G}_m \longrightarrow \mathcal{G}$ *be as in Theorem 3.7. Then*

(1) $N(F)^o$ *is a semi-direct product:*

$$N(F)^o = Z(w_F)^o \ltimes W(F) \ ,$$

  where

$$W(F) = \{g \in G \mid \lim_{t \longrightarrow 0} w_F(t)gw_F(t)^{-1} = e\}$$
$$= \text{unipotent radical of } N(F)^o \ ,$$
$$Z(w_F)^o = \text{connected component of centralizer of } w_F$$
$$= \text{a Levi component of } N(F)^o \ .$$

(2) $Z(w_F)^o$ *is generated by two commuting connected subgroups* $G_h(F)$ *and* $\widetilde{G}_\ell(F)$ *with finite intersection* A *(so that* $Z(w_F)^o \cong (G_h \times \widetilde{G}_\ell)/A$*) and*

$$Z(F)^o = \widetilde{G}_\ell(F) \ltimes W(F) \ .$$

(3) *When* $F = F_S$*, then* $G_h(F_S) = L_S$*; in general,* G_h *is semi-simple without compact factors and* $G_h/(\text{center})$ *is isomorphic to* $\text{Aut}(F)^o$*.*

Proof The first part is a simple decomposition which applies to any parabolic subgroup $P(w)$ for any one-parameter subgroup w, and is proved by decomposing $\mathfrak{g}$ into 0, $+$, and $-$ eigenspaces under $\text{Ad}\,w$ and exponentiating. To prove (2) and (3), we may assume $F = F_S$ (getting the general case by conjugating by $k \in K$). In the notation above, note that

$$\text{Lie}\,Z(w_S) = Z(\mathfrak{a}) + \sum_{\substack{\varphi \in \mathbb{R}^\Psi \\ \langle dw_s, \varphi \rangle = 0}} \mathfrak{g}^\varphi \ .$$

Moreover, the $\varphi \in {}_\mathbb{R}\Psi$ with $\langle dw_S, \varphi \rangle = 0$ are of two types:

$$\text{(a)} \qquad \varphi = \tfrac{1}{2}(\pm \gamma_i \pm \gamma_j), \pm \tfrac{1}{2}\gamma_i, \quad i, j \notin S,$$

$$\text{(b)} \qquad \varphi = \tfrac{1}{2}(\gamma_i - \gamma_j), \quad i, j \in S.$$

Clearly, $\mathfrak{l}_S \subset \text{Lie}\, Z(w_S)$. But also, since no roots of type (a) and of type (b) add up to a root, it follows that $\mathfrak{l}_S$ is an *ideal* in $\text{Lie}\, Z(w_S)$. This proves that we have a decomposition $Z(w_S)^o \cong (L_S \times \widetilde{G}_{\ell,S})/A$ as required. By Proposition 2.7, we know that

$$Z(\mathfrak{a}) = \mathfrak{a} \oplus m(\mathfrak{a}),$$

where

$$m(\mathfrak{a}) = \text{type (e)-factors in the decomposition of } \mathfrak{g} = Z(\mathfrak{a}) \cap \mathfrak{k}.$$

Note that if σ is the Cartan involution of $\mathfrak{g}$, then

$$\sigma(\mathfrak{g}^\varphi) = \mathfrak{g}^{-\varphi};$$

hence, as has already been noted earlier,

$$\sigma(\mathfrak{l}_S) = \mathfrak{l}_S.$$

Taking into account the fact that

$$\text{Lie}\, \widetilde{G}_{\ell,S} = \{x \in \text{Lie}\, Z(w_S) \mid [x, \mathfrak{l}_S] = 0\},$$

it follows that $\sigma(\text{Lie}\, \widetilde{G}_{\ell,S}) = \text{Lie}\, \widetilde{G}_{\ell,S}$. Since $\text{Lie}\, \widetilde{G}_{\ell,S}$ is normalized by $\mathfrak{a}$, we deduce the following explicit expression:

$$\text{Lie}\, \widetilde{G}_{\ell,S} = \sum_{\substack{\varphi = \frac{\gamma_i - \gamma_j}{2} \\ i,j \in S}} \mathfrak{g}^\varphi + \sum_{i \in S} \mathbb{R}x_i + (\text{ideal in } m(\mathfrak{a})).$$

(The middle factor comes from the fact that

$$\sum_{i \in S} \mathbb{R}x_i = \{x \in \mathfrak{a} \mid \gamma_i(x) = 0, \quad i \notin S\} = \{x \in \mathfrak{a} \mid [x, \mathfrak{l}_S] = (0)\}.)$$

Next, we check that $\widetilde{G}_\ell(F_S)$ acts identically on F_S. Since it commutes with L_S, which acts transitively on F_S, it suffices to check that

$$g c_S(o) = c_S(o) \text{ for all } g \in \widetilde{G}_{\ell,S},$$

or that

$$c_S^{-1} g c_S(o) = o \text{ for all } g \in \widetilde{G}_{\ell,S},$$

or that

$$\text{Ad}\, c_S(\text{Lie}\, \widetilde{G}_{\ell,S}) \subset \mathfrak{k}_\mathbb{C} + \mathfrak{p}_-.$$

But

$$\operatorname{Ad} c_S(x) = x, \text{ for all } x \in m(\mathfrak{a}) ,$$

$$\operatorname{Ad} c_S(\mathfrak{g}^\varphi) = \sum_{\varphi' \sim \frac{\gamma_i' - \gamma_j'}{2}} \mathfrak{g}^{\varphi'} \quad \text{if } \varphi = \tfrac{1}{2}(\gamma_i - \gamma_j) , \quad i, j \in S ,$$

$$\operatorname{Ad} c_S(x_i) = h_i \quad \text{if } i \in S ,$$

and hence

$$\operatorname{Ad} c_S(\operatorname{Lie} \widetilde{G}_{\ell,S}) \subset \sum_{\varphi' \sim \frac{\gamma_i' - \gamma_j'}{2}} \mathfrak{g}^{\varphi'} + \mathfrak{t}_{\mathbb{C}} + m(\mathfrak{a}) .$$

All these factors are in $\mathfrak{k}_{\mathbb{C}}$, so $\widetilde{G}_{\ell,S}$ centralizes F_S.

Next, we check that $W(F_S)$ centralizes F_S. Again, since $W(F_S)$ is normalized by L_S, it suffices to show that

$$g c_S(o) = c_S(o) , \quad \text{for all } g \in W(F_S) ,$$

or that

$$\operatorname{Ad} c_S(\operatorname{Lie} W(F_S)) \subset \mathfrak{k}_{\mathbb{C}} + \mathfrak{p}_- .$$

But

$$\operatorname{Ad} c_S(\operatorname{Lie} W(F_S)) = \sum_{\substack{\varphi \in \Psi \\ \langle dw_S', \varphi \rangle < 0}} \mathfrak{g}^\varphi .$$

These roots are:

(i) $\varphi \sim -\tfrac{1}{2}(\gamma_i' + \gamma_j')$ or $-\tfrac{1}{2}(\gamma_i')$, $i \in S$, any j ;
(ii) $\varphi \sim -\tfrac{1}{2}(\gamma_i' - \gamma_j')$, $i \in S$, $j \notin S$,

which are all compact, or negative non-compact. This proves that

$$\widetilde{G}_{\ell,S} \ltimes W(F_S) \subset Z(F_S)^o .$$

Since $L_S/(\text{center})$ acts faithfully on F_S, we must have equality here, and this finishes the proof.   □

The above proof gives us another useful piece of information: for all F, let

$$
\begin{array}{ccc}
\mathfrak{H} & \xrightarrow{\ f_F\ } & D \\
\cap & & \cap \\
\mathbb{P}^1 & \xrightarrow{\ f_F\ } & \check{D}
\end{array}
$$

be as in Theorem 3.3; in particular,

$$f_F(\mathrm{i}) = o ,$$

$$f_F(\infty) = o_F .$$

Define now (recall that s_o denotes the symmetry of D at o):

$$o_F^0 = f_F(0) = s_o(o_F) ,$$
$$F^0 = s_o(F) = \text{boundary component containing } o_F^0 .$$

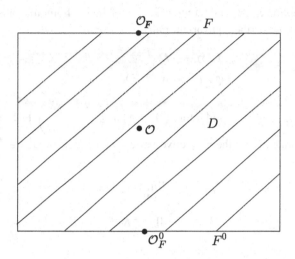

If $F = F_S$, then

$$o_{F_S}^0 \neq c_S^{-1}(o) .$$

In general, since $\begin{pmatrix} 1 & -i \\ 0 & 1 \end{pmatrix} \in SL(2, \mathbb{C})$ carries i to 0,

$$o_F^0 = \varphi_F \left(1, \begin{pmatrix} 1 & -i \\ 0 & 1 \end{pmatrix} \right) \cdot o .$$

Proposition 3.11 $\widetilde{G}_\ell(F)$ *fixes* o_F^0.

Proof Note that the symmetry s_o interchanges o_F^0 and the basepoint $o_F \in F$, hence, for $g \in \widetilde{G}_\ell(F)$,

$$go_F^0 = gs_o(o_F) = s_o(\sigma(g)(o_F)) = s_o(o_F) = o_F^0 ,$$

because σ preserves $\widetilde{G}_\ell \subset N(F)$. $\qquad\qquad\square$

3.4

The next topic we want to discuss is the natural projection of D onto one of its boundary components F:

$$\pi_F : D \longrightarrow F .$$

We begin by considering an analog of the linear projection $\pi : C \longrightarrow C_1$ of a cone onto one of its boundary components, discussed in Chapter II, Section 3. Please bear with us though – this will not immediately define π_F, as it is the "wrong projection."

Fix a basepoint $o \in D$ and let $w_F : \mathbb{G}_m \longrightarrow \mathscr{G}$ be the homomorphism associated to F and o, i.e., the equivariant pair

$$f : \mathfrak{H} \longrightarrow D ,$$
$$\varphi : U^1 \times \mathrm{SL}(2,\mathbb{R}) \longrightarrow G ,$$

with $f(\mathrm{i}) = o$, $f(\infty) \in F$, and let $w_F(t) = \varphi \left(1, \begin{pmatrix} t & 0 \\ 0 & t^{-1} \end{pmatrix} \right)$. Then $\mathbb{G}_m$ acts via w_F algebraically on the projective variety $\check{D}$; hence, for all $x \in \check{D}$, we may define:

$$p_F(x) = \lim_{t \longrightarrow 0} w_F(t) \cdot x .$$

For instance, $p_F(o) = o_F^0$. Now, for all $g \in N(F)_{\mathbb{C}} = P(w_F^{-1})_{\mathbb{C}}$, we calculate:

$$
\begin{aligned}
p_F(go) &= \lim_{t \longrightarrow 0} w_F(t) \cdot g \cdot o \\
&= \lim_{t \longrightarrow 0} (w_F(t) g w_F(t)^{-1})(w_F(t) \cdot o) \qquad\qquad (3.1) \\
&= \overline{g}(o_F^0) ,
\end{aligned}
$$

where $\overline{g} = \lim_{t \longrightarrow 0} w_F(t) g w_F(t)^{-1}$, which exists by assumption. Moreover, note that

$$N(F)_{\mathbb{C}} = Z(w_F)_{\mathbb{C}} \cdot W(F)_{\mathbb{C}} ,$$
$$N(F^0)_{\mathbb{C}} = Z(w_F)_{\mathbb{C}} \cdot W(F^0)_{\mathbb{C}} ,$$

and that $\overline{g}$ is just the projection of g into its $Z(w_F)_{\mathbb{C}}$-component. Therefore $\overline{g} \in N(F^0)_{\mathbb{C}}$, hence $\overline{g} \cdot o_F^0 \in F^0$. Now define the open subset $D(\check{F}) \subset \check{D}$ by

$$D(\check{F}) = \bigcup_{g \in N(F)_{\mathbb{C}}} g \cdot D = N(F)_{\mathbb{C}} \cdot o$$

(the last equality occurs because $N(F)$ acts transitively on D). By the first description, $D(\check{F})$ is open; by the second, it is the orbit of an algebraic group acting on a projective variety $\check{D}$, and hence it is a Zariski-open subset of a subvariety. Therefore $D(\check{F})$ is a Zariski-open subset of $\check{D}$. Now formula (3.1) shows that

(a) $p_F(D(\check{F})) \subset \check{F}^0$ and $p_F(D) \subset F^0$,

(b) $p_F : D(\check{F}) \longrightarrow \check{F}^0$ is equivariant for $N(F)_{\mathbb{C}}$

(not $N(F^0)_{\mathbb{C}}$!), where $N(F)_{\mathbb{C}}$ acts on $\check{F}^0$ by

$$N(F)_{\mathbb{C}} \longrightarrow Z(w_F)_{\mathbb{C}} \hookrightarrow N(F^0)_{\mathbb{C}} \longrightarrow \mathrm{Aut}\,(D_F)_{\mathbb{C}}\;;$$

hence p_F is continuous on $\mathrm{D}(\check{F})$.

But now a map defined like p_F is automatically a morphism of varieties whenever it is continuous. In fact, embed $\check{D} \subset \mathbb{P}^n_{\mathbb{C}}$ such that $\check{D} \not\subset$ hyperplane, and such that w_F acts on $\mathbb{P}^n$ too, via $(x_0,\dots,x_n) \longmapsto (t^{r_0}x_0,\dots,t^{r_n}x_n)$, where $r_0 \geq \cdots \geq r_n$. Let $r_{m-1} > r_m = r_n$. Let L_1, resp. L_2, be the linear spaces $x_0 = \cdots = x_{m-1} = 0$, resp. $x_m = \cdots = x_n = 0$. Then p_F on $\mathbb{P}^n \setminus L_2$ is the linear projection with center L_2 and image L_1. Moreover, $L_1 \cap L_2 = \emptyset$ and $p_F(L_2) \subset L_2$. Thus, if p_F is continuous in the classical topology on any connected $S \subset \mathbb{P}^n$, then either $S \subset L_2$ or $S \subset \mathbb{P}^n \setminus L_2$. It follows that $D(\check{F}) \subset \mathbb{P}^n \setminus L_2$.

Now, to define $\pi_F : D(\check{F}) \longrightarrow \check{F}$, we proceed as follows. Take the symmetries

$$s_o : \check{D} \longrightarrow \check{D}\,,$$
$$s_{o_F} : \check{F} \longrightarrow \check{F}\,.$$

Define

$$\pi_F(x) = s_{o_F}(s_o(p_F(x)))$$

(since $s_o(F^0) = F$, this makes sense). Then, for all $g \in N(F)_{\mathbb{C}}$, if $\bar{g}$ is its projection into $Z(w_F)_{\mathbb{C}}$ and $\bar{\bar{g}}$ is its projection into $\mathrm{Aut}\,(D_F)_{\mathbb{C}}$, then

$$\begin{aligned}
\pi_F(gx) &= s_{o_F}(s_o(p_F(gx)))\\
&= s_{o_F}(s_o(\bar{g}(p_Fx)))\\
&= s_{o_F}(\sigma(\bar{g})(s_o(p_Fx)))\\
&= s_{o_F}(\sigma_F(\bar{\bar{g}})(s_o(p_Fx)))\\
&= \bar{\bar{g}}(s_{o_F}(s_o(p_Fx)))\\
&= \bar{\bar{g}}\pi_F(x)\,.
\end{aligned}$$

Here $\sigma : G_{\mathbb{C}} \longrightarrow G_{\mathbb{C}}$ denotes the Cartan involution relative to the basepoint o, and $\sigma_F : \mathrm{Aut}\,(D_F)_{\mathbb{C}} \longrightarrow \mathrm{Aut}\,(D_F)_{\mathbb{C}}$ denotes the Cartan involution relative to the basepoint o_F. Therefore

(a') $\pi_F(D(\check{F})) \subset \check{F}$ and $\pi_F(D) \subset F$;

(b') π_F is equivariant for $N(F)_{\mathbb{C}}$ acting on $D(\check{F})$ and on $\check{F}$ in the natural way, i.e., restricting the action of $G_{\mathbb{C}}$ on $\check{D}$;

(c') π_F is a morphism of varieties.

As a corollary of (b'), we deduce the following intrinsic definition of π_F independent of the choice of basepoint:

(d') $\pi_F(x)$ = the point of $\check{F}$ with stabilizer equal to the image in $\mathrm{Aut}\,(D_F)_\mathbb{C}$ of $N(F)_\mathbb{C} \cap (\mathrm{Stab.\ of\ } x)$.

Sometimes π_F on D is called the *geodesic projection* of D onto F. The reason for this is as follows: for all $x \in D$, let $f_{F,x}$ be the unique symmetric holomorphic map defined by

$$f_{F,x} : \mathfrak{H} \longrightarrow D ,$$

$$i \longmapsto x ,$$

$$\infty \longmapsto \mathrm{pt.\ of\ } F .$$

Then $f_{F,x}(i\mathbb{R})$ is a geodesic in D through x and one checks easily – check it first for $x = o$, then use (b') – that

$$\pi_F(x) = f_{F,x}(\infty) = \lim_{t \to \infty} f_{F,x}(it). \tag{3.2}$$

Note that if, at the beginning of Subsection 3.4, we had interchanged the roles of F and F^0 and defined $p_{F^0} : D \longrightarrow F$ by

$$p_{F^0}(x) = \lim_{t \to \infty} w_F(t) \cdot x ,$$

this would look similar to (3.2). However, $w_F(\mathbb{R}) \cdot x$ is *not* a geodesic in general, and p_{F^0} is definitely *not* $N(F)$-equivariant! Incidentally, an alternative to our round-about approach via p_F would be to define π_F directly using (b'); this would require a careful root analysis via Proposition 2.7 to establish the key lemma:

$$N(F)_\mathbb{C} \cap (K_\mathbb{C} \cdot P_-) \subset K_{h,\mathbb{C}} \cdot P_{h,-} \cdot \tilde{G}_{\ell,\mathbb{C}} \cdot W_\mathbb{C} ,$$

where K_h, P_h are the subgroups of G_h analogous to K and P in G.

3.5

The final topic we want to discuss is that of *rational* boundary components. As in Subsection 2.5, say $D = \mathscr{G}(\mathbb{R})^o/K$, where $\mathscr{G}$ is an algebraic group defined over $\mathbb{Q}$.

Definition 3.12 A boundary component F of D is *rational* if its normalizer $N(F)$ is defined over $\mathbb{Q}$ (i.e., $N(F) = \mathscr{N}(F)(\mathbb{R}) \cap \mathscr{G}(\mathbb{R})^o$ for some algebraic subgroup $\mathscr{N}(F)$ defined over $\mathbb{Q}$).

To understand the effect of this definition, let $\mathscr{A} \subset \mathscr{G}$ be a maximal $\mathbb{Q}$-split torus, and let $\mathscr{B} \subset \mathscr{G}$ be a maximal $\mathbb{R}$-split torus with $\mathscr{A} \subset \mathscr{B}$. As in Subsection 2.5, let γ_i, $1 \leq i \leq r$, be the strongly orthogonal real roots and let $\{1, \ldots, r\} = I_0 \cup \cdots \cup I_s$ be the partition so that $\mathscr{A}$ is defined by $\gamma_i = 1$ for $i \in I_0$ and $\gamma_i = \gamma_j$ for $i, j \in I_k$, $k = 1, \ldots, s$. For $i \geq 1$, let β_i be the restriction to $\mathscr{A}$ of

γ_j, $j \in I_i$. For all $S \subset \{1,\ldots,r\}$, let F_S be the associated boundary component. Then we may state:

Lemma 3.13 *The boundary component F_S is rational if and only if $S = I_{i_1} \cup \cdots \cup I_{i_t}$, where $1 \leq i_1,\ldots,i_t \leq s$.*

Proof If S has the above property, then $w_S(\mathbb{G}_m)$ is a one-parameter subgroup of $\mathscr{A}$. Since $\mathscr{A}$ is split, w_S is defined over $\mathbb{Q}$, and hence so is P_S. By Proposition 3.6, this means F_S is rational. Note that if $\mathscr{G}$ is $\mathbb{Q}$-simple, then this construction gives all maximal $\mathbb{Q}$-rational parabolics containing $\mathscr{A}$. To prove the converse, we may assume that $\mathscr{G}$ is $\mathbb{Q}$-simple since $\mathscr{A}$ and $\mathscr{N}(F_S)$ are products of their intersections with the $\mathbb{Q}$-simple factors. Then $\mathscr{N}(F_S)$ is a maximal $\mathbb{Q}$-rational parabolic (see (2) in the following list), and hence S is of the desired form. $\square$

In general, rational boundary components behave just like ordinary boundary components. Thus one proves easily the following.

(1) If we decompose $\mathscr{G}$ into its $\mathbb{Q}$-simple factors as $\mathscr{G} = \mathscr{G}_1 \times \cdots \times \mathscr{G}_k$, and let $D = D_1 \times \cdots \times D_k$ be the corresponding decomposition, then a boundary component $F = F_1 \times \cdots \times F_k$ is rational if and only if the F_i are rational.

(2) If $\mathscr{G}$ is $\mathbb{Q}$-simple, then the association $F \longmapsto N(F)$ defines a bijection between the set of rational boundary components and the set of maximal $\mathbb{Q}$-parabolics of $\mathscr{G}$. If $\mathscr{G} = \mathscr{G}_1 \times \cdots \times \mathscr{G}_t$ over $\mathbb{R}$, then each rational F decomposes as $F = F_1 \times \cdots \times F_t$, with $\overline{F}_i \subsetneq D_i$; hence every maximal $\mathbb{Q}$-parabolic $\mathscr{P}$ decomposes as $\mathscr{P} = \mathscr{P}_1 \times \cdots \times \mathscr{P}_t$, with $\mathscr{P}_i \subset \mathscr{G}_i$ a maximal $\mathbb{R}$-parabolic.

(3) Every rational boundary component equals gF_S for some $g \in \mathscr{G}(\mathbb{Q})$ and some S (for which F_S is rational).

(4) If $s = \mathbb{Q}$-rank $\mathscr{G}$, there is a symmetric holomorphic map

$$
\begin{array}{ccc}
\mathfrak{H}^s & \xrightarrow{\ f_1\ } & D \\
\cap & & \cap \\
(\mathbb{P}^1)^s & \xrightarrow{\ f_2\ } & \check{D}
\end{array}
$$

associated to

$$\varphi : U^1 \times \mathrm{SL}(2,\mathbb{R})^s \longrightarrow G$$

such that $\varphi(1, \text{diagonal matrices})$ is a maximal $\mathbb{Q}$-split torus and the rational boundary components are the $\mathscr{G}(\mathbb{Q})$-transforms of the boundary components F_S containing

$$f_2(\ldots, \underbrace{i}_{i \notin S}, \ldots, \underbrace{\infty}_{i \in S}, \ldots).$$

(5) Jumping ahead and using the results of Section 4, we may ask, for F a rational boundary component, which of the factors of $\mathcal{N}(F)$ is $\mathbb{Q}$-rational. We will have (see Section 4) that the algebraic connected component of $\mathcal{N}(F)$ equals

$$[\mathcal{G}_h(F) \cdot \mathcal{G}_\ell(F) \cdot \mathcal{M}(F)] \cdot \mathcal{V}(F) \cdot \mathcal{U}(F) .$$

Then

(a) w_F is in a maximal $\mathbb{Q}$-split torus so it is rational; hence $\mathcal{G}_h \cdot \mathcal{G}_\ell \cdot \mathcal{M}$, and $\mathcal{U}$ and $\mathcal{V}$, as 0, 1, 2-eigenspaces for $\mathrm{Ad}w_F$, are $\mathbb{Q}$-rational;

(b) $\mathcal{G}_h \cdot \mathcal{M}$ as the centralizer of $\mathcal{U}$ in $\mathcal{G}_h \cdot \mathcal{G}_\ell \cdot \mathcal{M}$ is $\mathbb{Q}$-rational; hence $\mathcal{G}_\ell$ as the normal complement to $\mathcal{G}_h \cdot \mathcal{M}$ in $\mathcal{G}_h \cdot \mathcal{G}_l \cdot \mathcal{M}$ is $\mathbb{Q}$-rational.

(6) Because of (5), we may speak of rational boundary components of rational boundary components and we claim: given boundary components $F_1 < F_2$, with F_2 rational, then F_1 is rational as a boundary component of F_2 if and only if it is rational as a boundary component of D. In fact, our assumptions say that $N(F_1) \cap N(F_2)$ is a real parabolic, with $N(F_2)$ defined over $\mathbb{Q}$. We may assume $\mathcal{G}$ is $\mathbb{Q}$-simple. By the uniqueness of the expression of parabolics as an intersection of maximal parabolics, $N(F_1) \cap N(F_2)$ is defined over $\mathbb{Q}$ if and only if $N(F_1)$ is defined over $\mathbb{Q}$.

Note that if $G = \mathrm{Aut}(D)^o$, then we may have $\mathcal{G}(\mathbb{R})^o = G \cdot M_0$, with M_0 compact and acting trivially on D. From the point of view of algebraic groups over $\mathbb{Q}$, M_0 may be conjugate to simple factors of G, and hence not defined over $\mathbb{Q}$ and impossible to throw out. Then M_0 appears as an "extra" factor in each $\mathcal{N}(F)(\mathbb{R})$ and combines with the $M(F)$ to be introduced in Section 4. In particular, $\mathcal{G}_h(F) \cdot \mathcal{M}(F)$ is a semi-simple algebraic group over $\mathbb{Q}$ such that $F = [\mathcal{G}_h(F) \cdot \mathcal{M}(F)](\mathbb{R})^o/(\text{max. compact})$, i.e., we recover for F the presentation we have for D.

4 Siegel domains of the third kind

4.1

Let D be a bounded symmetric domain, and let F be a boundary component. The purpose of this section is to work out in considerable detail the structure of the group $N(F)$ and from this to realize D as a particular open subset of a rather simple ambient space $D(F)$: this is an abstract version of Piatetskii-Shapiro's models of D as "Siegel domains of third kind." We will briefly indicate at the end how to make the link with the more concrete Siegel domains.

First we study $N(F)$ more closely. As in Section 3, it is easiest to write things out explicitly for $N(F_S)$, and then to observe that the same things happen for every F by conjugating. In this way, we now want to introduce the *fundamental 5-factor decomposition* of $N(F)$.

Special case $F = F_S$ We have seen that

$$\mathfrak{n}(F_S) := \operatorname{Lie} N(F_S) = Z(\mathfrak{a}) \oplus \sum_{\substack{\varphi = \frac{\pm \gamma_i \pm \gamma_j}{2} \text{ or } \frac{\pm \gamma_i}{2} \\ i,j \notin S}} \mathfrak{g}^\varphi \oplus \sum_{\substack{\varphi = \frac{\gamma_i \pm \gamma_j}{2} \text{ or } \frac{\gamma_i}{2} \\ i \in S, \text{ any } j}} \mathfrak{g}^\varphi .$$

Under the homomorphism $w_S : \mathbb{G}_m \longrightarrow N(F_S)$, we may decompose $\operatorname{Lie} N(F_S)$ into the direct sum of three eigenspaces, where $\operatorname{Ad}(w_S(t))$ is multiplication by $1, t, t^2$, respectively:

$$\mathfrak{n}(F_S)_0 = Z(\mathfrak{a}) \oplus \sum_{\substack{\varphi = \frac{\pm \gamma_i \pm \gamma_j}{2} \text{ or } \frac{\pm \gamma_i}{2} \\ i,j \notin S}} \mathfrak{g}^\varphi \oplus \sum_{\substack{\varphi = \frac{\gamma_i - \gamma_j}{2} \\ i,j \in S}} \mathfrak{g}^\varphi ;$$

$$\mathfrak{n}(F_S)_1 = \sum_{\substack{\varphi = \frac{\gamma_i \pm \gamma_j}{2} \text{ or } \frac{\gamma_i}{2} \\ i \in S, j \notin S}} \mathfrak{g}^\varphi : \quad \text{call this } \mathfrak{v}(F_S) ;$$

$$\mathfrak{n}(F_S)_2 = \sum_{\substack{\varphi = \frac{\gamma_i + \gamma_j}{2} \\ i,j \in S}} \mathfrak{g}^\varphi : \quad \text{call this } \mathfrak{u}(F_S) .$$

These are the root spaces $\mathfrak{g}^\varphi$, where $\langle dw_S, \varphi \rangle = 0$, 1, or 2, respectively. Then $\mathfrak{u} + \mathfrak{v}$ is the maximal nilpotent ideal, so, defining as in Theorem 3.10

$$W(F_S) = \text{unipotent radical of } N(F_S) ,$$

we have

$$\mathfrak{u}(F_S) + \mathfrak{v}(F_S) = \operatorname{Lie} W(F_S) .$$

Define:

$U(F_S) -$ commutative subgroup of $W(F_S)$ with Lie algebra $\mathfrak{u}(F_S)$;

$V(F_S) = \exp \mathfrak{v}(F_S)$, a subset of G diffeomorphic to $\mathfrak{v}(F_S)$.

Then

$$W(F_S) \cong V(F_S) \times U(F_S) ,$$

where

$$U(F_S) \subset \text{center } W(F_S)$$

and

$$W(F_S)/U(F_S) \text{ is an abelian Lie group} \cong V(F_S) .$$

We also want to refine the decomposition of $Z(w_S)$ introduced in Theorem 3.10. If σ is the Cartan involution of $\mathfrak{g}$ then

$$\sigma(\mathfrak{g}^\varphi) = \mathfrak{g}^{-\varphi} \text{ for all } \varphi \in {}_{\mathbb{R}}\Psi \ ;$$

hence

$$\sigma([\mathfrak{g}^\varphi, \mathfrak{g}^{-\varphi}]) = [\mathfrak{g}^\varphi, \mathfrak{g}^{-\varphi}] \ ,$$

and

$$[\mathfrak{g}^\varphi, \mathfrak{g}^{-\varphi}] = [\mathfrak{g}^\varphi, \mathfrak{g}^{-\varphi}] \cap \mathfrak{a} \oplus \underbrace{[\mathfrak{g}^\varphi, \mathfrak{g}^{-\varphi}] \cap m(\mathfrak{a})}_{\text{call this } [\mathfrak{g}^\varphi, \mathfrak{g}^{-\varphi}]_{(e)}} \ .$$

Note that we have

$$\mathfrak{l}_S = \mathfrak{g}_h(F_S) = \sum_{\substack{\varphi = \frac{\pm \gamma_i \pm \gamma_j}{2} \text{ or } \frac{\pm \gamma_i}{2} \\ i, j \notin S}} (\mathfrak{g}^\varphi + [\mathfrak{g}^\varphi, \mathfrak{g}^{-\varphi}]_{(e)}) + \sum_{i \notin S} \mathbb{R}x_i \ .$$

Define analogously:

$$\mathfrak{g}_\ell(F_S) = \sum_{\substack{\varphi = \frac{\gamma_i - \gamma_j}{2} \\ i, j \in S}} (\mathfrak{g}^\varphi + [\mathfrak{g}^\varphi, \mathfrak{g}^{-\varphi}]_{(e)}) + \sum_{i \in S} \mathbb{R}x_i \ .$$

This is the subalgebra of $\operatorname{Lie} \widetilde{G}_\ell(F_S)$ generated by the $\mathfrak{g}^\varphi$ and the x_i. It is also an ideal in $\operatorname{Lie} \widetilde{G}_\ell(F_S)$, so $\operatorname{Lie} \widetilde{G}_\ell(F_S)$ is the sum of $\mathfrak{g}_\ell(F_S)$ and a compact ideal $m(F_S)$. Therefore, we have

$$\mathfrak{n}(F_S)_0 = \mathfrak{g}_h(F_S) \oplus \mathfrak{g}_\ell(F_S) \oplus m(F_S) \ .$$

Globally,

$$Z(w_S)^o = G_h(F_S) \cdot \underbrace{G_\ell(F_S) \cdot M(F_S)}_{\text{this is } \widetilde{G}_\ell(F_S)}$$

(which stands for the *direct product* of three groups modulo a finite subgroup), where

$\qquad G_h(F_S)$ is semi-simple, with no compact factors ,

$\qquad G_\ell(F_S)$ is reductive, with no compact factors ,

$\qquad M(F_S)$ is compact .

Thus, finally,

$$N(F_S)^o = [G_h(F_S) \cdot G_\ell(F_S) \cdot M(F_S)] \times V(F_S) \times U(F_S) \ .$$

General case We can use that $F = k \cdot F_S$, for some $k \in K$, $S \subset \{1, \ldots, r\}$, and hence $N(F) = kN(F_S)k^{-1}$, to get the same decomposition for $N(F)$. However, we may also characterize the decomposition intrinsically using the homomorphism w_F associated canonically to F. Then

$$\operatorname{Lie} N(F) = \text{sum of } 0, 1, 2\text{-eigenspaces for Ad} w_F(t)$$
$$= \operatorname{Lie} Z(w_F) \oplus \mathfrak{v}(F) \oplus \mathfrak{u}(F) .$$

Let $V(F) = \exp \mathfrak{v}(F)$, $U(F) = \exp \mathfrak{u}(F)$. Next, $Z(w_F)^o$ decomposes into $G_h(F)$ and $\widetilde{G}_\ell(F)$, as in Theorem 3.10, where $\widetilde{G}_\ell$ acts identically on F. Writing $\widetilde{G}_\ell$ as a product of its non-compact and compact factors, we obtain

$$N(F)^o = [G_h(F) \cdot G_\ell(F) \cdot M(F)] \times V(F) \times U(F) .$$

4.2

Next, we want to look deeper at the group-theoretic structure of $N(F)^o$. To state the first result, note that $\mathrm{d}\varphi_F$ maps the k'th eigenspace for Ad $\begin{pmatrix} t & 0 \\ 0 & t^{-1} \end{pmatrix}$ in $\mathrm{SL}(2, \mathbb{R})$ into the k'th eigenspace for Ad $w_F(t)$ in $\mathfrak{g}$. In particular

$$\omega_F := \mathrm{d}\varphi_F \left(0, \begin{pmatrix} 0 & 1 \\ 0 & 0 \end{pmatrix} \right) \in \mathfrak{u}(F) ,$$

and hence

$$\Omega_F := \varphi_F \left(1, \begin{pmatrix} 1 & 1 \\ 0 & 1 \end{pmatrix} \right) = \exp \omega_F \in U(F) .$$

With this definition, we have:

Theorem 4.1

(1) $[G_h(F) \cdot M(F)] \times W(F)$ *centralizes* $U(F)$ *and* $\mathfrak{u}(F)$.
(2) *The orbit of* Ω_F *by* $G_\ell(F)$,

$$C(F) = \{ g\Omega_F g^{-1} \mid g \in G_\ell(F) \} ,$$

is an open homogeneous cone in $U(F)$, *self-adjoint with respect to the positive-definite quadratic form*

$$\langle x, y \rangle = -B(x, \sigma(y))$$

on $\mathfrak{u}(F)$, *hence on* $U(F)$. *The centralizer of* Ω_F *in* $G_\ell(F)$ *is the maximal compact subgroup* $G_\ell(F) \cap K$, *hence*

$$C(F) \cong G_\ell(F)/G_\ell(F) \cap K .$$

Proof We show first that†

$$G_\ell(F) \cap K = \{g \in G_\ell(F) \mid \operatorname{Ad} g(\omega_F) = \omega_F\} \ ,$$

and that $\operatorname{Ad} M(F)$ fixes ω_F. In fact, note that

$$o_F^0 = \varphi_F \left(1, \begin{pmatrix} 1 & -i \\ 0 & 1 \end{pmatrix}\right) \cdot o$$

$$= \exp(-i\omega_F) \cdot o \ ;$$

hence, for all $g \in \tilde{G}_\ell(F) = G_\ell(F) \cdot M(F)$,

$$g \cdot o = g \exp(i\omega_F) \cdot \exp(-i\omega_F) \cdot o$$

$$= \exp(i\operatorname{Ad} g(\omega_F)) \cdot g o_F^0 \ .$$

By Proposition 3.11, $g o_F^0 = o_F^0$. So

$$g \cdot o = \exp(i\operatorname{Ad} g(\omega_F)) \cdot \exp(-i\omega_F) \cdot o$$

$$= \exp(i(\operatorname{Ad} g(\omega_F) - \omega_F)) \cdot o \ .$$

But nothing in $U_{\mathbb{C}}$ fixes o, so $g \in K$ if and only if $\operatorname{Ad} g(\omega_F) = \omega_F$. Next, count dimensions taking $F = F_S$. Using the fact that $\sigma(\mathfrak{g}^\varphi) = \mathfrak{g}^{-\varphi}$, we find

$$\dim G_\ell(F)/G_\ell(F) \cap K = \dim \{-1\text{-eigenspace of } \sigma \text{ in } \mathfrak{g}_\ell(F)\}$$

$$= |S| + \sum_{\substack{i>j \\ i,j \in S}} \dim \mathfrak{g}^{(\gamma_i - \gamma_j)/2} \ .$$

But, using Proposition 2.7 and the remarks following it, we have

$$\dim \mathfrak{g}^{(\gamma_i - \gamma_j)/2} = \# \text{ of } (b_{ij})\text{-factors in } \mathfrak{g} = \dim \mathfrak{g}^{(\gamma_i + \gamma_j)/2} \ , \text{ for } i > j \ ,$$

and moreover $\mathfrak{g}^{\gamma_i}$ is 1-dimensional and lies entirely in the (a_i)-factor, i.e., the i'th copy of $\mathfrak{sl}(2, \mathbb{R})$. Thus,

$$|S| + \sum_{\substack{i>j \\ i,j \in S}} \dim \mathfrak{g}^{(\gamma_i - \gamma_j)/2} = \sum_{\substack{i \geq j \\ i,j \in S}} \dim \mathfrak{g}^{(\gamma_i + \gamma_j)/2} = \dim \mathfrak{u}(F) \ .$$

Therefore,

$$g \longmapsto \operatorname{Ad} g(\omega_F)$$

defines an isomorphism of $G_\ell(F)/G_\ell(F) \cap K$ with an open subset $C(F)$ of $\mathfrak{u}(F)$. It follows from Chapter II, Proposition 1.10, that $C(F)$ is a homogeneous cone, self-adjoint for any inner product $\langle \cdot, \cdot \rangle$ such that $^t(\operatorname{Ad} g)^{-1} = \operatorname{Ad}(\sigma(g))$, or

$$\langle \operatorname{Ad} g(x), \operatorname{Ad} \sigma g(y) \rangle = \langle x, y \rangle \text{ for all } x, y \in \mathfrak{u}(F) \ .$$

---

† This argument is due to P. Deligne.

The inner product $-B(x, \sigma(y))$ has this property. This proves (2).

As for (1), by checking roots one sees

$$[\mathfrak{g}_h(F), \mathfrak{u}(F)] = [\mathfrak{v}(F) + \mathfrak{u}(F), \mathfrak{u}(F)] = (0) ,$$

and hence $G_h(F) \times W(F)$ centralizes $U(F)$. Finally, since

$$[\mathfrak{m}(F), \omega_F] = (0) ,$$

we have, for all $x \in \mathfrak{g}_\ell(F)$,

$$[\mathfrak{m}(F), [x, \omega_F]] = [[\mathfrak{m}(F), x], \omega_F] = (0) ,$$

hence

$$[\mathfrak{m}(F), \mathfrak{u}(F)] = (0) ,$$

and $M(F)$ centralizes $U(F)$. $\qquad\square$

Corollary 4.2 $\varphi_F(U^1) \subset G_h(F) \times M(F)$.

Proof Since $\varphi_F(U^1)$ and w_F commute, $\varphi_F(U^1) \subset G_h \times G_\ell \times M$ and, by Theorem 4.1, it suffices to show that φ_F centralizes $\mathfrak{u}(F)$. But if $F = F_S$, then

$$\mathfrak{u}(F) \subset \text{factors of type } (a_i), (b_{ij}) , \quad i, j \in S ,$$

and

$$\varphi_{F_S}(e^{i\theta}) = \varphi(e^{i\theta}; \dots, \underbrace{1}_{\text{if } i \in S}, \dots, \underbrace{e^{i\theta}}_{\text{if } i \notin S}, \dots) .$$

$\qquad\square$

The next theorem will show that $U(F)$ is the center of the unipotent radical $W(F)$ of the parabolic subgroup $N(F)$. We denote by J the operator on the vector space $\mathfrak{v}(F)$

$$J = \text{Ad}(\varphi_F(-\mathrm{i}, I)) .$$

Then, clearly, J defines a complex structure on $\mathfrak{v}(F)$: $\varphi_F(-1, -I) = h_U(-1) = u_0(1) = e$; hence

$$J^2 = \text{Ad}\,\varphi_F(-1, I) = \text{Ad}\,\varphi_F(1, -I) = w_F(-1) ,$$

which is $-\text{Id}$ on $\mathfrak{v}(F)$.

Theorem 4.3 $[v, Jv] \in \overline{C(F)} \setminus \{0\}$, *for all* $0 \neq v \in \mathfrak{v}(F)$.

We have the consequence:

Corollary 4.4 $U(F)$ *is the center of* $W(F)$. $\qquad\square$

Proof of Theorem 4.3 Since C is self-adjoint with respect to $\langle \cdot, \cdot \rangle$, it is sufficient to show that

$$v \neq 0 \Longrightarrow \langle [v, Jv], y \rangle > 0 \text{ for all } y \in C .$$

But, writing $y = \operatorname{Ad} g^{-1}(\omega_F)$ with $g \in G_\ell$, and noting that

$$\langle \operatorname{Ad}(\sigma(g))u, u' \rangle = \langle u, \operatorname{Ad}(g)^{-1}u' \rangle \text{ for all } u, u' \in U(F) ,$$

we obtain

$$
\begin{aligned}
\langle [v, Jv], y \rangle &= \langle [v, Jv], \operatorname{Ad} g^{-1}(\omega_F) \rangle = \langle \operatorname{Ad}\sigma(g)[v, Jv], \omega_F \rangle \\
&= \langle [\operatorname{Ad}\sigma(g)v, \operatorname{Ad}\sigma(g)Jv], \omega_F \rangle \\
&= \langle [\operatorname{Ad}\sigma(g)v, J\operatorname{Ad}\sigma(g)v], \omega_F \rangle ,
\end{aligned}
$$

where, for the last equality, we used $\varphi_F(U^1) \subset G_h \times M$, so $\varphi_F(U^1)$ and G_ℓ commute. Thus it suffices to show

$$v \neq 0 \Longrightarrow \langle [v, Jv], \omega_F \rangle > 0 .$$

Now,

$$
\begin{aligned}
\langle [v, Jv], \omega_F \rangle &= -B([v, Jv], \sigma(\omega_F)) \\
&= B(v, [\sigma(\omega_F), Jv]) \\
&= B(v, \sigma([\omega_F, \sigma Jv])) .
\end{aligned}
$$

So it suffices to prove

$$[\omega_F, \sigma Jv] = -v, \text{ for all } 0 \neq v \in \mathfrak{v}(F) ,$$

and use the fact that $B(x, \sigma(y))$ is negative-definite. But

$$
\begin{aligned}
\sigma \circ J &= \operatorname{Ad}\varphi_F \left(i, \begin{pmatrix} 0 & 1 \\ -1 & 0 \end{pmatrix} \right) \circ \operatorname{Ad}\varphi_F(-i, I) \\
&= \operatorname{Ad}\varphi_F \left(1, \begin{pmatrix} 0 & 1 \\ -1 & 0 \end{pmatrix} \right) ,
\end{aligned}
$$

and hence

$$
\begin{aligned}
[\omega_F, \sigma Jv] &= (\operatorname{Ad}\Omega_F - I) \left(\operatorname{Ad}\varphi_F \left(1, \begin{pmatrix} 0 & 1 \\ -1 & 0 \end{pmatrix} \right) v \right) \\
&= \operatorname{Ad}\varphi_F \left(1, \begin{pmatrix} -1 & 1 \\ -1 & 0 \end{pmatrix} \right) v - \operatorname{Ad}\varphi_F \left(1, \begin{pmatrix} 0 & 1 \\ -1 & 0 \end{pmatrix} \right) v .
\end{aligned}
$$

Now suppose we put together the $+1$ and -1-eigenspaces for $\operatorname{Ad} w_F(t)$: the $+1$-space is $\mathfrak{v}(F)$ and the whole space is given by

$$\mathfrak{h} = \{ x \in \mathfrak{g} \mid \operatorname{Ad} w_F(-1)x = -x \} ,$$

which is a subspace invariant under $\mathrm{SL}(2,\mathbb{R})$. Since $\begin{pmatrix} t & 0 \\ 0 & t^{-1} \end{pmatrix}$ has only ± 1-eigenspaces, $\mathfrak{h}$ is the sum of copies of the usual two-dimensional representation $\mathbb{R}^2$. Therefore we need only calculate in $\mathbb{R}^2$, with $v = \begin{pmatrix} 1 \\ 0 \end{pmatrix}$:

$$\begin{pmatrix} -1 & 1 \\ -1 & 0 \end{pmatrix}\begin{pmatrix} 1 \\ 0 \end{pmatrix} - \begin{pmatrix} 0 & 1 \\ -1 & 0 \end{pmatrix}\begin{pmatrix} 1 \\ 0 \end{pmatrix} = -\begin{pmatrix} 1 \\ 0 \end{pmatrix}.$$

$\square$

4.3

Next, we want to interpret the group-theoretic structure of $N(F)^o$ geometrically as a decomposition of D. First, by Proposition 3.6, $N(F)$, and hence $N(F)^o$, acts transitively on D, so that, using our more refined decomposition of $N(F)^o$, we can write

$$D \cong ([G_h(F) \cdot G_\ell(F) \cdot M(F)] \times W(F)) / [K_h(F) \cdot K_\ell(F) \cdot M(F)],$$

where

$$K_h(F) = G_h(F) \cap K = \text{ a maximal compact subgroup of } G_h(F),$$
$$K_\ell(F) = G_\ell(F) \cap K = \text{ a maximal compact subgroup of } G_\ell(F).$$

Therefore, firstly, there is an $N(F)^o$-equivariant mapping

$$\Phi_F : D \longrightarrow C(F) \text{ with } \Phi_F(o) = \Omega_F,$$

defined by the map of homogeneous spaces

$$([G_h \cdot G_\ell \cdot M] \times W) / [K_h \cdot K_\ell \cdot M]$$
$$\longrightarrow ([G_h \cdot G_\ell \cdot M] \times W) / ([G_h \cdot K_\ell \cdot M] \times W) \cong G_\ell/K_\ell.$$

Secondly, the whole domain D can be decomposed as a real manifold:

$$D \cong F \times C(F) \times W(F) \tag{4.1}$$

by

$$x \longmapsto (\pi_F(x), \Phi_F(x), w(x)),$$

where $w(x)$ is defined as follows. Let

$$\pi_F(x) = g_h(\pi_F(o)) \text{ with } g_h \in G_h,$$
$$\Phi_F(x) = g_\ell(\Omega_F) \text{ with } g_\ell \in G_\ell,$$

then

$$x = w(x) \cdot g_h \cdot g_\ell \cdot o \text{ with } w(x) \in W(F).$$

To understand the situation better, we introduce a new open subset $D(F) \subset \check{D}$.

Definition 4.5 Let $D(F) = U(F)_{\mathbb{C}} \cdot D = \bigcup_{g \in U(F)_{\mathbb{C}}} g \cdot D$.

Note that, since $U(F)$ is a normal subgroup of $N(F)^o$, it follows that $N(F)^o \cdot U(F)_{\mathbb{C}}$ is a subgroup of $G_{\mathbb{C}}$, which clearly acts transitively on $D(F)$. Recall the point:

$$o_F^0 = \exp(-i\omega_F) \cdot o \in D(F) \,.$$

Lemma 4.6 $K_h(F) \cdot G_\ell(F) \cdot M(F) = \{ g \in N(F)^o \cdot U(F)_{\mathbb{C}} \mid g o_F^0 = o_F^0 \}$.

Proof We saw in Proposition 3.11 that $G_\ell \cdot M$ fixes o_F^0. Moreover, K_h fixes o and K_h commutes with $\exp(-i\omega_F)$, so K_h fixes o_F^0. This shows "$\subset$." But, since $K_{\mathbb{C}} \cdot P_-$ is the stabilizer of o in $G_{\mathbb{C}}$, the RHS equals

$$N(F)^o \cdot U(F)_{\mathbb{C}} \cap \exp(-i\omega_F) \cdot K_{\mathbb{C}} P_- \cdot \exp(i\omega_F)$$

(intersection inside $G_{\mathbb{C}}$). A simple root calculation shows that the Lie algebra of this group equals the Lie algebra of the LHS. But considering $G_{\mathbb{C}}$ as the real points of $R_{\mathbb{C}/\mathbb{R}}\mathscr{G}_{\mathbb{C}}$, both $N(F)^o \cdot U(F)_{\mathbb{C}}$ and $K_{\mathbb{C}} \cdot P_-$ are the connected components of the real points of algebraic subgroups. So this intersection has finitely many components. Since $K_h \subset G_h$ is maximal compact, and $W(F) \cdot U(F)_{\mathbb{C}}$ is torsion-free, there is no group H with

$$N(F)^o \cdot U(F)_{\mathbb{C}} \supset H \underset{\text{(finite index)}}{\supsetneq} K_h \times G_\ell \times M \,.$$

$\square$

Therefore, if we take o_F^0 as basepoint of $D(F)$, we get an isomorphism:

$$\begin{aligned} (G_h(F)/K_h(F)) \times W(F) \cdot U(F)_{\mathbb{C}} &\xrightarrow{\sim} D(F) \,, \\ (g, w) &\longmapsto w \cdot g(o_F^0) \,, \end{aligned} \tag{4.2}$$

and hence

$$F \times V(F) \times U(F)_{\mathbb{C}} \cong D(F) \,.$$

The projection π_F to F is again just given by

$$N(F)^o \cdot U(F)_{\mathbb{C}}/K_h \cdot G_\ell \cdot M \longrightarrow N(F)^o \cdot U(F)_{\mathbb{C}}/K_h \cdot G_\ell \cdot M \cdot W \cdot U_{\mathbb{C}} \,,$$

$$\wr\| \qquad\qquad\qquad\qquad\qquad\qquad \wr\|$$

$$D(F) \qquad\qquad\qquad\qquad\qquad\qquad G_h/K_h$$

$$\wr\|$$

$$F$$

which is $N(F)^o \cdot U(F)_{\mathbb{C}}$-equivariant, and hence is the π_F defined in Section 3.4. Also $N(F)^o$ is a subgroup of $N(F)^o \cdot U(F)_{\mathbb{C}}$, so projection onto the *imaginary part* of the U-coordinate is given by

$$N(F)^o \cdot U(F)_{\mathbb{C}}/K_h \cdot G_\ell \cdot M \quad \longrightarrow \quad N(F)^o \cdot U(F)_{\mathbb{C}}/N(F)^o \ ,$$
$$\text{≀}\| \qquad\qquad\qquad\qquad \text{≀}\downarrow$$
$$D(F) \qquad\qquad\qquad\qquad\qquad U(F)$$

which is also $N(F)^o \cdot U(F)_{\mathbb{C}}$-equivariant, where $N(F)^o \cdot U(F)_{\mathbb{C}}$ acts on $U(F)$ via the vertical isomorphism in the preceding diagram, which amounts to letting $N(F)^o$ act on $U(F)$ by conjugation and letting $iU(F)$ act on $U(F)$ by translation. We call this Φ_F because we claim that we get a commutative diagram involving this projection:

$$
\begin{array}{ccc}
D & \xrightarrow{\ \Phi_F\ } & C(F) \\
\cap & & \cap \\
D(F) & \xrightarrow{\ \Phi_F\ } & U(F) \ .
\end{array}
$$

To see this, let $x = w \cdot g_h \cdot g_\ell \cdot o \in D$, with $w \in W$, $g_h \in G_h$, $g_\ell \in G_\ell$. Then $\Phi_F(x) = g_\ell(\Omega_F)$, and, since G_ℓ acts on $C(F)$ by conjugation, we obtain

$$\Phi_F(x) = g_\ell \exp(\omega_F) g_\ell^{-1} \ .$$

But

$$x = w \cdot g_h \cdot g_\ell \cdot \exp(i\omega_F) \cdot \exp(-i\omega_F) \cdot o$$
$$= w \cdot g_h \cdot g_\ell \cdot \exp(i\omega_F) \cdot g_\ell^{-1} \cdot o_F^0 \quad (\text{since } g_\ell o_F^0 = o_F^0).$$

Since $w \cdot g_h \in N(F)$, and $g_\ell \cdot \exp(i\omega_F) \cdot g_\ell^{-1} \in iU(F) \subset U(F)_{\mathbb{C}}$, the imaginary part of the $U(F)_{\mathbb{C}}$-coordinate of x is just $g_\ell \cdot \exp(\omega_F) \cdot g_\ell^{-1}$. In fact:

Lemma 4.7 $D = \{x \in D(F) \mid \Phi_F(x) \in C(F)\}$.

Proof Let $x = g \cdot \exp(iu) \cdot o_F^0$ be any element of $D(F)$, where $g \in N(F)^o$ and $u \in \mathfrak{u}(F)$. If $u \in C(F)$, then $u = \operatorname{Ad} h(\omega_F)$ with $h \in G_\ell$, and hence

$$x = g \cdot \left(h \exp(i\omega_F) h^{-1} \right) \cdot o_F^0$$
$$= g \cdot h \cdot \exp(i\omega_F) \cdot o_F^0 \quad (\text{since } h \cdot o_F^0 = o_F^0)$$
$$= g \cdot h \cdot o \in D \ .$$

$\square$

The idea of Siegel domains is that $D(F)$ is a much simpler complex manifold than D, and that D is defined inside $D(F)$ by the tube-domain-like requirement

$$\Phi_F(x) \in C(F) \ .$$

To see what $D(F)$ is, note that because of the isomorphism (4.2) above, $U(F)_{\mathbb{C}}$ acts freely on $D(F)$, making it into a principal homogeneous space over

$$D(F)' := D(F)/U(F)_{\mathbb{C}}$$
$$\cong N(F)^o \cdot U(F)_{\mathbb{C}}/K_h \cdot G_\ell \cdot M \cdot U(F)_{\mathbb{C}}$$
$$= N(F)^o/K_h \cdot G_\ell \cdot M \cdot U(F) \,.$$

In turn, $V(F)$ acts freely on $D(F)'$, making it into a principal homogeneous space over F:

$$
\begin{array}{c}
D(F) \\
\pi'_F \Big\downarrow \text{ fibers } U(F)_{\mathbb{C}} \\
D(F)' \\
p_F \Big\downarrow \text{ fibers } V(F) \\
F
\end{array}
$$

with π_F the composite map.

Since $U(F)_{\mathbb{C}}$ is a complex subgroup of $G_{\mathbb{C}}$, we have that $U(F)_{\mathbb{C}}$ acts holomorphically on $D(F)$, and $D(F)'$ has a complex structure making all the maps holomorphic. Note, however, that $V(F)$ has no natural complex structure and that $D(F)' \longrightarrow F$ is only real-analytically a principal homogeneous space with group $V(F)$.

This is as far as we need to carry this analysis of D and $D(F)$. The full picture, however, is the following.

(i) There is a holomorphic section of $D(F)$ over $D(F)'$, such that

$$D(F) \cong D(F)' \times U(F)_{\mathbb{C}} \,.$$

(ii) For each $x \in F$, there is a complex structure $J_x : V(F) \longrightarrow V(F)$ (with J_{o_F} being the J defined above) such that $(V(F), J_x)$ acts complex-analytically on $p_F^{-1}(x)$.

(iii) Altogether, $D(F)'$ is a complex vector bundle over F, which can be trivialized as

$$D(F)' \cong \mathbb{C}^k \times F \,,$$

such that each $v \in V(F)$ acts as

$$(x, a) \longmapsto (x + \lambda_v(a), a)$$

where $\lambda_v(a)$ is holomorphic in a and linear in v.

Via (i) and (iii), we get a *holomorphic* isomorphism:

$$D(F) \cong U(F)_{\mathbb{C}} \times \mathbb{C}^k \times F$$

(NB *not* the same as the more elementary group-theoretic isomorphism $D(F) \cong U(F)_{\mathbb{C}} \times V(F) \times F$ in (4.2) above).

(iv) In this product representation

$$\Phi_F(x,y,z) = \operatorname{Im} x - h_z(y,y) \,,$$

where h_z is a real-bilinear quadratic form

$$\mathbb{C}^k \times \mathbb{C}^k \longrightarrow U(F)$$

depending real-analytically on z.

Thus

$$D \cong \left\{ (x,y,z) \in U(F)_{\mathbb{C}} \times \mathbb{C}^k \times F \mid \operatorname{Im} x \in C(F) + h_z(y,y) \right\} \,.$$

For proofs, see Koranyi–Wolf [7] and Satake [11]. We summarize the essential maps that we will use in Section 5 below:

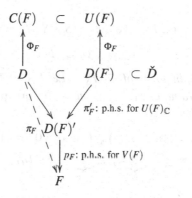

This whole diagram is $N(F)^o$-equivariant; all but D and $C(F)$ is $N(F)^o \cdot U(F)_{\mathbb{C}}$-equivariant.

An interesting topic that we have not seen explored is to investigate the third open subset of $\check{D}$:

$$D(\check{F}) = N(F)_{\mathbb{C}} \cdot D = \bigcup_{g \in N(F)_{\mathbb{C}}} g \cdot D \,.$$

Here, $D(\check{F})$ is a *Zariski-open* subset of $\check{D}$ (see Subsection 3.4) and appears to sit in a double fibration:

$$D(\check{F})$$

$$\Big\downarrow \text{p.h.s. for } U(F)_{\mathbb{C}}$$

$$D(\check{F})'$$

$$\Big\downarrow \text{cx. vector bundle and p.h.s. for } V(F)$$

$$\check{F}$$

in which $D(F)$ and $D(F)'$ are just the inverse images of $F \subset \check{F}$.

4.4

The final point of this section is to compare $N(F)$ and $N(F')$ and their decompositions when $F \subset \overline{F'}$. We formulate the situation in the following theorem.

Theorem 4.8 *Let $F \subset \overline{F'}$ be two boundary components of D. Then*

(i) $U(F) \supset U(F')$, $G_\ell(F) \supset G_\ell(F')$, $G_h(F) \subset G_h(F')$.

(ii) *$C(F')$ is a boundary component of $C(F)$. Moreover, fixing F, then the map $F' \longmapsto C(F')$ is an order-reversing bijection between the set of boundary components F' with $F \subset \overline{F'}$ of D and the set of boundary components of $C(F)$.*

(iii) *Now assume $F = F_S$, $F' = F_{S'}$, where $S \supset S'$. Then*

$$A_S = (A \cap G_\ell(F_S))^o$$

is a maximal split torus of $G_\ell(F_S)$. Let $U_D = \prod_{i \in S} \exp(\mathfrak{g}^{\gamma_j}) \subset U(F_S)$ and define coordinates $x_1, \ldots, x_d$ on U_D by $x_i(\exp \mathfrak{g}^{\gamma_j}) = 0$ if $i \neq j$ and $x_i(\Omega_{F_S}) = 1$. Then

$$A_S \cdot \Omega_{F_S} = \{(x_1, \ldots, x_d) \in U_D \mid x_i > 0 \text{ for all } i = 1, \ldots, d\},$$

$$\Omega_{F_{S'}} = (\ldots, \underbrace{0}_{i \in S \setminus S'}, \ldots, \underbrace{1}_{i \in S'}, \ldots) \in U_D,$$

and $C(F_{S'})$ is the boundary component of $C(F_S)$ whose intersection with U_D is given by

$$\{(x_1, \ldots, x_n) \mid x_i = 0 \text{ if } i \in S \setminus S', \ x_i > 0 \text{ if } i \in S'\}.$$

(iv) *Finally, if $D = \mathcal{G}(\mathbb{R})^o / K$, with $\mathcal{G}$ defined over $\mathbb{Q}$, and F is a rational boundary component, then the subgroups $U(F)$, $G_\ell(F) \subset N(F)$ are defined over $\mathbb{Q}$; in particular, $U(F)$ comes from a $\mathbb{Q}$-vector space and for all F' with $F \subset \overline{F'}$,*

$$\begin{bmatrix} F' \text{ is a rational boundary} \\ \text{component of } D \end{bmatrix} \iff \begin{bmatrix} C(F') \text{ is a rational} \\ \text{boundary component of } C(F) \end{bmatrix} .$$

Proof By Proposition 3.4, any pair $F \subset \overline{F'}$ is conjugate by some $k \in K$ to a pair $F_S \subset \overline{F}_{S'}$, where $S \supset S'$. For such pairs, (i) is immediate by the explicit formulae at the beginning of this section. Now, $A \subset G_\ell(F_S) \cdot G_h(F_S)$ and is a maximal split torus in G. So $A = A_\ell \cdot A_h$, where $A_\ell = (A \cap G_\ell(F_S))^o$ and $A_h = (A \cap G_h(F_S))^o$ are maximal split tori in $G_\ell(F_S)$ and $G_h(F_S)$. In fact, $A_\ell = \exp\left(\sum_{i \in S} \mathbb{R} x_i\right)$. To investigate about the situation in $U(F)$, let us identify $U(F)$ and $\mathfrak{u}(F)$ via exp, and calculate instead in $\mathfrak{u}(F)$. First of all, ω_{F_S} is the natural basepoint in

$$\sum_{i \in S} \mathfrak{g}^{\gamma_i} .$$

(In the notation of Theorem 2.4, it is $\sum_{i \in S} \frac{-y_i + ih_i}{2}$.) Since A_ℓ acts on $\mathfrak{g}^{\gamma_i}$ via the character e^{γ_i}, it is a simple calculation to describe $A_S \cdot \omega_{F_S}$ as in part (iii) of the theorem. And if $S \supset S'$, then, by definition, $\omega_{F_{S'}}$ is just the projection of ω_{F_S} into the subspace

$$\sum_{i \in S'} \mathfrak{g}^{\gamma_i} .$$

Now, by definition,

$$C(F_{S'}) = \{ \operatorname{Ad} g(\omega_{F_{S'}}) \mid g \in N(F_{S'}) \} .$$

But as $N(F_{S'}) = G_\ell(F_{S'}) \cdot (\text{centralizer of } U(F_{S'}))$, in fact for any group H with

$$G_\ell(F_{S'}) \subset H \subset N(F_{S'}),$$

we have

$$C(F_{S'}) = \{ \operatorname{Ad} g(\omega_{F_{S'}}) \mid g \in H \} .$$

For instance, since $G_\ell(F_S) \supset G_\ell(F'_S)$, it follows that $H = N(F_{S'}) \cap G_\ell(F_S)$ will do. But let

$$w_{S'} : \mathbb{G}_m \longrightarrow \mathscr{A}$$

be the homomorphism defined above, relative to S'. Then $N(F_{S'}) = P(w_{S'}^{-1})$, so H is the parabolic subgroup $P(w_{S'}^{-1})_{G_\ell(F_S)}$ of $G_\ell(F_S)$ defined by $w_{S'}^{-1}$, i.e.,

$$H = \left\{ g \in G_\ell(F_S) \mid \lim_{t \to 0} w_{S'}(t) g w_{S'}(t)^{-1} \text{ exists} \right\} .$$

But $w_{S'}$ is a one-parameter subgroup of $G_\ell(F_S)/(\text{center}) = \operatorname{Aut}(C(F_S))^o$ (via Ad). Acting on the basepoint ω_{F_S},

$$\lim_{t \to 0} \operatorname{Ad} w_{S \setminus S'}(t) \omega_{F_S} = \omega_{F_{S'}} .$$

Note that the parabolic subgroups $P(w_{S'}^{-1})_{G_\ell(F_S)}$ and $P(w_{S \backslash S'})_{G_\ell(F_S)}$ of $G_\ell(F_S)$ agree, since w_S is central in G_ℓ. Therefore, by Chapter II, Proposition 3.4, the boundary component of $C(F_S)$ through $\omega_{F_{S'}}$ is the orbit of $\omega_{F_{S'}}$ under $P(w_{S \backslash S'})_{G_\ell(F_S)} = P(w_{S'}^{-1})_{G_\ell(F_S)}$ in $\mathfrak{u}(F_S)$, which is just $C(F_{S'})$, namely the locus of points $\operatorname{Ad} g(\omega_{F_{S'}})$, with $g \in H$.

To prove the second part of (ii), we no longer normalize F and F', but instead we may assume G is simple because, in the general case, everything breaks up into a product. The first step is to check that, for any two boundary components F, F',

$$F \subset \overline{F'} \text{ or } F' \subset \overline{F} \Longleftrightarrow N(F) \cap N(F') \text{ is parabolic.}$$

Ordering the roots $\sum u_i \gamma_i$ lexicographically, the corresponding minimal real parabolic $P \supset A$ is given by

$$\operatorname{Lie} P = Z(\mathfrak{a}) + \sum_{\text{all } i,j} \mathfrak{g}^{\frac{\gamma_i + \gamma_j}{2}} + \sum_{i<j} \mathfrak{g}^{\frac{\gamma_i - \gamma_j}{2}} + \sum_i \mathfrak{g}^{\frac{\gamma_i}{2}} \, ;$$

and the maximal real parabolics containing P are P_S for

$$S = \{1\}, \{1,2\}, \dots, \{1,2,\dots,r\} \, .$$

To prove implication $\Longrightarrow$, if for instance $F \subset \overline{F'}$, we may assume $F = F_S$, $F' = F_{S'}$, with $S \supset S'$; applying the Weyl group, we may even assume $S = \{1,\dots,s\}$, $S' = \{1,\dots,s'\}$, with $s \geq s'$. Thus $N(F_S) \cap N(F_{S'}) \supset P$. Conversely, if $N(F) \cap N(F')$ contains a parabolic, by conjugating we can assume it contains P. Then $N(F) = P_{\{1,\dots,s\}}$ and $N(F') = P_{\{1,\dots,s'\}}$ for some s, s'. Thus $F = F_{\{1,\dots,s\}}$ and $F' = F_{\{1,\dots,s'\}}$ and, depending on whether $s > s'$ or $s < s'$, we get $F \subset \overline{F'}$ or $F' \subset \overline{F}$. Therefore, fixing F, and using the unique representation of parabolics as intersections of maximal parabolics, we find a bijection:

$$\{F' \mid F \subset \overline{F'} \text{ or } F' \subset \overline{F}\} \cong \{P \subset N(F) \text{ maximal real parabolic}\} \, .$$

But now $N(F) \cong G_\ell(F) \cdot G_h(F) \cdot M(F) \cdot W(F)$, so the maximal real parabolics $P \subset N(F)$ are of the form

$$P = P_\ell \cdot G_h(F) \cdot M(F) \cdot W(F) \, , \quad P_\ell \text{ maximal parabolic in } G_\ell(F) \, ,$$

or

$$P = G_\ell(F) \cdot P_h \cdot M(F) \cdot W(F) \, , \quad P_h \text{ maximal parabolic in } G_h(F) \, .$$

Using our standard models again, it is easy to see that $F \subset \overline{F'}$ corresponds to the first type and $F' \subset \overline{F}$ to the second; i.e., we get bijections:

$$\{F' \mid F \subset \overline{F'}\} \cong \{P \subset G_\ell(F) \text{ maximal real parabolic}\} \, ,$$
$$\{F' \mid F' \subset \overline{F}\} \cong \{P \subset G_h(F) \text{ maximal real parabolic}\} \, .$$

But if G is simple then so is $G_\ell(F)$ (and $G_h(F)$): for our purposes, it is enough to check this for $G_\ell(F_S)$. Then, for every permutation σ of $\{1,\dots,r\}$ which preserves S, there is an element $w_\sigma \in \mathrm{Norm}\,(A)$, such that $w_\sigma \gamma_i w_\sigma^{-1} = \gamma_{\sigma(i)}$; hence $w_\sigma w_S = w_S w_\sigma$ and $w_\sigma \in N(F_S)$. Hence w_σ projects to give an element of the Weyl group of the root system of $G_\ell(F_S)$, and it follows that the Weyl group of the root system of $G_\ell(F_S)$ is the full permutation group of S, and hence $G_\ell(F_S)$ is simple.

Therefore there is a bijection between maximal real parabolics in $G_\ell(F)$ and boundary components of $C(F)$. Altogether we have a bijection from the F' with $F \subset \overline{F'}$ to the boundary components C' of $C(F)$ defined by

$$N_G(F') \cap G_\ell(F) = N_{G_\ell(F)}(C') \ .$$

In fact, this C' is just $C(F')$ because $N_G(F') \cap G_\ell(F) \subset N_{G_\ell(F)}(C(F'))$ and $N_{G_\ell(F)}(C')$ is maximal. This proves (ii).

Finally, if we have a $\mathbb{Q}$-structure and F is $\mathbb{Q}$-rational, then $U(F)$ is the center of the unipotent radical of F, and hence is defined over $\mathbb{Q}$, and $G_\ell(F)$ is the factor of $Z(w_F)$ which acts non-trivially on $U(F)$, and hence is defined over $\mathbb{Q}$. Then, for all $\overline{F'} \supset F$,

F' is $\mathbb{Q}$-rational $\Longleftrightarrow N(F')$ is defined over $\mathbb{Q}$

$\Longleftrightarrow N(F') \cap G_\ell(F)$ is defined over $\mathbb{Q}$

$\Longleftrightarrow N_{G_\ell(F)}(C(F'))$ is defined over $\mathbb{Q}$

$\Longleftrightarrow C(F')$ is a $\mathbb{Q}$-rational boundary comp. of $C(F)$.

$\square$

Appendix: Connected components

We need to know in Section 5 that the key diagram of Section 4, namely

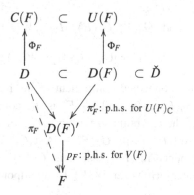

is, in fact, $N(F)$-equivariant, in addition to being $N(F)^o$-equivariant. To check this, note that $U(F)$, being the center of the unipotent radical of $N(F)^o$, is invariant under all *outer* automorphisms of $N(F)^o$, and hence is normal in $N(F)$. Therefore conjugation by $N(F)$ carries $U(F)_\mathbb{C}$ into itself, and hence $N(F)$ maps $D(F)$ into itself. Since $D(F)' \cong D(F)/U(F)_\mathbb{C}$, we may define the action of $N(F)$ on $D(F)'$ to make π'_F equivariant. But π_F was seen to be $N(F)$-equivariant in Subsection 3.4, and hence p_F is also $N(F)$-equivariant. Next define the action of $N(F)$ on $U(F)$ to be conjugation. Then the equivariance of Φ_F follows from:

Lemma 4.9 *Let $G \subset G'$ be groups, G normal in G'. Let $f : X \longrightarrow Y$ be a G-equivariant map of G'-spaces. If*

  (i) *G is transitive on X, and*
 (ii) *there is a point $o \in X$ such that $f(o)$ is the only point of Y fixed by* Stab$_G(o)$,

then f is G'-equivariant.

Proof Easy.   □

Apply the lemma with $G = N(F)^o, G' = N(F), X = D(F), Y = U(F), f = \Phi_F$, $o = \omega_F$, noting that $o^0_F \in U(F)$ is the only point fixed by K_ℓ. Finally, in view of the fact that the action of $N(F)$ on $D(F)$ maps D into itself, it follows that the action of $N(F)$ on $U(F)$ maps $C(F)$ into itself.

A few remarks on connected components may be in order.

(1) By definition, G is the connected component both of $\mathrm{Aut}\,(D)$ and of $\mathscr{G}(\mathbb{R})$,
(2) $N(F)$ is the full subgroup of G fixing F, but may be of finite index in the corresponding subgroup of $\mathrm{Aut}\,(D)$ or in the real points of the algebraic subgroup $\mathscr{N}(F) \subset \mathscr{G}(F)$.
(3) By definition, we made $G_h(F), G_\ell(F), M(F), W(F)$ connected, getting

$$N(F)^o = [G_h \cdot G_\ell \cdot M] \times W(F) \,.$$

Then $G_h(F)$ is the connected component of the real points of the algebraic subgroup $\mathscr{G}_h(F)$ of $\mathscr{G}$, and $G_\ell(F)$ is the connected component of the real points of the algebraic subgroup $\mathscr{G}_\ell(F)$ of $\mathscr{G}$. There are projections $G_h(F) \to \mathrm{Aut}\,(F)$, resp. $G_\ell(F) \to \mathrm{Aut}\,(C(F))$, which identify the connected components $\mathrm{Aut}\,(F)^o$, resp. $\mathrm{Aut}\,(C(F))^o$, with $G_h(F)/(\text{center})$, resp. $G_\ell(F)/(\text{center})$. However, $W(F)$, being nilpotent, equals $\mathscr{W}(F)(\mathbb{R})$.

5 Statement of the Main Theorem

Let $\mathscr{G}$ be a connected semi-simple linear algebraic group defined over $\mathbb{Q}$ such that $G = \mathscr{G}(\mathbb{R})^o$ is the connected component of the group of automorphisms of a hermitian symmetric domain D.

For a boundary component F of D, we get the associated groups

$$N(F), W(F), U(F) \qquad \text{(see Section 4)} ;$$

if F is rational, these are algebraic groups defined over $\mathbb{Q}$.

Let $\Gamma \subset G$ be an arithmetic group, i.e., $\Gamma \subset \mathscr{G}(\mathbb{Q})$, and, for any faithful rational representation $\rho : \mathscr{G} \longrightarrow \mathrm{GL}(n)$, the subgroup $\rho(\Gamma)$ of $\mathrm{GL}(n, \mathbb{Q})$ is commensurable with $\rho(G) \cap \mathrm{GL}(n, \mathbb{Z})$. If F is a rational boundary component of D, we let

$$\Gamma_F = \Gamma \cap N(F) ,$$

$\Gamma'_F = $ subgroup of elements of Γ_F acting trivially

by conjugation on $\mathfrak{u}(F)$,

$\overline{\Gamma}_F = $ group of automorphisms of $\mathfrak{u}(F)$ induced by Γ_F ;

these map $C(F)$ into itself ,

$$U(F)_{\mathbb{Z}} = \Gamma \cap U(F) .$$

Note that we have an exact sequence of groups:

$$1 \to \Gamma'_F \to \Gamma_F \to \overline{\Gamma}_F \to 1 .$$

Let $\{\sigma_\alpha\}$ be a $\overline{\Gamma}_F$-admissible polyhedral decomposition of $C(F) \subset U(F)$ (see Chapter II, Definition 4.10).

We now construct a "partial compactification in the direction F" of $D/U(F)_{\mathbb{Z}}$,

$$(D/U(F)_{\mathbb{Z}})_{\{\sigma_\alpha\}} .$$

Reconsider the fiber bundle from Section 4:

$$D(F)(= U(F)_{\mathbb{C}} \cdot D)) \xrightarrow{\pi_1} D(F)'(= D(F)/U(F)_{\mathbb{C}}) ,$$

where π_1 makes $D(F)$ into a principal fiber bundle over $D(F)'$ with structure group $U(F)_{\mathbb{C}}$.

From this, we get a quotient fiber bundle $D(F)/U(F)_{\mathbb{Z}} \xrightarrow{\overline{\pi}_1} D(F)'$, where $\overline{\pi}_1$ is a principal fiber bundle under the algebraic torus (over $\mathbb{C}$)

$$T(F) = U(F)_{\mathbb{C}}/U(F)_{\mathbb{Z}} .$$

The chosen collection $\{\sigma_\alpha\}$ defines an equivariant embedding (see Chapter I, Section 1)

$$T(F) \subset T(F)_{\{\sigma_\alpha\}} .$$

We form

$$(D(F)/U(F)_{\mathbb{Z}}) \times^{T(F)} T(F)_{\{\sigma_\alpha\}} \ ;$$

this is the fiber bundle over $D(F)'$ associated to $\bar{\pi}_1$ with fiber $T(F)_{\{\sigma_\alpha\}}$. Finally, define:

$$(D/U(F)_{\mathbb{Z}})_{\{\sigma_\alpha\}} = \text{interior of closure of } D/U(F)_{\mathbb{Z}} \text{ in}$$

$$(D(F)/U(F)_{\mathbb{Z}}) \times^{T(F)} T(F)_{\{\sigma_\alpha\}} \ .$$

Note that, since $\{\sigma_\alpha\}$ is a $\bar{\Gamma}_F$-invariant collection, the group Γ_F is still acting on $(D/U(F)_{\mathbb{Z}})_{\{\sigma_\alpha\}}$; we will see later that, in fact, $\Gamma_F/U(F)_{\mathbb{Z}}$ acts properly discontinuously on $(D/U(F)_{\mathbb{Z}})_{\{\sigma_\alpha\}}$.

Moreover, recall from Section 4 that we have a map

$$\Phi : D(F) \longrightarrow U(F)$$

such that

(a) Φ is equivariant for $N(F) \cdot U(F)_{\mathbb{C}}$, where $N(F)$ acts on $U(F)$ by inner automorphisms and $ia \in iU(F)$ acts on $U(F)$ by translation by a ;
(b) $\Phi^{-1}(C(F)) = D$;
(c) Φ induces a trivialization of the $U(F)_{\mathbb{C}}$-bundle π_1 in the imaginary direction:

$$D(F)/U(F) \xrightarrow[(\Phi,\pi_1)]{\sim} U(F) \times D(F)' \ .$$

Therefore, denoting as in Chapter I, Section 1 by $T(F)_c$ the maximal compact torus in $T(F)$, i.e., $U(F)/U(F)_{\mathbb{Z}}$, we see that Φ induces a trivialization in the "absolute value direction" of the $T(F)$-bundle $\bar{\pi}_1$;
(d) $[D(F)/U(F)_{\mathbb{Z}}]/T(F)_c \xrightarrow{\sim} U(F) \times D(F)'$.

Moreover, Φ extends continuously to maps (which we also call Φ):

$$(D(F)/U(F)_{\mathbb{Z}}) \times^{T(F)} T(F)_{\{\sigma_\alpha\}} \quad \xdashrightarrow{\Phi} \quad U(F)_{\{\sigma_\alpha\}}$$
$$\cup \qquad\qquad\qquad\qquad\qquad\qquad \cup$$
$$D(F)/U(F)_{\mathbb{Z}} \qquad\qquad \xrightarrow{\Phi} \qquad U(F)$$

by the definition

$$\Phi(x,y) = \Phi(x) \cdot \text{ord}(y) \ .$$

Here "$\cdot$" denotes the action of $U(F)$ on the partial compactification $U(F)_{\{\sigma_\alpha\}}$, and this definition is justified by the second equivariance assertion in (a). As in Section 1, define

$$C(F)'' = \text{interior of closure of } C(F) \text{ in } U(F)_{\{\sigma_\alpha\}} \ .$$

The quotient

$$\left((D(F)/U(F)_{\mathbb{Z}}) \times^{T(F)} T(F)_{\{\sigma_\alpha\}} \right) \Big/ T(F)_c$$

is a fiber bundle over $D(F)'$ with fiber $U(F)_{\{\sigma_\alpha\}}$. It is, again via Φ, just the product

$$U(F)_{\{\sigma_\alpha\}} \times D(F)' .$$

It therefore follows easily from (b) that $(D/U(F)_{\mathbb{Z}})_{\{\sigma_\alpha\}}$ can also be described as $\Phi^{-1}(C(F)'')$ as follows:

$$
\begin{array}{ccc}
(D/U(F)_{\mathbb{Z}})_{\{\sigma_\alpha\}} = \Phi^{-1}(C(F)'') & \xrightarrow{\;\Phi\;} & C(F)'' \\
\cap & & \cap \\
(D(F)/U(F)_{\mathbb{Z}}) \times^{T(F)} T(F)_{\{\sigma_\alpha\}} & \xrightarrow{\;\Phi\;} & U(F)_{\{\sigma_\alpha\}} .
\end{array}
$$

These maps will be useful later in the study of $(D/U(F)_{\mathbb{Z}})_{\{\sigma_\alpha\}}$.

The following definition describes the simplicial data on which our compactification will depend.

Definition 5.1 A Γ-*admissible collection* of polyhedra $\{\sigma_\alpha\}$ is a collection of polyhedra

$$\{\sigma_\alpha^F\} \subset \overline{C(F)} ,$$

one for every rational boundary component F of D, which are $\overline{\Gamma}_F$-admissible and which satisfy the following two compatibility conditions:

(1) if $F_1 = \gamma \cdot F_2$ with $\gamma \in \Gamma$, then

$$\left\{ \sigma_\alpha^{F_1} \right\} = \left\{ \gamma \cdot \sigma_\alpha^{F_2} \right\}$$

via the natural isomorphism

$$\gamma : C(F_2) \xrightarrow{\;\sim\;} C(F_1) ;$$

(2) if $F_1 \subset \overline{F}_2$, then

$$\left\{ \sigma_\alpha^{F_2} \right\} \text{ is exactly the set of cones } \sigma_\alpha^{F_1} \cap \overline{C(F_2)}$$

(recall that $\overline{C(F_2)} = \overline{C(F_1)} \cap U(F_1)$ (see Subsection 4.4)).

We can now formulate our main theorem.

Theorem 5.2 (Main Theorem I) *Let $\mathscr{G}$ be a semi-simple algebraic group defined over $\mathbb{Q}$ such that $G = \mathscr{G}(\mathbb{R})^\circ$ is the connected component of the automorphism group of a hermitian symmetric domain D. Let $\Gamma \subset G$ be an arithmetic group. Let $\{\sigma_\alpha\}$ be a Γ-admissible collection of polyhedra. Then there exists* *a unique Hausdorff analytic variety $\overline{D/\Gamma}$ containing D/Γ as an open dense subset and such that, for every rational boundary component F of D, there are open analytic morphisms π_F making the following diagram commutative:*

$$
\begin{array}{ccc}
D/U(F)_{\mathbb{Z}} & \hookrightarrow & (D/U(F)_{\mathbb{Z}})_{\{\sigma_\alpha^F\}} \\
\downarrow & & \Big| \pi_F \\
& & \downarrow \\
D/\Gamma & \hookrightarrow & \overline{D/\Gamma}
\end{array}
$$

and such that every point of $\overline{D/\Gamma}$ is in the image of one of the maps π_F. Furthermore, $\overline{D/\Gamma}$ is a compact algebraic space.

Proof of uniqueness Set $\widetilde{D/\Gamma} = \bigsqcup_F (D/U(F)_{\mathbb{Z}})_{\{\sigma_\alpha^F\}}$. Then $\overline{D/\Gamma}$ is the quotient of $\widetilde{D/\Gamma}$ by an equivalence relation which is described by a *closed* graph (closed because $\overline{D/\Gamma}$ is a Hausdorff space):

$$\Lambda \subset \widetilde{D/\Gamma} \times \widetilde{D/\Gamma}.$$

But because the π_F are open, Λ is just the closure of the equivalence relation defined on $\bigsqcup_F D/U(F)_{\mathbb{Z}}$ by the action of Γ; this shows that $\overline{D/\Gamma}$ is unique. $\square$

The fact that $\overline{D/\Gamma}$ is an algebraic space comes from the following by-product of the proof of the Main Theorem.

Proposition 5.3 *There exists a natural morphism from $\overline{D/\Gamma}$ to the Baily–Borel "minimal" compactification $(D/\Gamma)^*$, inducing the identity morphism on D/Γ.*

Now, $(D/\Gamma)^*$ is a projective algebraic variety; hence this proposition implies that $\overline{D/\Gamma}$ is a Moishezon space, i.e., an algebraic space over $\mathbb{C}$.

Note also that, to prove the proposition, it suffices to construct a continuous map from $\overline{D/\Gamma}$ to $(D/\Gamma)^*$: the fact that this map is holomorphic will then follow from the Riemann extension theorem.

To construct $\overline{D/\Gamma}$, we shall construct the equivalence relation Λ explicitly and show in Section 6 that Λ is closed.

We use the following straightforward lemma.

Lemma 5.4 *Let F and F' be two rational boundary components such that $F' \subset \overline{F}$. Then*

(1) *$U(F') \subset U(F) \subset N(F')$ and $T(F') \subset T(F)$;*
(2) *$T(F)_{\{\sigma_\alpha^{F'}\}} \cong T(F) \times^{T(F')} T(F')_{\{\sigma_\alpha^{F'}\}}$, an open subset of $T(F)_{\{\sigma_\alpha^F\}}$;*

(3) *the quotient by the action of* $U(F)_{\mathbb{Z}}$ *on the left factor of*

$$(D(F')/U(F')_{\mathbb{Z}}) \times^{T(F')} T(F')_{\{\sigma_{\alpha}^{F'}\}}$$

is canonically isomorphic to an open subset of

$$(D(F)/U(F)_{\mathbb{Z}}) \times^{T(F)} T(F)_{\{\sigma_{\alpha}^{F'}\}} \, ,$$

which is an open subset of

$$(D(F)/U(F)_{\mathbb{Z}}) \times^{T(F)} T(F)_{\{\sigma_{\alpha}^{F}\}} \, ;$$

(4) *this induces an étale map*

$$(D/U(F')_{\mathbb{Z}})_{\{\sigma_{\alpha}^{F'}\}} \longrightarrow (D/U(F)_{\mathbb{Z}})_{\{\sigma_{\alpha}^{F}\}} \, .$$

$\square$

Now introduce the following equivalence relation on $\widetilde{D/\Gamma}$. Let

$$x_1 \in (D/U(F_1)_{\mathbb{Z}})_{\{\sigma_{\alpha}^{F_1}\}} \text{ and } x_2 \in (D/U(F_2)_{\mathbb{Z}})_{\{\sigma_{\alpha}^{F_2}\}} \, .$$

Then $x_1 \sim x_2$ if and only if

(1) there is a rational boundary component F and an element $\gamma \in \Gamma$, such that

$$F_1 \subset \overline{F} \quad \text{and} \quad \gamma \cdot F_2 \subset \overline{F} \, ;$$

(2) there is a point $x \in (D/U(F)_{\mathbb{Z}})_{\{\sigma_{\alpha}^{F}\}}$ that

   (a) projects to x_1 via the canonical map defined by Lemma 5.4,

$$(D/U(F)_{\mathbb{Z}})_{\{\sigma_{\alpha}^{F}\}} \longrightarrow (D/U(F_1)_{\mathbb{Z}})_{\{\sigma_{\alpha}^{F_1}\}} \, ,$$

   (b) projects to $\gamma \cdot x_2$ via the canonical map

$$(D/U(F)_{\mathbb{Z}})_{\{\sigma_{\alpha}^{F}\}} \longrightarrow (D/U(\gamma \cdot F_2)_{\mathbb{Z}})_{\{\sigma_{\alpha}^{\gamma \cdot F_2}\}} \, .$$

The transitivity condition of this relation is an easy consequence of Lemma 5.4 plus the following result.

Lemma 5.5 *Let*

$$x \in (D/U(F)_{\mathbb{Z}})_{\{\sigma_{\alpha}^{F}\}} \, .$$

Among all rational boundary components F' such that there is some $x' \in (D/U(F')_{\mathbb{Z}})_{\{\sigma_{\alpha}^{F'}\}}$ projecting to $x \in (D/U(F)_{\mathbb{Z}})_{\{\sigma_{\alpha}^{F}\}}$ via the canonical map

$$(D/U(F')_{\mathbb{Z}})_{\{\sigma_{\alpha}^{F'}\}} \longrightarrow (D/U(F)_{\mathbb{Z}})_{\{\sigma_{\alpha}^{F}\}} \, ,$$

there is a maximal one F_x, i.e., $F' \subset \overline{F}_x$.

We will call this boundary component F_x the *boundary component associated with x* or say that x *belongs to* the F_x-stratum.

Proof To the point x there is associated, in a canonical way, a polyhedron $\sigma_\alpha \subset U(F)$: it is the unique polyhedron such that, via the map Φ defined above,

$$\Phi(x) = z + \infty \cdot \sigma_\alpha$$

(in the notation of Chapter I).

Let $C_x \subset C(F)$ be the smallest boundary component such that $\sigma_\alpha \subset \overline{C}_x$. This defines a rational boundary component F_x such that

$$F \subset \overline{F}_x \, , \, C_x = C(F_x) \, ;$$

see Subsection 4.4. It is immediate that there exists a point in $(D/U(F_x)_\mathbb{Z})_{\{\sigma_\alpha^{F_x}\}}$ projecting to x via the canonical map

$$(D/U(F_x)_\mathbb{Z})_{\{\sigma_\alpha^{F_x}\}} \longrightarrow (D/U(F)_\mathbb{Z})_{\{\sigma_\alpha^F\}} \, .$$

This argument also shows that F_x is the maximal element with the required properties. $\square$

6 Proof of the Main Theorem

6.1

Let $\mathscr{G}$ be a connected semi-simple algebraic group defined over $\mathbb{Q}$ such that $D = \mathscr{G}(\mathbb{R})^o/K$ is a bounded symmetric domain. Let D^* be the union of D and its rational boundary components. In Chapter II, Subsection 4.1, we defined Siegel sets $\mathfrak{S}_\omega \subset D$, associated to a minimal $\mathbb{Q}$-parabolic subgroup $\mathscr{P} \subset \mathscr{G}$ and a relatively compact subset $\omega \subset \mathscr{P}(\mathbb{R})^o$. Our first goal is to analyze the "Satake topology" on D^*, introduced by Satake (see Baily and Borel [1], §4.8), using Siegel sets.

The *Satake topology* is a topology on D^* which is different from the topology induced via the Harish-Chandra embedding from the vector space topology on $\mathfrak{p}_+$. To define it, take any arithmetic subgroup $\Gamma \subset \mathscr{G}(\mathbb{Q})$ and choose any fundamental set for Γ:

$$\Omega = C \cdot \mathfrak{S}_\omega \, ,$$

where $C \subset \mathscr{G}(\mathbb{Q})$ is a finite subset and where $\mathfrak{S}_\omega$ is a Siegel set with respect to a minimal $\mathbb{Q}$-parabolic. A fundamental system of neighborhoods of $x \in D^*$ is, by definition, given by all subsets $U \subset D^*$ such that

$$\Gamma_x \cdot U = U \, , \text{ where } \Gamma_x = \{\gamma \in \Gamma \mid \gamma \cdot x = x\} \, ,$$

and such that $\gamma \cdot U \cap \overline{\Omega}$ is a neighborhood of $\gamma \cdot x$ in $\overline{\Omega}$ (for the topology on $\overline{\Omega}$ induced from $\overline{D} \subset \mathfrak{p}_+$), whenever $\gamma \cdot x \in \overline{\Omega}$.

The topology is characterized by the following theorem.

Theorem 6.1 *The Satake topology is the unique topology on D^* having the following properties:*

(i) *it induces the natural topology on D and on the closure $\overline{\mathfrak{S}}_\omega$ of any Siegel set $\mathfrak{S}_\omega$;*

(ii) *the group $\mathcal{G}(\mathbb{Q})$ acts continuously on D^*;*

(iii) *if $x, x' \in D^*$ are not equivalent with respect to the action of an arithmetic group $\Gamma \subset \mathcal{G}(\mathbb{Q})$, then there exist neighborhoods U of x and U' of x' such that $\Gamma \cdot U \cap U' = \emptyset$;*

(iv) *let $\Gamma \subset \mathcal{G}(\mathbb{Q})$ be an arithmetic group. For every $x \in D^*$, there exists a fundamental set of neighborhoods $\{U\}$ of x such that*

$$\gamma \cdot U = U \text{ if } \gamma \cdot x = x \,,$$
$$\gamma \cdot U \cap U = \emptyset \text{ if } \gamma \cdot x \neq x \,.$$

The Baily–Borel compactification $(D/\Gamma)^*$ of D/Γ is the quotient

$$(D/\Gamma)^* = D^*/\Gamma \,,$$

equipped with the quotient topology. Baily and Borel [1] proved:

Theorem 6.2 *The compactification $(D/\Gamma)^*$ is a compact Hausdorff space containing D/Γ as an open dense subset. The compactification $(D/\Gamma)^*$ is a finite union of subspaces of the form*

$$F/\Gamma_F \,,$$

where F is a rational boundary component.

The closure of F/Γ_F in $(D/\Gamma)^*$ is the union of F/Γ_F and of subspaces $F'/\Gamma_{F'}$, of strictly smaller dimension.

We wish to describe explicitly the Satake topology in terms of the coordinates on D given by its presentation as a Siegel domain, and, in particular, we want to show that a fundamental system of open sets can be given using the concept of rational core of a cone (see Chapter II, Subsection 5.1). The result is as follows.

Theorem 6.3 *Let F be a rational boundary component of D and let $x \in F$. Let*

$C_0(F) \subset C(F)$ *be a rational core. Let* $U \subset D$ *be an open set. Then*

$$
\begin{bmatrix}
\text{there exists a} \\
\text{neighborhood } V \text{ of } x \in D^* \\
\text{in the Satake topology} \\
\text{such that } U \supset V \cap D
\end{bmatrix}
\Longleftrightarrow
\begin{bmatrix}
\text{there exists a} \\
\text{neighborhood } E \text{ of } x \in F \\
\text{and } 1 \le t < \infty \text{ , such that} \\
U \supset \pi_F^{-1}(E) \cap \Phi_F^{-1}(tC_0)
\end{bmatrix} .
$$

Proof Fix an arithmetic group $\Gamma \subset \mathcal{G}(\mathbb{Q})$, and let $\Gamma_x = \{\gamma \in \Gamma \mid \gamma x = x\}$. On the LHS, we can assume V is Γ_x-invariant; on the RHS, we can assume C_0 is $\overline{\Gamma}_F$-invariant and E is Γ_x-invariant. Thus we are comparing Γ_x-invariant open subsets of D, namely $V \cap D$ and $\pi_F^{-1}(E) \cap \Phi_F^{-1}(C_0)$, so we may as well assume U is Γ_x-invariant. Similarly, if $\mathcal{G} = \mathcal{G}_1 \times \mathcal{G}_2$ over $\mathbb{Q}$, and this corresponds to $D = D_1 \times D_2$, then the Satake topology on D is the product of the Satake topology on D_1 and on D_2; hence, both the sets $V \cap D$ and the sets

$$
\pi_F^{-1}(E) \cap \Phi_F^{-1}(C_0)
$$

can be assumed to be products. Therefore we may as well assume U is a product too, and thus the result for D_1 and D_2 implies it for D. So we may assume $\mathcal{G}$ is $\mathbb{Q}$-simple. We start on the left:

$$
U \supset V \cap D \Longleftrightarrow \text{ for all Siegel sets } \mathfrak{S}_\omega \subset D \text{ with } x \in \overline{\mathfrak{S}}_\omega,
$$
$$
\overline{U \cap \mathfrak{S}_\omega} \text{ is a neighborhood of } x \text{ in } \overline{\mathfrak{S}}_\omega .
$$

Here we mean any Siegel set $\mathfrak{S}_\omega$ with respect to any minimal $\mathbb{Q}$-parabolic $\mathcal{P}$. This is straightforward.

Next we need a lemma about when $x \in \overline{\mathfrak{S}}_\omega$.

Lemma 6.4 *If* $\mathfrak{S}_\omega$ *is a Siegel set with respect to* $\mathcal{P}$, *and* F *is a boundary component of* D, *then*

$$
\overline{\mathfrak{S}}_\omega \cap F \ne \emptyset \Longleftrightarrow \mathcal{P} \subset \mathcal{N}(F) \text{ and } F \text{ is rational .}
$$

Proof Let $\mathcal{A} \subset \mathcal{P}$ be a maximal $\mathbb{Q}$-split torus, $A = \mathcal{A}(\mathbb{R})^o$ and A^+ the positive piece, see Chapter II, Definition 4.1. Let K be a maximal compact subgroup such that $\operatorname{Lie} K \perp \operatorname{Lie} A$, and let p be fixed by K. Then it suffices to show

$$
\overline{A^+ p} \cap F \ne \emptyset \Longleftrightarrow \mathcal{P} \subset \mathcal{N}(F) , \ F \text{ rational },
$$

because, if $\omega \subset \mathcal{P}(\mathbb{R})$ is compact with $\overline{\mathfrak{S}}_\omega = \omega \cdot \overline{A^+ p}$, then ω leaves fixed every F such that $\mathcal{P} \subset \mathcal{N}(F)$. Let $s = \dim A$, and let

$$
\tfrac{1}{2}(\beta_1 - \beta_2), \tfrac{1}{2}(\beta_2 - \beta_3), \dots, \tfrac{1}{2}(\beta_{s-1} - \beta_s), \beta_s, \text{ or } \tfrac{1}{2}\beta_s
$$

be the simple positive roots (ordering defined by $\mathcal{P}$) as usual. We constructed in Subsection 3.5 symmetric holomorphic maps

$$
\begin{array}{ccc}
\mathfrak{H}^s & \xrightarrow{\ f_1\ } & D \\
\cap & & \cap \\
(\mathbb{P}^1)^s & \xrightarrow{\ f_2\ } & \check{D}
\end{array}
$$

associated to

$$\varphi : U^1 \times \mathrm{SL}(2,\mathbb{R})^s \to G$$

with $\varphi(1, \text{diag. matrices}) = A$ and with

$$x = \varphi\left(1, \begin{pmatrix} t_1 & 0 \\ 0 & t_1^{-1} \end{pmatrix}, \ldots, \begin{pmatrix} t_s & 0 \\ 0 & t_s^{-1} \end{pmatrix}\right) \in A ,$$

such that $\beta_i(x) = t_i^2$. Then A^+ is the image of elements

$$\begin{pmatrix} t_1 & 0 \\ 0 & t_1^{-1} \end{pmatrix}, \ldots, \begin{pmatrix} t_s & 0 \\ 0 & t_s^{-1} \end{pmatrix} ,$$

where $t_1 \geq t_2 \geq \cdots \geq t_s \geq 1$. Then

$$\overline{A^+ p} = f_2\left(\{(ix_1, \ldots, ix_s) \mid \infty \geq x_1 \geq \cdots \geq x_s \geq 1\}\right) ,$$

i.e., $\overline{A^+ p}$ meets the boundary components $F_1, \ldots, F_s$, where

$$f_2(\infty, i, \ldots, i) \in F_1, f_2(\infty, \infty, i, \ldots, i) \in F_2, \ldots, f_2(\infty, \infty, \ldots, \infty) \in F_s .$$

Now, if

$$w_i(t) = \varphi(1, \underbrace{\begin{pmatrix} t & 0 \\ 0 & t^{-1} \end{pmatrix}, \ldots, \begin{pmatrix} t & 0 \\ 0 & t^{-1} \end{pmatrix}}_{i \text{ factors}}, I, \ldots, I) ,$$

then all but the i'th simple root vanishes on w_i; hence, $\mathscr{P} = \mathscr{P}(w_1^{-1}) \cap \cdots \cap \mathscr{P}(w_r^{-1})$ expresses $\mathscr{P}$ as an intersection of maximal $\mathbb{Q}$-parabolics, and it is immediate that $\mathscr{P}_i = \mathscr{N}(F_i)$. $\qquad\square$

Next, write out the algebraic connected component of $\mathscr{N}(F)$ as

$$\widetilde{\mathscr{G}}_h(F) \cdot \mathscr{G}_\ell(F) \cdot \mathscr{W}(F) ,$$

where all these factors are defined over $\mathbb{Q}$, $\widetilde{\mathscr{G}}_h(F)(\mathbb{R})^o = G_h(F) \cdot M(F)$ in the notation of Section 4, and $\mathscr{W}$ is the unipotent radical. Then a minimal $\mathbb{Q}$-parabolic $\mathscr{P} \subset \mathscr{N}(F)$ is equal to

$$\mathscr{P} = \widetilde{\mathscr{P}}_h \cdot \mathscr{P}_\ell \cdot \mathscr{W} ,$$

where $\widetilde{\mathscr{P}}_h \subset \widetilde{\mathscr{G}}_h$ and $\mathscr{P}_\ell \subset \mathscr{G}_\ell$ are minimal $\mathbb{Q}$-parabolics. Then, for any base-point $o \in D$, a cofinal collection of Siegel sets for $\mathscr{P}$ is given as

$$\mathfrak{S}_\omega = (\omega_W \cdot \omega_h \cdot \omega_\ell) \cdot A^+ \cdot o ,$$

where

$$\omega_h \subset \widetilde{P}_h = \widetilde{\mathscr{P}}_h(\mathbb{R})^o \ , \ \omega_\ell \subset P_\ell = \mathscr{P}_\ell(\mathbb{R})^o \ , \ \omega_W \subset W = \mathscr{W}(\mathbb{R})$$

are compact, and where $\mathscr{A} \subset \mathscr{P}$ is a maximal $\mathbb{Q}$-split torus conjugate to the original maximal split torus in $\mathscr{P}$, with Lie $\mathscr{A} \perp \mathrm{Stab}\,(o)$, and where

$$A^+ = \left\{ g \in A = \mathscr{A}(\mathbb{R})^o \ \middle| \ \begin{array}{c} \beta(g) \geq 1 \text{ for all positive roots } \beta \text{ of } A \\ \text{i.e., roots occurring in Lie } \mathscr{P} \end{array} \right\} .$$

Note that $A = A_h \cdot A_\ell$, where $A_h \subset \widetilde{P}_h$ and $A_\ell \subset P_\ell$ are conjugates of the maximal $\mathbb{Q}$-split tori in $\widetilde{\mathscr{P}}_h$ and $\mathscr{P}_\ell$. Moreover, the root system of A (written multiplicatively, i.e., as a subset $\Phi \subset \mathrm{Hom}\,(A, \mathbb{R}_{>0})$) is given by

$$\begin{array}{ccc} \beta_i^{\pm 1}, & \beta_i^{\pm 1/2} \beta_j^{\pm 1/2} & (\text{plus possibly } \beta_i^{\pm 1/2}) , \\ 1 \leq i \leq s & 1 \leq i < j \leq s & 1 \leq i \leq s \end{array}$$

with simple positive roots

$$\beta_1^{1/2} \beta_2^{-1/2}, \beta_2^{1/2} \beta_3^{-1/2}, \ldots, \beta_{s-1}^{1/2} \beta_s^{-1/2}, \text{ and } \beta_s \text{ or } \beta_s^{1/2} ,$$

and with

$$A_h = \{ g \in A \mid \beta_1(g) = \cdots = \beta_u(g) = 1 \}$$
$$A_\ell = \{ g \in A \mid \beta_{u+1}(g) = \cdots = \beta_s(g) = 1 \} .$$

Our next step is the following lemma.

Lemma 6.5 *Assume ω_ℓ is large enough so that $x \in \mathrm{Int}_F (\omega_\ell \cdot \pi_F(o))$. Let $U_n \subset G_h(F)$ be sets such that $U_n \cdot \pi_F(o)$ is a fundamental system of neighborhoods of $x \in F$, for $n = 1, 2, \ldots$. Then the intersection with $\mathfrak{S}_\omega$ of a fundamental system of neighborhoods of $x \in \overline{\mathfrak{S}}_\omega$ is given by*

$$S_{n,t} = \omega_W \cdot U_n \cdot \omega_\ell \cdot A_{\ell,t} \cdot o ,$$

where

$$A_{\ell,t} = \{ x \in A_\ell \mid \beta_1(x) \geq \beta_2(x) \geq \cdots \geq \beta_u(x) \geq t \} .$$

Proof Recall from Section 3 that we can map $\mathrm{SL}(2, \mathbb{R})^u \longrightarrow G$, taking diagonal matrices $\delta = \left(\begin{pmatrix} t_1 & 0 \\ 0 & t_1^{-1} \end{pmatrix}, \ldots \right)$ to elements $x \in A_\ell$ with $\beta_i(x) = t_i^2$, and from this obtain a homomorphism

$$\varphi : G_h(F) \times \mathrm{SL}(2, \mathbb{R})^u \longrightarrow G$$

plus a symmetric holomorphic map for φ:

$$
\begin{array}{ccc}
F \times \mathfrak{H}^u & \xrightarrow{\ f_1\ } & D \\
\cap & & \cap \\
\check{F} \times (\mathbb{P}^1)^u & \xrightarrow{\ f_2\ } & \check{D} .
\end{array}
$$

Consider the map

$$
g : \omega_W \times \omega_\ell \times \overline{F} \times A'_{\ell,1} \longrightarrow \overline{D} ,
$$

where

$$
A'_{\ell,t} = \{ \underline{t} = (t_1, \dots, t_u) \mid \infty \geq t_1 \geq t_2 \geq \cdots \geq t_u \geq t^{1/2} \} ,
$$

given by

$$
g(u, v, a, i\underline{t}) = u \cdot v \cdot f_2(a, i\underline{t}).
$$

Note that $\operatorname{Im} g$ is compact, hence closed, and that $\mathfrak{S}_\omega \subset \operatorname{Im} g$. Therefore, $\overline{\mathfrak{S}}_\omega \subset \operatorname{Im} g$, which is why we have introduced g. On the other hand, $f_2^{-1}(x) = \{(x, \infty, \dots, \infty)\}$, and hence

$$
g^{-1}(x) = \omega_W \times \omega_\ell \times \{x\} \times \{(\infty, \dots, \infty)\} .
$$

Since ω_W and ω_ℓ are compact, any open subset of $\omega_W \times \omega_\ell \times \overline{F} \times A'_{\ell,1}$ containing $g^{-1}(x)$ contains

$$
\omega_W \times \omega_\ell \times \{U_n \cdot \pi_F(o)\} \times A'_{\ell,t}
$$

for some n and t; and, since g is a proper map, the images $S_{n,t}$ by g of these sets are a fundamental system of neighborhoods in $\operatorname{Im} g$ of x. But, for n large, these are subsets of $\overline{\mathfrak{S}}_\omega$, hence the lemma follows. $\qquad\square$

Corollary 6.6 *Let $U \subset D$ be an open subset. Then there exists an open neighborhood V of $x \in D^*$ in the Satake topology with $U \supset V \cap D$ if and only if, for all minimal $\mathbb{Q}$-parabolics $\mathscr{P}_\ell \subset \mathscr{G}_\ell$ and all compact subsets $\omega_W \subset W$ and $\omega_\ell \subset P_\ell$,*

$$
U \supset S_{n,t}
$$

for some $n \geq 1$, $t \geq 1$, where $S_{n,t}$ are defined as in Lemma 6.5 above. $\qquad\square$

We now use the product representation

$$
D \cong F \times C(F) \times W(F)
$$

introduced in Subsection 4.3. In this representation

$$
S_{n,t} = (U_n \cdot \pi_F(o)) \times (\omega_\ell \cdot A_{\ell,t} \cdot \Omega_F) \times \omega_W ;
$$

here Ω_F is the basepoint of $C(F)$. If $E_n = U_n \cdot \pi_F(o)$, then, by definition, the

E_n are a fundamental set of neighborhoods of x in F. Now U is Γ_x-invariant, and hence $U \supset S_{n,t}$ implies

$$U \supset E_n \times (\overline{\Gamma}_F \cdot \omega_\ell \cdot A_{\ell,t} \cdot \Omega_F) \times (\Gamma \cap W) \cdot \omega_W .$$

If ω_W is large enough, $(\Gamma \cap W) \cdot \omega_W = W$, so there is no condition on the third factor. Moreover, if we take a finite union of the sets in the middle factor, one for each $\overline{\Gamma}_F$-conjugacy class of minimal $\mathbb{Q}$-parabolics $\mathscr{P}_\ell \subset \mathscr{G}_\ell$, while taking the ω_ℓ large enough, then

$$\bigcup_i \overline{\Gamma}_F \cdot \omega_\ell^{(i)} \cdot A_\ell^{(i)} \cdot \Omega_F = C(F) ,$$

while

$$\bigcup_i \overline{\Gamma}_F \cdot \omega_\ell^{(i)} \cdot A_{\ell,t}^{(i)} \cdot \Omega_F = t \cdot C_0(F) ,$$

where $C_0(F)$ lies between two rational cores of $C(F)$. Thus the condition on U is equivalent to

$$U \supset \pi_F^{-1}(E_n) \cap \Phi_F^{-1}(t C_0(F)) \text{ for some } n \geq 1, t \geq 1 .$$

This concludes the proof of Theorem 6.3.   □

In fact, if one adjoins to $\pi_F^{-1}(E) \cap \Phi_F^{-1}(t C_0(F))$ a suitable set of elements of the rational boundary components F' with $\overline{F'} \supset F$, then one can prove that these extended sets, as F, E, and t vary (E is open and relatively compact in F), form a *basis of open sets for the Satake topology on D^**. However, we do not need this additional fact.

Combining the above proof with the results of Chapter II, Subsection 4.3, we find:

Corollary 6.7 *Every Siegel set $\mathfrak{S}_\omega \subset D$ is covered by a finite number of sets of the form*

$$E \times (C_0 \cap \sigma_\alpha^F) \times \omega_W \subset F \times C(F) \times W(F) \cong D ,$$

where $E \subset F$ and $\omega_W \subset W(F)$ are compact, $C_0 \subset C(F)$ is a rational core, and σ_α^F is one of the polyhedra in our decomposition of $C(F)$.   □

6.2

The next step is to define a map

$$f : \widetilde{D/\Gamma} = \bigsqcup_F (D/U(F)_\mathbb{Z})_{\{\sigma_\alpha^F\}} \longrightarrow (D/\Gamma)^* .$$

Let $x \in (D/U(F)_{\mathbb{Z}})_{\{\sigma_\alpha^F\}}$ and let F_x be its associated rational boundary component; hence there exists a point $x' \in (D/U(F_x)_{\mathbb{Z}})_{\{\sigma_\alpha^{F_x}\}}$ projecting to x via the canonical map

$$(D/U(F_x)_{\mathbb{Z}})_{\{\sigma_\alpha^{F_x}\}} \longrightarrow (D/U(F)_{\mathbb{Z}})_{\{\sigma_\alpha^F\}} \; .$$

We then define the image via f of x as the image of x' via the succession of maps

$$(D/U(F_x)_{\mathbb{Z}})_{\{\sigma_\alpha\}} \xrightarrow{\pi_{F_x}} F_x \subset D^* \longrightarrow (D/\Gamma)^* \; .$$

The following facts are immediate.

(1) The definition of $f(x)$ is independent of the choice of x'.

(2) The restriction $f|_D$ is just the natural projection from D to $D/\Gamma \subset (D/\Gamma)^*$.

(3) Two points in $\widetilde{D/\Gamma}$ that are equivalent under the equivalence relation introduced in Section 5 have the same image in $(D/\Gamma)^*$. Indeed, $F_{\gamma \cdot x} = \gamma \cdot F_x$, for all $\gamma \in \Gamma$.

(4) $f|_{(D/U(F)_{\mathbb{Z}})_{\{\sigma_\alpha\}}}$ factors as

$$(D/U(F)_{\mathbb{Z}})_{\{\sigma_\alpha\}} \xrightarrow{f_F} D^*/U(F)_{\mathbb{Z}} \longrightarrow (D/\Gamma)^* \; .$$

Proposition 6.8 *If $(D/\Gamma)^*$ and $D^*/U(F)_{\mathbb{Z}}$ are given the Satake topology, the map*

$$f : \widetilde{D/\Gamma} \longrightarrow (D/\Gamma)^*$$

and the maps f_F through which it factors are continuous.

Proof It is clearly sufficient to show that f_F is continuous. We use the following elementary lemma.

Lemma 6.9 *Let $\overline{X}$ and $\overline{Y}$ be two topological spaces which are second countable, $\overline{Y}$ also being a T_3-space, and let $X \subset \overline{X}$, resp. $Y \subset \overline{Y}$, be an open and dense, resp. open, subset. Let*

$$f : \overline{X} \longrightarrow \overline{Y}$$

be a map such that $f(X) \subset Y$. Assume that, for every $x \in \overline{X}$ and $y = f(x) \in \overline{Y}$, and for any neighborhood V of y, $f^{-1}(V \cap Y)$ contains a set of the form $U \cap X$, where U is a neighborhood of x. Then f is continuous.

Proof We have to show that if

$$x_1, x_2, \ldots \longrightarrow x \; ,$$

then

$$y_1 = f(x_1), y_2 = f(x_2), \ldots \longrightarrow y = f(x) .$$

Suppose first that $x_i \in X \subset \overline{X}$. Let V be any neighborhood of y and choose a neighborhood U of x as in the hypotheses of the lemma. Then, for large enough i, we have

$$x_i \in U \cap X .$$

Hence we get $y_i \in f(U \cap X) \subset V$ for $i \gg 0$, i.e., $y_i \longrightarrow y$.

Now consider the general case. Since $X \subset \overline{X}$ is dense, we may find convergent sequences

$$x_{i1}, x_{i2}, \ldots \longrightarrow x_i \text{ with } x_{ij} \in X .$$

Let V be an arbitrary open neighborhood of y and let V_1 be a neighborhood of y such that

$$V_1 \subset \overline{V}_1 \subset V .$$

(We use the fact that $\overline{Y}$ is a T_3-space.) Denote by U, resp. U_1, the neighborhoods of x whose existence is guaranteed by hypothesis. Then, for $i \gg 0$, we have $x_i \in U_1$. Since U_1 is a neighborhood of x_i for $i \gg 0$, we have

$$x_{ij} \in U_1 \text{ for } i \gg 0, j > J(i) .$$

This shows that $y_{ij} \in V_1$ for $i \gg 0$, $j > J(i)$. Since, by the first part of the proof, the sequence y_{ij} converges to y_i we conclude that $y_i \in \overline{V}_1 \subset V$ for $i \gg 0$. This finishes the proof. □

To verify the assumptions of the lemma for f_F, we use induction on codim F. Therefore we may assume $f_{F'}$ continuous for all F' with $\overline{F'} \supset F$.

Then to check f_F itself is continuous, we need only verify the assumption for x in the F-stratum of $(D/U(F)_{\mathbb{Z}})_{\{\sigma_\alpha^F\}}$. Let $y = f_F(x) \in F$. Then a fundamental system of neighborhoods of y in the Satake topology, intersected with D, is given, according to Theorem 6.3, by the sets $\pi_F^{-1}(E) \cap \Phi_F^{-1}(C_0)$, where $E \subset F$ is a neighborhood of x and where C_0 is a rational core of $C(F)$. But π_F extends to a continuous map

$$\overline{\pi}_F : (D/U(F)_{\mathbb{Z}})_{\{\sigma_\alpha^F\}} \to F ,$$

and Φ_F extends to a continuous map

$$\overline{\Phi}_F : (D/U(F)_{\mathbb{Z}})_{\{\sigma_\alpha^F\}} \longrightarrow C(F)'' .$$

(Recall that $C(F)''$ is the interior of the closure of $C(F)$ in $U(F)_{\{\sigma_\alpha^F\}}$.) Moreover, C_0 contains some cylindrical set $a + C(F) \subset C(F)$, and

$$(a + C(F))'' := \text{interior of closure of } a + C(F) \text{ in } U(F)_{\{\sigma_\alpha^F\}}$$

is an open subset of $C(F)''$ containing all points of $U(F)_{\{\sigma_\alpha^F\}}$ in the F-stratum, i.e., points $x + \infty \cdot \sigma_\alpha^F$, where σ_α^F meets $C(F)$ (or $\sigma_\alpha^F \not\subset \partial C(F)$). Therefore $\pi_F^{-1}(E) \cap \Phi_F^{-1}(C_0)$ (inverse image in $D/U(F)_{\mathbb{Z}}$) contains

$$D/U(F)_{\mathbb{Z}} \cap \underbrace{\overline{\pi_F^{-1}}(E) \cap \overline{\Phi_F^{-1}}((a+C(F))'')}_{\text{inverse image in } (D/U(F)_{\mathbb{Z}})_{\{\sigma_\alpha^F\}}},$$

and the latter terms are neighborhoods of x in $(D/U(F)_{\mathbb{Z}})_{\{\sigma_\alpha^F\}}$. $\qquad\square$

6.3

We now show that the equivalence relation Λ is closed. Since any pair (y, z) of equivalent points in $\widetilde{D/\Gamma}$ is a limit of equivalent points (y_i, z_i), where

$$y_i \in (D/U(F_1)_{\mathbb{Z}}) \subset \widetilde{D/\Gamma},$$

$$z_i \in (D/U(F_2)_{\mathbb{Z}}) \subset \widetilde{D/\Gamma},$$

it suffices to show that, for any two sequences $y_i, z_i \in D$, where $y_i = \gamma_i z_i$ with $\gamma_i \in \Gamma$, such that

$$y_i \bmod U(F_1)_{\mathbb{Z}} \longrightarrow \bar{y} \in (D/U(F_1)_{\mathbb{Z}})_{\{\sigma_\alpha^{F_1}\}},$$

$$z_i \bmod U(F_2)_{\mathbb{Z}} \longrightarrow \bar{z} \in (D/U(F_2)_{\mathbb{Z}})_{\{\sigma_\alpha^{F_2}\}},$$

it follows that $(\bar{y}, \bar{z}) \in \Lambda$. But, by Proposition 6.8, the images of $\bar{y}$ and $\bar{z}$ in $(D/\Gamma)^*$ are equal, and hence lie in the same stratum F/Γ_F of $(D/\Gamma)^*$. By the definition of f, it follows that $\overline{\delta_1 F} \supset F_1$, $\overline{\delta_2 F} \supset F_2$, for some $\delta_1, \delta_2 \in \Gamma$, and where $\bar{y}$ lies in the $\delta_1 F$-stratum, and $\bar{z}$ lies in the $\delta_2 F$-stratum. Replacing $y_i, \bar{y}$ by $\delta_1^{-1} y_i, \delta_1^{-1} \bar{y}$ and $z_i, \bar{z}$ by $\delta_2^{-1} z_i, \delta_2^{-1} \bar{z}$, we can assume $\overline{F} \supset F_1$ and $\overline{F} \supset F_2$.

Next, lift $\bar{y}, \bar{z}$ to points $y^*, z^* \in (D/U(F)_{\mathbb{Z}})_{\{\sigma_\alpha^F\}}$. Since

$$(D/U(F)_{\mathbb{Z}})_{\{\sigma_\alpha^F\}} \longrightarrow (D/U(F_i)_{\mathbb{Z}})_{\{\sigma_\alpha^{F_i}\}}$$

is obtained by dividing by $U(F_i)_{\mathbb{Z}}$ plus an open immersion, it follows that

$$y^* = \lim_{i \to \infty} \lambda_i \cdot y_i, \quad \lambda_i \in U(F_1)_{\mathbb{Z}},$$

$$z^* = \lim_{i \to \infty} \mu_i \cdot z_i, \quad \mu_i \in U(F_2)_{\mathbb{Z}}.$$

So, replacing y_i by $\lambda_i \cdot y_i$, and z_i by $\mu_i \cdot z_i$, and $\bar{y}$ by y^*, and $\bar{z}$ by z^*, it suffices to prove the assertion in the special case:

$$F_1 = F_2 : \text{ call this just } F;$$

$$\bar{y}, \bar{z} \in F\text{-stratum of } (D/U(F)_{\mathbb{Z}})_{\{\sigma_\alpha^F\}}.$$

Now consider the following continuous maps:

$$(D/U(F)_{\mathbb{Z}})_{\{\sigma_\alpha^F\}} \xrightarrow{f_F} D^*/U(F)_{\mathbb{Z}} \longrightarrow (D/\Gamma)^*$$
$$\cup \qquad\qquad\qquad \cup$$
$$F \qquad\longrightarrow\quad F/\Gamma_F\,.$$

Since $\bar{y}, \bar{z}$ have the same image in $(D/\Gamma)^*$, their images in $D^*/U(F)_{\mathbb{Z}}$ differ by $\gamma \in \Gamma_F$. Again, replacing $z_i, \bar{z}$ by $\gamma^{-1}z_i, \gamma^{-1}\bar{z}$, we may assume that $\bar{y}$ and $\bar{z}$ have the same image $p \in F \subset D^*/U(F)_{\mathbb{Z}}$. Let $U_1 \subset D^*$ be a neighborhood of p in the Satake topology such that $\Gamma_p \cdot U_1 = U_1$ and $\gamma U_1 \cap U_1 = \emptyset$ if $\gamma \notin \Gamma_p$. Then

$$U_2 = f_F^{-1}(U_1/U(F)_{\mathbb{Z}})$$

is an open subset of $(D/U(F)_{\mathbb{Z}})_{\{\sigma_\alpha^F\}}$ containing $\bar{y}$ and $\bar{z}$. Therefore y_i mod $U(F)_{\mathbb{Z}}$ and z_i mod $U(F)_{\mathbb{Z}}$ lie in U_2 if $i \gg 0$. In other words, the points y_i, z_i in D lie in U_1 if $i \gg 0$. Now $y_i = \gamma_i z_i$, for $\gamma_i \in \Gamma$, so, by the assumption on U_1, we have $\gamma_i \in \Gamma_p$, hence $\gamma_i \in \Gamma_F$. This reduces us to the following assertion.

Proposition 6.10 *The action of* $\Gamma_F/U(F)_{\mathbb{Z}}$ *on* $(D/U(F)_{\mathbb{Z}})_{\{\sigma_\alpha^F\}}$ *is properly discontinuous. Consequently,* Γ_F*-equivalence is a closed equivalence relation in* $(D/U(F)_{\mathbb{Z}})_{\{\sigma_\alpha^F\}} \times (D/U(F)_{\mathbb{Z}})_{\{\sigma_\alpha^F\}}$.

Proof We use the double fibration:

$$(D/U(F)_{\mathbb{Z}})_{\{\sigma_\alpha^F\}} \xrightarrow{\pi_F'} D(F)' \xrightarrow{p_F} F\,.$$

Recall that $U(F) \subset W(F) \subset N(F)$ are all defined over $\mathbb{Q}$; hence, the quotient group by which $N(F)$ acts on $U(F)$, namely the image of

$$N(F) \longrightarrow \mathrm{Aut}(U(F))\,,$$

is also defined over $\mathbb{Q}$. Let $N(F)' = \mathrm{Ker}(N(F) \longrightarrow \mathrm{Aut}(U(F)))$; this is defined over $\mathbb{Q}$. Finally, $N(F)'/W(F)$ is defined over $\mathbb{Q}$ and equals G_h up to compact factors. Therefore all the following are arithmetic, hence discrete subgroups:

$$\Gamma_F \subset N(F)\,,$$
$$\Gamma_F' \subset N(F)'\,,$$
$$\Gamma \cap W(F) = W(F)_{\mathbb{Z}} \subset W(F)\,,$$
$$\Gamma \cap U(F) = U(F)_{\mathbb{Z}} \subset U(F)\,,$$
$$\bar{\Gamma}_F \subset \mathrm{Aut}(U(F))\,,$$
$$\Gamma_F'/W(F)_{\mathbb{Z}} \subset G_h(F) \cdot (\text{comp. factors})\,.$$

We use the following elementary lemma.

Lemma 6.11 *Let X, Y be two Hausdorff topological spaces acted on by groups G, resp. H, and let*

$$\phi : G \longrightarrow H ,$$
$$f : X \longrightarrow Y$$

be a homomorphism and an equivariant continuous map. If H acts properly discontinuously on Y and $\operatorname{Ker} \phi$ acts properly discontinuously on X, then G acts properly discontinuously on X. ☐

We deduce:

(1) $\Gamma'_F / W(F)_{\mathbb{Z}}$ acts properly discontinuously on F;

hence, since p_F is a principal fiber bundle with structure group $V(F)$,

(2) $\Gamma'_F / U(F)_{\mathbb{Z}}$ acts properly discontinuously on $D(F)'$ by the lemma;

hence,

(3) $\Gamma'_F / U(F)_{\mathbb{Z}}$ acts properly discontinuously on $(D/U(F)_{\mathbb{Z}})_{\{\sigma_\alpha\}}$.

Next we make use of the Γ_F-equivariant map

$$\Phi_F : (D/U(F)_{\mathbb{Z}})_{\{\sigma_\alpha^F\}} \longrightarrow C(F)'' .$$

Since, by the method of proof of Theorem 1.4, $\overline{\Gamma}_F$ acts properly discontinuously on $C(F)''$, the lemma plus (3) above give the proposition.

☐

This proves that Λ is closed. Let $\overline{D/\Gamma}$ be the quotient of $\widetilde{D/\Gamma}$ by the equivalence relation Λ. The above proof shows that $\overline{D/\Gamma}$ is locally isomorphic at all points of the F-stratum to

$$(D/U(F)_{\mathbb{Z}})_{\{\sigma_\alpha^F\}} / (\Gamma_F / U(F)_{\mathbb{Z}}) .$$

Since this is an analytic space modulo a properly discontinuous group action, $\overline{D/\Gamma}$ has an analytic structure, and the maps

$$\pi_F : (D/U(F)_{\mathbb{Z}})_{\{\sigma_\alpha^F\}} \longrightarrow \overline{D/\Gamma}$$

are open. This, together with the fact that the maps f_F are continuous, also shows that the natural map from $\overline{D/\Gamma}$ to $(D/\Gamma)^*$ is continuous.

It remains to check that $\overline{D/\Gamma}$ is compact. Now, D/Γ it is covered by the images of finitely many Siegel sets hence, by Corollary 6.7, is covered by the images of finitely many sets of the form

$$S = E \times (C_0 \cap \sigma_\alpha) \times \omega_W \subset F \times C(F) \times W(F) \cong D .$$

We claim that the closures of the images of these sets in $(D/U(F)_{\mathbb{Z}})_{\{\sigma_\alpha^F\}}$ are already compact, and hence so are the closures of their images in $\overline{D/\Gamma}$. To see this, recall that, dividing by the compact group $T(F)_c$,

$$(D/U(F)_{\mathbb{Z}})_{\{\sigma_\alpha\}}/T(F)_c \cong C(F)'' \times D(F)'$$
$$\cong F \times C(F)'' \times V(F)\,.$$

Now, $C_0 \cap \sigma_\alpha$ has compact closure in $C(F)''$, and this decomposition is related to our previous one by

$$
\begin{array}{ccc}
D & \cong & F \times C(F) \times V(F) \times U(F) \\
\downarrow & & \downarrow \\
D/U(F) & \cong & F \times C(F) \times V(F) \\
\cap & & \cap \\
(D/U(F)_{\mathbb{Z}})_{\{\sigma_\alpha^F\}}/T(F)_c & \cong & F \times C(F)'' \times V(F)\,,
\end{array}
$$

so the closure of the image of our set S in $(D/U(F)_{\mathbb{Z}})_{\{\sigma_\alpha^F\}}/T(F)_c$ is compact, hence the same happens in $(D/U(F)_{\mathbb{Z}})_{\{\sigma_\alpha^F\}}$.

7 An intrinsic form of the Main Theorem

Recall from Borel [2], §17.1, the concept of a *neat subgroup* $\Gamma \subset \mathscr{G}(\mathbb{C})$: this is a subgroup consisting of elements g such that, for one faithful representation,

$$\rho : \mathscr{G} \longrightarrow \mathrm{GL}(n)$$

(and hence for all representations ρ), the group inside $\mathbb{C}^*$ generated by the eigenvalues of $\rho(g)$ is torsion-free. Such a Γ is itself torsion-free. But, even more, for every pair of subgroups

$$\mathscr{H}_1 \subset \mathscr{H}_2 \subset \mathscr{G}\,,$$

with $\mathscr{H}_1$ normal in $\mathscr{H}_2$, the group

$$\Gamma \cap \mathscr{H}_2(\mathbb{C})/\Gamma \cap \mathscr{H}_1(\mathbb{C})$$

is also torsion-free.

We now return to the set-up of Sections 5 and 6: $\mathscr{G}$ is semi-simple, defined over $\mathbb{Q}$, and $G = \mathscr{G}(\mathbb{R})^o$ is the connected group of automorphisms of a bounded symmetric domain D; $\Gamma \subset \mathscr{G}(\mathbb{Q}) \subset G$ is an arithmetic subgroup. *We will assume throughout this section that Γ is also neat.* Since by [2], Proposition 17.6, any Γ has a neat subgroup $\Gamma_1 \subset \Gamma$ of finite index, this is not too restrictive. It has the effect that not only does Γ act freely on D, hence $\Gamma \xrightarrow{\sim} \pi_1(D/\Gamma)$, but that also for all rational boundary components F, the group $\mathrm{Im}(\Gamma_F \longrightarrow \mathrm{Aut}(F))$ acts freely on F, and $\overline{\Gamma}_F$ acts freely on $C(F)$, and $\Gamma_F/U(F)_{\mathbb{Z}}$ acts freely on $(D/U(F)_{\mathbb{Z}})_{\{\sigma_\alpha^F\}}$; hence

$$(D/U(F)_{\mathbb{Z}})_{\{\sigma_\alpha^F\}} \longrightarrow \overline{D/\Gamma}$$

is étale.

Borel [3] showed that the Baily–Borel compactification $(D/\Gamma)^*$ enjoys the following property:

Every holomorphic map

$$f : (\mathring{\Delta})^k \times \Delta^{n-k} \longrightarrow D/\Gamma$$

extends to a holomorphic map

$$f^* : \Delta^n \longrightarrow (D/\Gamma)^* .$$

Here, as usual, Δ is the unit disc, and $\mathring{\Delta} = \Delta \setminus \{0\}$. Letting $\exp : \mathfrak{H} \longrightarrow \mathring{\Delta}$ be the universal covering as usual, such an f lifts to an equivariant map $\tilde{f}$ and a homomorphism ϕ:

$$\phi : \mathbb{Z}^k \longrightarrow \Gamma ,$$

$$\tilde{f} : \mathfrak{H}^k \times \Delta^{n-k} \longrightarrow D .$$

Let

$$\gamma_i = \phi(\ldots,0,\underbrace{1}_{\text{place } i},0,\ldots) , \tag{7.1}$$

i.e., γ_i is the image under ϕ of the covering map $z \longmapsto z + 1$ in the i'th factor $\mathfrak{H}$.

Now, f will not in general extend to

$$\overline{f} : \Delta^n \longrightarrow \overline{D/\Gamma} ,$$

and we will analyze here when it does. Enlarge $\mathfrak{H}$ to

$$\mathfrak{H}^* = \mathfrak{H} \cup \{i\infty\} ,$$

where a fundamental system of neighborhoods of $i\infty$ is given by

$$W_c = \{z \in \mathfrak{H} \mid \operatorname{Im} z > c\} \cup \{i\infty\} .$$

Extend the map $\exp : \mathfrak{H} \to \mathring{\Delta}$ to $\exp : \mathfrak{H}^* \to \Delta$ by $\exp(i\infty) = 0$.

As a preliminary remark, we have

Proposition 7.1 *The map f^* above lifts to a continuous map*

$$\tilde{f}^* : (\mathfrak{H}^*)^k \times \Delta^{n-k} \longrightarrow D^* ,$$

extending the map $\tilde{f} : \mathfrak{H}^k \times \Delta^{n-k} \longrightarrow D$, where here we put on D^* the Satake topology (see Subsection 6.1).

Proof Let $x \in (\mathfrak{H}^*)^k \times \Delta^{n-k}$, let $\exp(x)$ be its image in $\Delta^k \times \Delta^{n-k}$, and let $P = f^*(\exp x)$. Let $Q \in D^*$ lie over P, and let U_1 be a neighborhood of Q in the Satake topology such that, for all $\gamma \in \Gamma$, we have $\gamma U_1 = U_1$ if $\gamma Q = Q$, but $\gamma U_1 \cap U_1 = \emptyset$ if $\gamma Q \neq Q$. Let U_2 be the image of U_1 in $(D/\Gamma)^*$. Then $\exp(x)$ has a neighborhood V_2 such that $f^*(V_2) \subset U_2$. Let V_1 be the component of $\exp^{-1}(V_2)$ containing x. Then

$$\tilde{f}\left(V_1 \cap (\mathfrak{H}^k \times \Delta^{n-k})\right) \subset \bigcup_{\gamma \in \Gamma} \gamma U_1 .$$

We may assume V_1 is a product of sets of the type W_c and discs in $\mathbb{C}$, and hence that $V_1 \cap (\mathfrak{H}^k \times \Delta^{n-k})$ is still connected. Therefore

$$\tilde{f}\left(V_1 \cap (\mathfrak{H}^k \times \Delta^{n-k})\right) \subset \gamma \cdot U_1$$

for some $\gamma \in \Gamma$. We then define

$$\tilde{f}^*(x) = \gamma \cdot Q .$$

It is easy to check that this $\tilde{f}^*$ is continuous.   □

Consider the point $P = \tilde{f}^*(i\infty, \ldots, i\infty, 0, \ldots, 0)$ in D^*. Let F be the unique boundary component containing it. Then

$$\operatorname{Im} \tilde{f}^* \subset D \cup \bigcup_{F_1 \supset F} F_1 .$$

Note that, by continuity, $\gamma_i P = P$, for all $i = 1, \ldots, k$, hence $\gamma_i \in \Gamma_F$.

The next result was suggested to us by P. Deligne; it generalizes a result of Y. Namikawa [9].

Theorem 7.2 *In the notation introduced above, $\gamma_i \in \overline{C(F)} \cap U(F)_{\mathbb{Z}}$ for all $i = 1, \ldots, k$; hence, $\tilde{f}$ induces a holomorphic map*

$$f^0 : (\mathring{\Delta})^k \times \Delta^{n-k} \longrightarrow D/U(F)_{\mathbb{Z}} ,$$

and the following statements are equivalent:

(i) *all γ_i are in one and the same $\sigma_\alpha^F \subset C(F)$ for some α;*
(ii) *f^0 extends to a holomorphic map*

$$\overline{f}^0 : \Delta^n \longrightarrow (D/U(F)_{\mathbb{Z}})_{\{\sigma_\alpha^F\}} ;$$

(iii) *f extends to a holomorphic map*

$$\overline{f} : \Delta^n \longrightarrow \overline{D/\Gamma} .$$

Proof We prove first that $\gamma_i \in U(F)_{\mathbb{Z}}$: in fact, restricting f to the i'th factor $\mathring{\Delta}$, holding all other variables constant, we get a map

$$f_i : \mathring{\Delta} \longrightarrow D/\Gamma$$

that extends to $\Delta \longrightarrow (D/\Gamma)^*$, and hence has no essential singularity at $0 \in \Delta$. Since $\dim \Delta = 1$ and $\overline{D/\Gamma}$ is compact, f_i also extends to a map

$$\overline{f}_i : \Delta \longrightarrow \overline{D/\Gamma} .$$

Also $\overline{f}_i(0)$ maps to a point of $(D/\Gamma)^*$ which is in the image by $D^* \longrightarrow (D/\Gamma)^*$ of a stratum F_1, where $\overline{F}_1 \supset F$. Therefore $\overline{f}_i(0)$ is in the image of the map

$$(D/U(F)_\mathbb{Z})_{\{\sigma_\alpha^F\}} \longrightarrow \overline{D/\Gamma} .$$

Since this map is étale, it follows that $\overline{f}_i$ lifts, in a small neighborhood $\Delta' \subset \Delta$ of 0, to

$$\overline{f}_i^0 : \Delta' \longrightarrow (D/U(F)_\mathbb{Z})_{\{\sigma_\alpha^F\}} ,$$

and hence f_i lifts to

$$f_i^0 : \mathring{\Delta}' = \Delta' \setminus \{0\} \longrightarrow D/U(F)_\mathbb{Z} .$$

But $\gamma_i \in \mathrm{Im}\left(f_{i,*}^0 : \pi_1(\mathring{\Delta}') \longrightarrow \pi_1(D/U(F)_\mathbb{Z})\right)$, so $\gamma_i \in U(F)_\mathbb{Z}$.

In particular, we know now that f^0 is defined. Next, we show that f^0 always extends to what we will call here a *semi-proper meromorphic map* $\overline{f}^0$ from Δ^n to $(D/U(F)_\mathbb{Z})_{\{\sigma_\alpha^F\}}$; i.e., a many-valued map whose graph

$$\mathrm{Gr}(\overline{f}^0) \subset \Delta^n \times (D/U(F)_\mathbb{Z})_{\{\sigma_\alpha^F\}}$$

is a closed analytic subset, mapping properly to Δ^n, and restricting to f^0 on $(\mathring{\Delta})^k \times \Delta^{n-k}$. To see this, consider the following diagram:

$$
\begin{array}{ccccc}
\Delta^n \times (D/U(F)_\mathbb{Z})_{\{\sigma_\alpha^F\}} & \supset & \overline{\mathrm{Gr}(f^0)} & \supset & \mathrm{Gr}(f^0) \\
\downarrow{\scriptstyle \text{étale}} & & \downarrow & & \downarrow{\scriptstyle \wr} \\
\Delta^n \times \overline{D/\Gamma} & \supset & \overline{\mathrm{Gr}(f)} & \supset & \mathrm{Gr}(f) \\
\downarrow{\scriptstyle \substack{\text{proper,}\\ \text{birational}}} & & \downarrow & & \downarrow{\scriptstyle \wr} \\
\Delta^n \times (D/\Gamma)^* & \supset & \mathrm{Gr}(f^*) & \supset & \mathrm{Gr}(f) \\
& & \downarrow{\scriptstyle \wr} & & \downarrow{\scriptstyle \wr} \\
& & \Delta^n & \supset & (\mathring{\Delta})^k \times \Delta^{n-k}
\end{array}
$$

Since f^* exists, it follows that $\mathrm{Gr}(f^*)$ is a closed analytic set isomorphic to Δ^n. Since $\overline{D/\Gamma} \longrightarrow (D/\Gamma)^*$ is proper and birational, the closure $\overline{\mathrm{Gr}(f)}$ of $\mathrm{Gr}(f)$ is also analytic, and maps properly to $\mathrm{Gr}(f^*)$, hence to Δ^n. Since $(D/U(F)_\mathbb{Z})_{\{\sigma_\alpha^F\}} \longrightarrow \overline{D/\Gamma}$ is étale, the closure $\overline{\mathrm{Gr}(f^0)}$ is étale over $\overline{\mathrm{Gr}(f)}$ and in particular is analytic. Since $\mathrm{Gr}(f^0) = \mathrm{Gr}(f)$ is open dense in both $\overline{\mathrm{Gr}(f^0)}$ and $\overline{\mathrm{Gr}(f)}$, in fact $\overline{\mathrm{Gr}(f^0)}$ is just an open subset of $\overline{\mathrm{Gr}(f)}$.

Finally, $\overline{\mathrm{Gr}(f^0)} = \overline{\mathrm{Gr}(f)}$. To see this, let $x_i \in \mathfrak{H}^k \times \Delta^{n-k}$ be a sequence with $x_i \longrightarrow x \in (\mathfrak{H}^*)^k \times \Delta^{n-k}$, such that $f(\exp x_i) \longrightarrow y \in \overline{D/\Gamma}$. But $\tilde{f}(x_i) \longrightarrow \tilde{f}^*(x)$ and $\tilde{f}^*(x) \in F_1$ for some rational boundary component F_1 with $\overline{F}_1 \supset F$. Then the image of y in $(D/\Gamma)^*$ is in the image of F_1, so y lifts to a point z in the F_1-stratum of $(D/U(F_1)_\mathbb{Z})_{\{\sigma_\alpha^{F_1}\}}$; we may choose z so that z maps to $\tilde{f}^*(x) \bmod U(F_1)_\mathbb{Z}$ via $(D/U(F_1)_\mathbb{Z})_{\{\sigma_\alpha^{F_1}\}} \longrightarrow D^*/U(F_1)_\mathbb{Z}$. Now, since $(D/U(F_1)_\mathbb{Z})_{\{\sigma_\alpha^{F_1}\}} \longrightarrow \overline{D/\Gamma}$ is étale, there is a unique sequence $z_i \in D/U(F_1)_\mathbb{Z}$ converging to z, with z_i over $f(\exp x_i)$. In fact, $\{z_i\}$ must be equal to $\gamma \cdot \tilde{f}(x_i) \bmod U(F_1)_\mathbb{Z}$, for some $\gamma \in \Gamma$. But then, in $D^*/U(F_1)_\mathbb{Z}$,

$$
\begin{array}{ccc}
\mathrm{Im}\,(\gamma \cdot f(x_i)) & \overset{i \to \infty}{\longrightarrow} & \mathrm{Im}\,(z) = \tilde{f}^*(x) \bmod U(F_1)_\mathbb{Z} \\
\| & & \\
\gamma \cdot \mathrm{Im}\,(\tilde{f}(x_i)) & \overset{i \to \infty}{\longrightarrow} & \gamma \cdot \tilde{f}^*(x) \bmod U(F_1)_\mathbb{Z}\,,
\end{array}
$$

i.e., $\gamma \cdot \tilde{f}^*(x) = \tilde{f}^*(x) \bmod U(F_1)_\mathbb{Z}$. Thus $\gamma \in \Gamma_{F_1}$. Replace z by $\gamma^{-1}z$; then $f^0(\exp x_i) = \tilde{f}(x_i) \bmod U(F_1)_\mathbb{Z}$ converges to z, so $(\exp(x), z) \in \overline{\mathrm{Gr}(f^0)}$, as required.

The equivalence of (ii) and (iii) is now clear: since $\mathrm{Gr}(\overline{f^0}) \overset{\sim}{\longrightarrow} \mathrm{Gr}(\overline{f})$, we see that $\overline{f}$ is single-valued if and only if $\overline{f^0}$ is single-valued. To bring in (i), use the following maps:

$$
(D/U(F)_\mathbb{Z})_{\{\sigma_\alpha^F\}} \subset (D(F)/U(F)_\mathbb{Z})_{\{\sigma_\alpha^F\}}
$$
$$
\pi_1 \Big\downarrow \; T(F)-\text{bundle}
$$
$$
D(F)'
$$
$$
\Big\downarrow
$$
$$
F
$$

For all $x \in \Delta^n$, $\overline{f}(x)$ is a connected compact analytic set. But F is a bounded domain and $D(F)'$ is an affine bundle over F; so $D(F)'$ contains no positive-dimensional compact analytic sets. Thus $\overline{\pi}_1 \circ \overline{f}$ is single-valued. We can now apply:

Lemma 7.3 *Let*

(1) $g : \Delta^n \longrightarrow Y$ *be an analytic map;*

(2) $\pi : E \longrightarrow Y$ *be a topologically trivial principal analytic fiber bundle with fiber an algebraic torus T; let $T \subset X_{\{\sigma_\alpha\}}$ be a torus embedding and let $\pi : E_{\{\sigma_\alpha\}} \longrightarrow Y$ be the associated fiber bundle;*

(3) $h : (\mathring{\Delta})^k \times \Delta^{n-k} \longrightarrow E$ be a lifting of g, which extends to a semi-proper meromorphic map $\Delta^n \longrightarrow E_{\{\sigma_\alpha\}}$.

Then

$$\left[\begin{array}{c} h \text{ extends to} \\ \overline{h} : \Delta^n \longrightarrow E_{\{\sigma_\alpha\}} \end{array} \right] \Longleftrightarrow \left[\begin{array}{c} \text{the monodromy elements} \\ \gamma_i \in \pi_1(T) = \mathrm{Ker}\,(\pi_1(E) \longrightarrow \pi_1(Y)) \\ \text{all lie in one } \sigma_\alpha \end{array} \right]$$

and, in this case, h extends to $\overline{h} : \Delta^n \to E_{\sigma_\alpha}$, where α is as in the assertion on the right. Here the γ_i are the images of the generators in $\pi_1((\mathring{\Delta})^k)$.

Proof Firstly, the result is local on Δ^n, so it is also local on Y. So we may assume $E \cong T \times Y$. Following h by projection onto T, it suffices to prove the lemma when Y is a point and $E = T$. Since h is semi-proper meromorphic, for all $\beta \in M(T) = \mathrm{Hom}\,(T, \mathbb{G}_m)$, we have that $\beta \circ h$ is a meromorphic function on Δ^n and

$$h \text{ extends to } \overline{h} : \Delta^n \to E_{\sigma_\alpha} \Longleftrightarrow \text{ for all } \beta \in \check{\sigma}_\alpha \cap M(T),$$

$$\beta \circ h \text{ is holomorphic on } \Delta^n .$$

But for all $\beta \in M(T)$, $\beta \circ h$ has zeroes and poles only on the coordinate hyperplanes $H_i \subset \Delta^n$, so we may write

$$(\beta \circ h) = \sum_{i=1}^{k} \ell_i(\beta) \cdot H_i ,$$

where ℓ_i is a linear function from $M(T)$ to $\mathbb{Z}$. But then $\arg(\beta \circ h)$ changes by $2\pi \ell_i(\beta)$ when going in a loop around H_i; i.e., $\arg \beta$ changes by $2\pi \ell_i(\beta)$ when traversing the loop γ_i in T. But, identifying $\pi_1(T)$ with $N(T)$, hence with $\mathrm{Hom}\,(M(T), \mathbb{Z})$, $\arg \mathfrak{X}^\beta$ in fact changes by $2\pi \langle \gamma, \beta \rangle$ in a loop homologous to γ. Thus

$$\ell_i(\beta) = \langle \gamma_i, \beta \rangle .$$

Therefore

$$\beta \circ h \text{ is holomorphic on } \Delta^n \Longleftrightarrow \langle \gamma_i, \beta \rangle \geq 0 \text{ for all } i ,$$

hence

$$h \text{ extends to } \overline{h} : \Delta^n \longrightarrow E_{\sigma_\alpha} \Longleftrightarrow \gamma_i \in \check{\sigma}_\alpha = \sigma_\alpha , \text{ for all } i .$$

Now, if h extends to $\overline{h} : \Delta^n \longrightarrow E_{\{\sigma_\alpha\}}$, then $\overline{h}(0, \dots, 0) \in E_{\sigma_\alpha}$, for some σ_α, and hence for all $\beta \in \check{\sigma}_\alpha$, the function $\beta \circ h$ is holomorphic on Δ^n, i.e., by what precedes, $\langle \gamma_i, \beta \rangle \geq 0$ for all i, and all $\beta \in \check{\sigma}_\alpha$; hence $\gamma_i \in \check{\sigma}_\alpha = \sigma_\alpha$ for all i. $\square$

The only question now is why $\gamma_i \in \overline{C(F)}$ in all cases. But this follows because f_i = restriction of f to i'th $\mathring{\Delta}$ always extends, so, by (iii) $\Rightarrow$ (i), we have that $\gamma_i \in$ some σ_α, hence $\gamma_i \in \overline{C(F)}$.   □

The theorem easily extends to the more general setting in which the inclusion

$$(\mathring{\Delta})^k \times \Delta^{n-k} \subset \Delta^n$$

is replaced by

$$U \cap T \subset U,$$

where U is a nice neighborhood of a point in the closed orbit of X_σ, with $T \subset X_\sigma$ a torus embedding. The condition that $U \cap T \longrightarrow D/\Gamma$ extends to $U \longrightarrow \overline{D/\Gamma}$ is that the image of a certain natural map

$$\sigma \longrightarrow \overline{C(F)}$$

should lie in some σ_α.

The above theorem in the special case $k = n = 1$ says: starting with any $f : \mathring{\Delta} \longrightarrow D/\Gamma$, we lift it to an equivariant $\tilde{f} : \mathfrak{H} \to D$. Then the monodromy γ is in $\overline{C(F)} \cap U(F)_{\mathbb{Z}}$. If we alter the lifting $\tilde{f}$ of f to $\delta \cdot \tilde{f}$, then γ changes to $\delta \gamma \delta^{-1}$. Thus f alone determines canonically an element

$$\mu(f) \in \Sigma_{\mathbb{Z}} = \left(\bigcup_F \overline{C(F)} \cap U(F)_{\mathbb{Z}} \right) \Big/ \Gamma .$$

(Here the union is over all rational boundary components F, and the $\overline{C(F)} \cap U(F)_{\mathbb{Z}}$ are considered as subsets of G; the action of Γ is through conjugation; μ is short for *monodromy*.) Let us clarify the meaning of $\Sigma_{\mathbb{Z}}$. First of all, inside G itself, let

$$|\widetilde{\Sigma}| = \bigcup_F \overline{C(F)} = \bigsqcup_F C(F) ,$$
$$\widetilde{\Sigma}_{\mathbb{Z}} = \Gamma \cap |\widetilde{\Sigma}| = \bigsqcup_F C(F) \cap U(F)_{\mathbb{Z}} .$$

If $\{\sigma_\alpha\}$ is a Γ-admissible collection of polyhedra as in Section 5, and $\widetilde{V}_\alpha = \text{res}_{\text{Int}\,\sigma_\alpha}$ (linear functions on $U(F)$), then $\widetilde{\Sigma} = (|\widetilde{\Sigma}|, \{\sigma_\alpha\}, \{\widetilde{V}_\alpha\})$ is a (infinite) conical polyhedral complex in the sense of Chapter I, Section 3, and $\widetilde{\Sigma}_{\mathbb{Z}}$ defines an integral structure on $\widetilde{\Sigma}$; namely, on each σ_α, let $\widetilde{L}_\alpha$ be the linear functions which are integral on $\widetilde{\Sigma}_{\mathbb{Z}}$.

Since Γ is neat, Γ acts freely on $|\widetilde{\Sigma}|$. And, for every Γ-admissible collection $\{\sigma_\alpha\}$, this action permutes the strata $\text{Int}\,\sigma_\alpha$ and preserves the spaces $\widetilde{V}_\alpha$ of linear functions, so we may form a quotient conical polyhedral complex:

$$|\Sigma| = |\widetilde{\Sigma}|/\Gamma \,,$$
$$S_\alpha = \mathrm{Im}\,(\mathrm{Int}\,\sigma_\alpha) \,,$$
$$V_\alpha = \widetilde{V}_\alpha \,.$$

As a topological space with piecewise-linear structure, $|\Sigma|$ is independent of $\{\sigma_\alpha\}$. And $\Sigma_{\mathbb{Z}}$ is a subset of $|\Sigma|$ and defines an integral structure on the conical polyhedral complex $\Sigma = (|\Sigma|, \{S_\alpha\}, \{V_\alpha\})$, by taking L_α to be the linear functions which are integral on $\Sigma_{\mathbb{Z}}$. Note that, whereas there are infinitely many σ_α, Σ has only a finite number of strata S_α, and, in fact, if $\mathbb{R}_{>0}$ acts by homotheties on the cones of Σ, then $|\Sigma|/\mathbb{R}_{>0}$ is compact. Heuristically, $|\Sigma|/\mathbb{R}_{>0}$ should be thought of as the set of all directions of approach to ∞ in D/Γ, as described by ratios of exponents.

Definition 7.4 A structure $\{S_\alpha, V_\alpha\}$ of conical polyhedral complexes on $|\Sigma|$ is *admissible* if it is induced by a Γ-admissible collection $\{\sigma_\alpha\}$.

Note that an admissible structure $\{S_\alpha, V_\alpha\}$ on $|\Sigma|$ comes from *only one* collection $\{\sigma_\alpha\}$: in fact, we can recover the σ_α by looking at

$$C(F) \hookrightarrow |\widetilde{\Sigma}| \longrightarrow |\Sigma|$$

and (i) taking inverse images of the S_α, (ii) taking their connected components, and (iii) closing them up in $\overline{C(F)}$.

Next, define, in analogy with the definition for a toroidal embedding in Chapter I, Section 3,

$$R.S.(D/\Gamma) = \left\{\varphi : \Delta \longrightarrow (D/\Gamma)^* \text{ analytic} \mid \varphi(\mathring{\Delta}) \subset D/\Gamma\right\} \,.$$

Then, as above, we get a map

$$\mu : R.S.(D/\Gamma) \longrightarrow \Sigma_{\mathbb{Z}}$$

by lifting and considering the monodromy. We can now formulate the Main Theorem I (Theorem 5.2) in a more intrinsic way.

Theorem 7.5 (Main Theorem II) *Let $\mathscr{G}$, G, D, and Γ be as in Theorem 5.2, where Γ is neat, and define the piecewise-linear topological space $|\Sigma|$ as above. Then there is a map*

$$\left\{ \begin{array}{c} \text{admissible conical} \\ \text{polyhedral stuctures} \\ \Sigma = \{|\Sigma|, S_\alpha, V_\alpha\} \} \end{array} \right\} \longrightarrow \left\{ \begin{array}{c} \text{toroidal embeddings } D/\Gamma \subset \overline{D/\Gamma} \\ \text{without monodromy, where } \overline{D/\Gamma} \\ \text{is a compact algebraic space} \end{array} \right\}$$

such that, if $\Sigma(\overline{D/\Gamma})$ is the conical polyhedral complex with integral structure *associated by the theory of Chapter I, Section 3, to the toroidal embedding on the right, there is a unique isomorphism φ making the diagram*

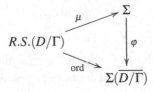

commute. In particular, there is a bijection between the set of strata of $\overline{D/\Gamma}$ and the set of strata $\{S_\alpha\}$ of Σ. Furthermore, if $\{S_\alpha^{(1)}\}$ is a subdivision of $\{S_\alpha^{(2)}\}$, then $\overline{D/\Gamma}^{(1)}$ dominates $\overline{D/\Gamma}^{(2)}$.

Proof Since $D/\Gamma \subset \overline{D/\Gamma}$ is locally like $D/U(F)_{\mathbb{Z}} \subset (D/U(F)_{\mathbb{Z}})_{\{\sigma_\alpha^F\}}$, it is certainly a toroidal embedding. Also

$$(D/U(F)_{\mathbb{Z}})_{\{\sigma_\alpha^F\}}/(\Gamma_F/U(F)_{\mathbb{Z}}) \longrightarrow \overline{D/\Gamma}$$

is bijective on the strata of type F. Morever, denoting, as in the proof of Proposition 6.10, by Γ_F' the intersection $\Gamma \cap N(F)'$ and by $\overline{\Gamma}_F$ the image of Γ in $\mathrm{Aut}(U(F))$, $\Gamma_F'/U(F)_{\mathbb{Z}}$ fixes every stratum of $(D/U(F)_{\mathbb{Z}})_{\{\sigma_\alpha\}}$, while $\overline{\Gamma}_F$ permutes without fixed points the strata of type F in $(D/U(F)_{\mathbb{Z}})_{\{\sigma_\alpha\}}$. It follows that the strata of $\overline{D/\Gamma}$ are all of the form

$$Y_\alpha/(\Gamma_F'/U(F)_{\mathbb{Z}}) ,$$

where

$$Y_\alpha \subset (D/U(F)_{\mathbb{Z}})_{\{\sigma_\alpha^F\}}$$

is a stratum corresponding to a σ_α meeting $C(F)$. The main point here is that $\Gamma_F'/U(F)_{\mathbb{Z}}$ acts on $\overline{Y}_\alpha$ so as to fix each stratum $Y_\beta \subset \overline{Y}_\alpha$. Since $D/U(F)_{\mathbb{Z}} \subset (D/U(F)_{\mathbb{Z}})_{\{\sigma_\alpha^F\}}$ is a toroidal embedding without self-intersection, it follows that, even though $D/\Gamma \subset \overline{D/\Gamma}$ may have self-intersection, it is without monodromy (in the sense of Chapter I, Section 3). Moreover, the polyhedral complex $C(F)^*$ associated to $D/U(F)_{\mathbb{Z}} \subset (D/U(F)_{\mathbb{Z}})_{\{\sigma_\alpha^F\}}$ is given by

$$\left(\bigcup_\alpha \sigma_\alpha^F = C(F) \cup \bigcup_{C'} C' ; \{\sigma_\alpha^F\} ; \text{restr. to } \sigma_\alpha \text{ of linear functions on } U(F) \right) .$$

Here C' runs through the rational boundary components of C. The equivalence relation Λ is generated by

(i) the isomorphisms

$$(D/U(F)_{\mathbb{Z}})_{\{\sigma_\alpha^F\}} \xrightarrow{\ \gamma\ } (D/U(\gamma F)_{\mathbb{Z}})_{\{\sigma_\alpha^{\gamma F}\}}, \text{ for } \gamma \in \Gamma ,$$

(ii) the étale maps

$$(D/U(F_1)_{\mathbb{Z}})_{\{\sigma_\alpha^{F_1}\}} \longrightarrow (D/U(F)_{\mathbb{Z}})_{\{\sigma_\alpha^F\}} ,$$

whenever $\overline{F}_1 \supset F$.

So the polyhedral complex $\Sigma(\overline{D/\Gamma})$ is obtained by dividing $\bigsqcup_F C(F)^*$ by

(i) the isomorphisms $\gamma : C(F)^* \longrightarrow C(\gamma F)^*$
(ii) the inclusions, whenever $\overline{F}_1 \supset F$,

$$C(F_1)^* \hookrightarrow C(F)^* .$$

Modulo the latter we get $\widetilde{\Sigma}$, and modulo the former we get Σ; this yields

$$\varphi : \Sigma \xrightarrow{\sim} \Sigma(\overline{D/\Gamma}) .$$

The fact that the diagram in the theorem commutes is implied by the commutativity of the following analogous diagram:

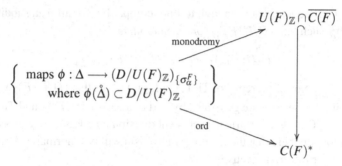

The commutativity here is a consequence of Proposition 3.3 in Chapter I. The uniqueness assertion about φ follows from the fact that the image of ord, modulo homotheties, is dense. Finally, if $\{S_\alpha^{(1)}\}$ is a subdivision of $\{S_\alpha^{(2)}\}$, then $\{\sigma_\alpha^{(1)}\}$ is a subdivision of $\{\sigma_\alpha^{(2)}\}$; hence there are vertical maps in the diagram

$$
\begin{array}{c}
(D/U(F)_{\mathbb{Z}})_{\{\sigma_\alpha^{(1),F}\}} \\
D/U(F)_{\mathbb{Z}} \hookrightarrow \quad\quad\quad | \\
\quad\quad\quad\quad\quad \downarrow \\
(D/U(F)_{\mathbb{Z}})_{\{\sigma_\alpha^{(2),F}\}}
\end{array}
$$

which induce a map in the vertical direction:

$$
\begin{array}{c}
\overline{D/\Gamma}^{(1)} \\
D/\Gamma \hookrightarrow \quad | \\
\quad\quad \downarrow \\
\overline{D/\Gamma}^{(2)}
\end{array}
$$

□

Finally, let us show that there are plenty of non-singular compactifications among those constructed. Recall that we assume throughout this section that Γ is neat.

Corollary 7.6 *For every Γ-admissible collection $\{\sigma_\alpha\}$ of polyhedra, there exists exists a subdivision $\{\sigma'_\alpha\}$ such that the morphism $\pi : \overline{D/\Gamma}' \longrightarrow \overline{D/\Gamma}$ is projective and $\overline{D/\Gamma}'$ is smooth.*

Proof Because Γ is neat, the morphism π_F in the diagram

$$
\begin{array}{ccc}
D/U(F)_{\mathbb{Z}} & \lhook\joinrel\longrightarrow & (D/U(F)_{\mathbb{Z}})_{\{\sigma_\alpha^F\}} \\
\downarrow & & \downarrow{\scriptstyle \pi_F} \\
D/\Gamma & \lhook\joinrel\longrightarrow & \overline{D/\Gamma}
\end{array}
$$

is étale. Consequently, it is enough to find, compatibly for all F, a subdivision $\sigma_\alpha'^F$ of σ_α^F such that $(D/U(F)_{\mathbb{Z}})_{\{\sigma_\alpha'^F\}}$ is smooth and

$$
(D/U(F)_{\mathbb{Z}})_{\{\sigma_\alpha'^F\}} \longrightarrow (D/U(F)_{\mathbb{Z}})_{\{\sigma_\alpha^F\}}
$$

is projective. Now Theorem 4 of TE I , Ch.I, §1, tells us that $(D/U(F)_{\mathbb{Z}})_{\{\sigma_\alpha'^F\}}$ is smooth if all cones $\sigma_\alpha'^F$ are generated by part of a basis of $U(F)_{\mathbb{Z}}$, and Theorem 11 of TE I , Ch.I, §1, constructs such a subdivision by an inductive procedure. It is also shown there that the resulting morphism will be a normalized blow-up and is, in particular, projective. $\square$

References

[1] W. Baily and A. Borel, Compactification of arithmetic quotients of bounded symmetric domains, *Annals of Math.* **84** (1966), 442–528.

[2] A. Borel, *Introduction aux Groupes Arithmétiques*. Paris: Hermann, 1969.

[3] A. Borel, Some metric properties of arithmetic quotients of symmetric spaces and an extension theorem, *J. Diff. Geom.* **6** (1972), 543–560.

[4] R. Gunning and H. Rossi, *Analytic Functions*. Englewood Cliffs, NJ: Prentice Hall, 1965.

[5] Harish-Chandra, Representations of semi-simple Lie groups VI, *Am. J. Math.* **78** (1956), 564–628.

[6] S. Helgason, *Differential Geometry and Symmetric Spaces*. New York: Academic Press, 1962.

[7] A. Koranyi and J. Wolf, Generalized Cayley transformations of bounded symmetric domains, *Am. J. Math.* **87** (1965), 899–939.

[8] R. P. Langlands, The dimension of spaces of automorphic forms, *Am. J. Math.* **85** (1963), 99–125.

[9] Y. Namikawa, On the canonical holomorphic map from the moduli space of stable curves to the Igusa monoidal transform, *Nagoya Math. J.* **52** (1973), 197–259.

[10] I. Satake, On the arithmetic of tube domains (blowing up of the point at infinity), *Bull. Am. Math. Soc.* **79** (1973), 1076–1094.

[11] I. Satake, *Realization of Symmetric Domains as Siegel Domains of the Third Kind*. Lecture notes, Berkeley, 1972.

[12] J. Wolf, Fine structure of hermitian symmetric spaces, in *Symmetric Spaces*, eds. W. Boothby and G. Weiss. New York: Marcel Dekker, 1972, pp. 271–357.

[13] J. Wolf, On the classification of hermitian symmetric spaces, *J. Math. Mechanics*, **13** (1964), 489–495.

IV

Further developments

This chapter is divided into two sections.

The first section is an immediate application of the construction of $\overline{D/\Gamma}$ in the previous chapter. With explicit local coordinates, we are able to give a criterion for the holomorphic extension of higher-order differential forms to cusps. We have also computed the local codimension of the space of "extendable forms" in the space of cusp forms in the case of Hilbert modular surfaces, from which one can compute the plurigenera of Hilbert modular surfaces.

The second section concerns a criterion for the projectivity of $\overline{D/\Gamma}$. We shall show that certain $\overline{D/\Gamma}$'s are obtained by blowing up Baily–Borel's compactification. Our method follows essentially that of Igusa [5], but the situation is much clearer now, since $\overline{D/\Gamma}$ has already been constructed.

1 Extension of differential forms to the cusps

1.1

Following the notations of the previous chapter, let

$\mathscr{G} = $ a connected semi-simple linear algebraic group

of hermitian type defined over $\mathbb{Q}$,

$G = \mathscr{G}(\mathbb{R})^{o}$,

$K = $ its maximal compact subgroup,

$\Gamma = $ an arithmetic subgroup of G,

$D = G/K$, a bounded symmetric domain in $\mathbb{C}^{N}$,

$\overline{D/\Gamma} = $ the compactification of D/Γ corresponding to

a Γ-admissible collection of polyhedra $\{\sigma_{\alpha}\}$.

Our purpose is to investigate when a top differential form extends to all "cusps" $\overline{D/\Gamma} \setminus (D/\Gamma)$.

Let f be an automorphic form of weight ℓ with respect to Γ:

$$f(\gamma z)j(\gamma,z)^{\ell} = f(z) \text{ for all } \gamma \in \Gamma, \ z \in D,$$

where $j(\gamma,z)$ is the jacobian of the map $\gamma : D \longrightarrow D$ at z, i.e., if

$$\omega = \mathrm{d}z_1 \wedge \cdots \wedge \mathrm{d}z_N, \text{ then } \gamma^*\omega = j(\gamma,z)\omega.$$

The differential form $f\omega^{\otimes \ell}$ is invariant under Γ, and hence can be considered as a form† in $\Omega^N(D/\Gamma)^{\otimes \ell}$, and every element in $\Omega^N(D/\Gamma)^{\otimes \ell}$ is of this type.

Our problem is: when does $f\omega^{\otimes \ell}$ extend to $\overline{D/\Gamma}$?

Let us first recall how $\overline{D/\Gamma}$ was constructed. By Chapter III, Theorem 5.2 (Main Theorem I), we have the following diagram:

$$
\begin{array}{ccc}
D/U(F)_{\mathbb{Z}} & \lhook\joinrel\longrightarrow & (D/U(F)_{\mathbb{Z}})_{\{\sigma_\alpha^F\}} \\
\downarrow & & \downarrow{\scriptstyle \pi_F} \\
D/\Gamma & \lhook\joinrel\longrightarrow & \overline{D/\Gamma}
\end{array}
$$

for every rational boundary component F.

Furthermore, if Γ is neat [1] (comp. also Chapter III, Section 7), and if each σ_α^F can be generated by a part of a basis of $U(F)_{\mathbb{Z}}$, then π_F is unramified and $\overline{D/\Gamma}$ is non-singular (see Chapter III, Corollary 7.6).

Assume this is the case.

Hence, to see if $f\omega^{\otimes \ell}$ extends, it is enough to check if it extends for

$$D/U(F)_{\mathbb{Z}} \hookrightarrow (D/U(F)_{\mathbb{Z}})_{\{\sigma_\alpha^F\}}$$

for every F. In fact, we only need to check this for $D/U(F)_{\mathbb{Z}} \hookrightarrow (D/U(F)_{\mathbb{Z}})_{\sigma_\alpha^F}$ for each top-dimensional simplicial cone σ_α^F.

Let us fix a rational boundary component F, and embed D into $D(F)$:

$$D(F) = U(F)_{\mathbb{C}} \times \mathbb{C}^m \times F,$$
$$D = \{(z,u,t) \in D(F) \mid \operatorname{Im} z - h_t(u,u) \in C(F)\}.$$

Introduce local coordinates (z_i), (u_j), (t_k) for each component; we may take ω as

$$\omega = \bigwedge_i \mathrm{d}z_i \wedge \bigwedge_j \mathrm{d}u_j \wedge \bigwedge_k \mathrm{d}t_k.$$

Now, f has an expansion as a Fourier–Jacobi series:

$$f = \sum_{\rho \in L^*} \varphi_\rho(u,t) \exp(2\pi \mathrm{i}\langle \rho, z \rangle),$$

† Assume there are no singularities in D/Γ.

where $\langle\,,\,\rangle$ is a positive-definite inner product on $U(F)$ for which $C(F)$ is self-adjoint, as defined in Chapter II, Section 1, and

$$L^* = \{\rho \in U(F) \mid \langle\rho,x\rangle \in \mathbb{Z}, \text{ for all } x \in U(F)_{\mathbb{Z}}\}\,.$$

This series is convergent in some cylindrical set

$$S(K,r) = \{(z,u,t) \in D(F) \mid t \in K, \operatorname{Im}z - h_t(u,u) - r \in C(F)\}\,,$$

where K is a compact set in F.

If none of the $\mathbb{Q}$-simple components of $\mathscr{G}$ is isomorphic to $\mathrm{SL}_2/\mathbb{Q}$, then we have "Koecher's theorem":

$$\varphi_\rho \neq 0 \text{ only for } \rho \in L^* \cap \overline{C(F)}\,,$$

(see Baily [1], p. 299).

We now express $f\omega^{\otimes\ell}$ in terms of the local coordinates in $(D/U(F)_{\mathbb{Z}})_{\sigma_\alpha^F}$.

Recall that $(D/U(F)_{\mathbb{Z}})_{\sigma_\alpha^F}$ is the interior of the closure of $(D/U(F)_{\mathbb{Z}})$ in $(D/U(F)_{\mathbb{Z}}) \times^{T(F)} T(F)_{\sigma_\alpha^F}$, and $T(F) = U(F)_{\mathbb{C}}/U(F)_{\mathbb{Z}}$. And, by our assumption,

$$\sigma_\alpha^F = \sum \mathbb{R}_{\geq 0}P_i, \text{ where } \{P_1,\ldots,P_n\} \text{ is a basis for } U(F)_{\mathbb{Z}}\,.$$

Let $\{Q_1,\ldots,Q_n\} \subset L^*$ be the dual basis of $\{P_1,\ldots,P_n\}$, $\check\sigma_\alpha = \sum \mathbb{R}_{\geq 0}Q_i$. Then $T(F)_{\sigma_\alpha^F}$ is isomorphic to $\mathbb{C}^n$ with coordinates $\{w_r\}$:

$$w_r = \exp(2\pi\mathrm{i}\langle Q_r,z\rangle)\,.$$

In terms of (w_r), (u_j), (t_k), which are coordinates on $(D/U(F)_{\mathbb{Z}})_{\sigma_\alpha^F}$, we may write

$$f = \sum_\rho \varphi_\rho(u,t)\exp(2\pi\mathrm{i}\langle\rho,z\rangle)$$

$$= \sum_\rho \varphi_\rho(u,t)\exp(2\pi\mathrm{i}\langle\sum\rho_rQ_r,z\rangle) \text{ (where } \rho_r = \langle\rho,P_r\rangle)$$

$$= \sum_\rho \varphi_\rho(u,t)\prod_r w_r^{\rho_r}\,,$$

$$\omega = \bigwedge_i \mathrm{d}z_i \wedge \bigwedge_j \mathrm{d}u_j \wedge \bigwedge_k \mathrm{d}t_k$$

$$= \text{constant}\cdot\bigwedge_r \mathrm{d}w_r \wedge \bigwedge_j \mathrm{d}u_j \wedge \bigwedge_k \mathrm{d}t_k \frac{1}{\prod_r w_r}\,.$$

Hence,

$$f\omega^{\otimes\ell} = \text{const.}\sum_\rho \varphi_\rho \prod_r w_r^{\rho_r}\frac{1}{(\prod_r w_r)^\ell}\left(\bigwedge_r \mathrm{d}w_r \wedge \bigwedge_j \mathrm{d}u_j \wedge \bigwedge_k \mathrm{d}t_k\right)^{\otimes\ell}\,.$$

Now, $(D/U(F)_{\mathbb{Z}})_{\sigma_\alpha^F} \setminus (D/U(F)_{\mathbb{Z}})$ is given by $\bigcup_r \{w_r = 0\}$, and hence $f\omega^{\otimes \ell}$ extends to $(D/U(F)_{\mathbb{Z}})_{\{\sigma_\alpha^F\}}$ if and only if

$$\varphi_\rho \neq 0 \text{ implies } \langle \rho, P_r \rangle \geq \ell \text{ for all } r .$$

Since $\{P_r\}$ form a basis for $U(F)_{\mathbb{Z}}$ and $\{\sigma_\alpha^F\}$ is admissible, there exists, for every $P \in U(F)_{\mathbb{Z}} \cap \overline{C(F)}$, an α such that P can be expressed as follows:

$$P = \sum a_r P_r \text{ with } a_r \in \mathbb{Z}_{\geq 0}, \text{ where } \sigma_\alpha^F = \sum \mathbb{R}_{\geq 0} \cdot P_r .$$

If $P \neq 0$, then

$$\langle \rho, P \rangle = \sum a_r \langle \rho, P_r \rangle \geq (\sum a_r)\ell \geq \ell .$$

This proves

Theorem 1.1 *Let D be a bounded symmetric domain in $\mathbb{C}^N$, let Γ be a neat arithmetic subgroup of* $\mathrm{Aut}(D)^o$, *let f be an automorphic form of weight ℓ with respect to Γ, i.e., $f\omega^{\otimes \ell} \in \Omega^N(D/\Gamma)^{\otimes \ell}$, and let $\overline{D/\Gamma}$ be the compactification of D/Γ corresponding to a Γ-admissible collection of polyhedra $\{\sigma_\alpha^F\}$, where each σ_α^F can be generated by a part of a basis of $U(F)_{\mathbb{Z}}$. Then $f\omega^{\otimes \ell}$ extends to $\overline{D/\Gamma}$ if and only if, for every rational boundary component F, in the Fourier expansion of f at F, namely*

$$f = \sum_{\rho \in U(F)_{\mathbb{Z}}^*} \varphi_\rho^F(u,t) \exp(2\pi \mathrm{i}\langle \rho, z \rangle) ,$$

φ_ρ^F *satisfies*

$$\varphi_\rho^F \neq 0 \Longrightarrow \langle \rho, P \rangle \geq \ell$$

for all non-zero $P \in U(F)_{\mathbb{Z}} \cap \overline{C(F)}$. □

Now, on any complex manifold U of dimension N, let us call sections of $(\Omega_U^N)^{\otimes \ell}$ *ℓ-fold top differentials*. Recall the general fact that, if V_1, V_2 are bimeromorphic compact analytic manifolds, and ω_1, ω_2 are meromorphic ℓ-fold top differentials on V_1, V_2 which correspond to each other, then ω_1 is regular, i.e., without poles, if and only if ω_2 is regular. Therefore, for any non-compact analytic manifold U of the form $\overline{U} \setminus Z$, with $\overline{U}$ a compact manifold, and Z a closed analytic subset, and for any holomorphic ℓ-fold top differential η on U, the condition that η extend to $\overline{U}$ is *independent* of the choice of $\overline{U}$. Let us call such η simply *extendable forms*. We see indeed from the above theorem that the criterion for $f\omega^{\otimes \ell}$ to extend to $\overline{D/\Gamma}$ is independent of the choice of $\{\sigma_\alpha^F\}$. Thus we may rephrase the conclusion as:

Corollary 1.2 *The ℓ-fold top differential $f\omega^{\otimes \ell}$ on D/Γ is extendable if and*

only if, for every F, we have that $\varphi_\rho^F \neq 0$ implies $\langle \rho, P \rangle \geq \ell$, for all $0 \neq P \in$
$U(F)_\mathbb{Z} \cap \overline{C(F)}$. $\qquad\qquad\qquad\qquad\qquad\qquad\qquad\qquad\qquad\qquad\qquad\qquad$ □

1.2

Let $\Gamma' \subset \Gamma$ be an arithmetic subgroup of G, and define $U(F)'_\mathbb{Z} = U(F) \cap \Gamma'$.

Apply the previous theorem to Γ', but keep f as an automorphic form with respect to Γ. Then $f\omega^{\otimes\ell}$ extends to $\overline{D/\Gamma'}$ if and only if we have

$$\varphi_\rho^F \neq 0 \Longrightarrow \langle \rho, P \rangle \geq \ell \text{ for all } F \text{ and all } 0 \neq P \in U(F)'_\mathbb{Z} \cap \overline{C(F)} \ . \qquad (1.1)$$

Assume that, for some positive integer q, $U(F)'_\mathbb{Z} \subseteq qU(F)_\mathbb{Z}$ for all F. Then (1.1) is a consequence of the following condition:

$$\varphi_\rho^F \neq 0 \Longrightarrow \langle \rho, P \rangle \geq \tfrac{\ell}{q} \text{ for all } F \text{ and all } 0 \neq P \in U(F)'_\mathbb{Z} \cap \overline{C(F)} \ . \qquad (1.2)$$

Now start with a cusp form of weight ℓ with respect to Γ and choose $q \geq \ell$; then f will satisfy (1.2) via the following proposition.

Proposition 1.3 *If f is a cusp form on D of weight ℓ with respect to Γ, then, for every rational boundary component F of D,*

$$\varphi_\rho^F \neq 0 \Longrightarrow \langle \rho, P \rangle \geq 1, \text{ for all non-zero } P \in U(F)_\mathbb{Z} \cap \overline{C(F)} \ .$$

Proof Since f is a cusp form, $\varphi_0^F = 0$ for every boundary component F.

Assume $\varphi_\tau^F \neq 0$ for some $\tau \in L^* \cap \overline{C(F)}$, with

$$\langle \tau, P \rangle = 0 \text{ for all } 0 \neq P \in U(F)_\mathbb{Z} \cap \overline{C(F)} \ .$$

Define

$$H_\tau = \{ x \in U(F) \mid \langle \tau, x \rangle = 0 \} \ .$$

Then $H_\tau \cap \overline{C(F)}$ defines a rational boundary component of $C(F)$. By Chapter III, Theorem 4.8, there is a rational boundary component F', with

$$F \subset \overline{F}' \text{ and } \overline{C(F')} = H_\tau \cap \overline{C(F)} \ .$$

Consider the Fourier expansion of f at F':

$$f = \sum_{\rho' \in L^* \cap \overline{C(F')}} \varphi_{\rho'}^{F'} (u', t') \exp(2\pi i \langle \rho', z' \rangle) \ .$$

Decompose $U(F)_\mathbb{C}$ into

$$U(F)_\mathbb{C} = U(F')_\mathbb{C} \oplus \mathbb{C}^k$$
$$z = z' + z'' \ .$$

Then

$$\varphi_0^{F'} = \sum_{\substack{\rho \in L^* \cap \overline{C(F')} \\ \langle \rho, C(F') \rangle = 0}} \varphi_\rho^F(u,t) \exp(2\pi i \langle \rho, z'' \rangle) \,,$$

which would not be identically zero because $\varphi_\tau^F \neq 0$. But this contradicts the fact that f is a cusp form. $\qquad\square$

We recall that, for any compact analytic manifold V of dimension N, the *Kodaira dimension* of V is defined to be

$$\kappa(V) = \mathrm{tr.deg}_{\mathbb{C}} \left(\bigoplus_{\ell=0}^\infty \Gamma(V, (\Omega_V^N)^{\otimes \ell}) \right) - 1 \,.$$

Then $-1 \leq \kappa(V) \leq N$, and V is said to be *of general type* if $\kappa(V) = N$. Equivalently, this means that, for some ℓ, there are $N+1$ ℓ-fold top differentials $\omega_0, \ldots, \omega_N$ such that $\omega_1/\omega_0, \ldots, \omega_N/\omega_0$ are algebraically independent meromorphic functions on V. If U is a non-compact manifold of type $\overline{U} \setminus X$, with $\overline{U}$ a compact analytic manifold and X a closed analytic subset, then we make the same definitions but using extendable ℓ-fold top differentials on U. Note that, in both cases, the Kodaira dimension is a biholomorphic invariant.

Theorem 1.4 *There exists* $\Gamma' \subset \Gamma$ *such that* D/Γ' *is of general type.*

Proof Let $f_0, \ldots, f_N$ be modular forms of weight ℓ such that $f_1/f_0, \ldots, f_N/f_0$ are algebraically independent, and let f be a cusp form of weight m with respect to Γ. Then $f \cdot f_0, \ldots, f \cdot f_N$ are cusp forms of weight $\ell + m$.

Let $\Gamma' \subset \Gamma$ be an arithmetic subgroup of G such that, for all rational boundary components F, we have

$$(\ell + m + 1) \cdot U(F)_{\mathbb{Z}} \supset U(F)'_{\mathbb{Z}} \,.$$

Then $f \cdot f_i$ all extend to sections of $(\Omega^N)^{\otimes(\ell+m)}$ over $\overline{D/\Gamma'}$. Hence the Kodaira dimension of $\overline{D/\Gamma'}$ is equal to N. $\qquad\square$

1.3

Let

$$K = \text{a real quadratic number field},$$
$$\mathcal{O} = \text{its ring of integers},$$
$$D = \mathfrak{H} \times \mathfrak{H} = \text{the product of upper half-planes}$$
$$\Gamma = \mathrm{SL}(2, \mathcal{O})/\{\pm I\} \,.$$

Then $\gamma = \begin{pmatrix} a & b \\ c & d \end{pmatrix} \in \Gamma$ acts on $z = (z_1, z_2) \in D$ by

$$\gamma z = \left(\frac{az_1 + b}{cz_1 + d}, \frac{a'z_2 + b'}{c'z_2 + d'} \right) ,$$

where $x \mapsto x'$ is the conjugation automorphism of K.

Let f be an automorphic form of weight ℓ:

$$f(\gamma z) = (cz_1 + d)^{2\ell}(c'z_2 + d')^{2\ell} f(z) .$$

Then f has a Fourier expansion:

$$f(z) = \sum_{\alpha \in L^* \cap \overline{V}} a_\alpha \exp(2\pi i \langle \alpha, z \rangle) ,$$

where

$$V = \text{the positive quadrant in } \mathbb{R}^2,$$
$$\langle \alpha, z \rangle = \alpha_1 z_1 + \alpha_2 z_2, \text{ for } \alpha = (\alpha_1, \alpha_2), z = (z_1, z_2),$$
$$L = \{ a \in \mathbb{R}^2 \mid z \longmapsto z + a \text{ lies in } \Gamma \},$$
$$L^* = \{ \alpha \in \mathbb{R}^2 \mid \langle \alpha, a \rangle \in \mathbb{Z} \text{ for all } a \in L \},$$

and this series is convergent for $\operatorname{Im} z_1 \operatorname{Im} z_2 \gg 0$.

Theorem 1.1 implies that $f(dz_1 \wedge dz_2)^{\otimes \ell}$ extends to $\overline{D/\Gamma}$ if and only if

$$a_\alpha = 0 \text{ for all } \alpha \text{ such that } \langle \alpha, a \rangle < \ell, \text{ for some } 0 \neq a \in L \cap \overline{V} .$$

We are going to study the number of those terms modulo the unit action. More precisely, if we identify L with $\mathcal{O}$, and define

$$(x, y) = xy + x'y' \text{ for } x, y \in K ,$$

then

$$L^* = \{ \rho \in K \mid (\rho, x) \in \mathbb{Z}, \text{ for all } x \in \mathcal{O} \} .$$

It is known as the *complementary module* of L. If we fix a base for L:

$$L = \mathbb{Z} \cdot 1 + \mathbb{Z} \cdot W ,$$

then

$$L^* = \mathbb{Z} \cdot \frac{1}{W - W'} + \mathbb{Z} \cdot \frac{W}{W - W'} .$$

Let C be the cone of totally positive numbers in K. Also fix some positive integer ℓ. Define

$$\varphi(\alpha) = \min_{0 \neq x \in L \cap \overline{C}} (\alpha, x) = \min_{x \in L \cap C} (\alpha, x) ,$$
$$\mathcal{S} = \{ \alpha \in L^* \cap \overline{C} \mid \varphi(\alpha) < \ell \} .$$

The group of totally positive units U_+ in $\mathcal{O}$ acts on $\mathcal{S}$ via $\varepsilon : \alpha \mapsto \varepsilon\alpha$.†

The following theorem was conjectured by Hirzebruch.

Theorem 1.5 *The set $\mathcal{S}/U_+$ has*

$$\tfrac{1}{2}\ell(\ell-1)\sum_{k=1}^{r}(b_k-2)$$

elements, where $(b_1,\ldots,b_r)$ is the cycle associated to the continued fraction expansion of W.

Before proving the theorem, we first construct a decomposition of C^* such that φ is linear on each sector.

By taking the convex hull of $L\cap C$, we obtain an admissible decomposition of C. Write each vertex V_k as $V_k = p_k - q_k W$ with $p_k, q_k \in \mathbb{Z}$. Assume $W > 1 > W'$. Then we have

$$p_{k-1}q_k - q_{k-1}p_k = 1 , \tag{1.3}$$

$$p_{k-1}q_{k+1} - q_{k-1}p_{k+1} = b_k \geq 2 , \tag{1.4}$$

$$V_k > V_{k+1}, V'_{k+1} > V'_k , \tag{1.5}$$

$$V_{k+1} + V_{k-1} = b_k V_k , \tag{1.6}$$

$$\lim_{k\to\infty} \frac{p_k}{q_k} = W, \ \lim_{k\to-\infty} \frac{p_k}{q_k} = W' . \tag{1.7}$$

The union of r consecutive sectors forms a fundamental domain with respect to the action of totally positive units (see Chapter I, Section 5, or Hirzebruch [4]).

For each k, define

$$W_k = \frac{p_{k+1} - p_k}{W - W'} - \frac{q_{k+1} - q_k}{W - W'}W' \in L^* .$$

By (1.5), $W_k > 0$ and $W'_k > 0$, hence $W_k \in L^* \cap C$. Furthermore,

$$(W_k, V_k) = \left(\frac{p_{k+1} - p_k}{W - W'} - \frac{q_{k+1} - q_k}{W - W'}W' \right)(p_k - q_k W)$$

$$+ \left(-\frac{p_{k+1} - p_k}{W - W'} - \frac{q_{k+1} - q_k}{W - W'}W \right)(p_k - q_k W')$$

$$= p_k(q_{k+1} - q_k) - q_k(p_{k+1} - p_k)$$

$$= 1, \text{ by } (1.3) .$$

---

† The action induced by Γ is actually $\alpha \mapsto \varepsilon^2\alpha$, which will change our number by a factor of 1 or 2.

Similarly,

$$(W_{k-1}, V_k) = 1 .$$

Define

$$C_k^* = \mathbb{R}_{>0} W_k + \mathbb{R}_{\geq 0} W_{k-1} .$$

Note from (1.3) and (1.4) that

$$\frac{p_{k+1} - p_k}{q_{k+1} - q_k} \geq \frac{p_k - p_{k-1}}{q_k - q_{k-1}} ,$$

with equality holding only when

$$b_k = 2 ,$$

in which case $W_k = W_{k-1}$ and C_k^* is only a ray.

The collection of C_k^* with $b_k \geq 3$ covers C^* because

$$\frac{a}{W - W'} - \frac{bW}{W - W'} \in C^* \Longleftrightarrow W > \frac{a}{b} > W' ,$$

and the fact that

$$\lim_{k \longrightarrow \infty} \frac{p_k - p_{k-1}}{q_k - q_{k-1}} = W , \quad \lim_{k \longrightarrow -\infty} \frac{p_k - p_{k-1}}{q_k - q_{k-1}} = W', \text{ by (1.7) .}$$

Lemma 1.6 $\overline{C_k^*} = \{\alpha \in \overline{C^*} \mid \varphi(\alpha) = \min_{x \in L \cap C}(\alpha, x) = (\alpha, V_k)\}.$

Proof Start with an element on the left. Now,

$$\alpha \in \overline{C_k^*} \Longrightarrow \alpha = a W_{k-1} + b W_k , \text{ for some } a, b \geq 0$$
$$\Longrightarrow (\alpha, V_k) = a + b \leq (\alpha, x) , \text{ for all } x \in L \cap C .$$

Conversely, suppose $\varphi(\alpha) = (\alpha, V_k) \neq 0$. Since $\{\mathbb{R}_{>0} W_i + \mathbb{R}_{>0} W_{i-1}\}$ covers C^*, we have $\alpha \subset \mathbb{R}_{>0} W_i + \mathbb{R}_{>0} W_{i-1}$ for some i. Assume that $i \neq k$. Without loss of generality, let us consider the case $i < k$. Then $\alpha = a W_{i-1} + b W_i$ for some $a, b > 0$; hence

$$a(W_{i-1}, V_k) + b(W_i, V_k) = (\alpha, V_k) \leq (\alpha, V_i) = a + b ,$$

and

$$(W_{i-1}, V_k) = 1 .$$

But we know that

$$(W_{i-1}, V_{i-1}) = (W_{i-1}, V_i) = 1 .$$

Hence V_k lies on the straight line joining V_{i-1} and V_i. By convexity, all the vertices $V_{i-1}, V_i, \ldots, V_k$ lie on the same line, and we get

$$\overline{C_i^*} = \overline{C_{i+1}^*} = \cdots = \overline{C_{k-1}^*} \subset \overline{C_k^*} ,$$

hence $\alpha \in \overline{C_k^*}$.   □

The collection of r consecutive C_k^* forms a fundamental domain in C^* with respect to the action of totally positive units.

Define $N_k = \#\{\alpha \in L^* \cap C_k^* \mid \varphi(\alpha) < \ell\}$, so that

$$\#(\mathscr{S}/U^+) = \sum_{b_k \geq 3} N_k \; .$$

Proof of Theorem 1.5 It suffices to prove

$$N_k = \tfrac{1}{2}\ell(\ell-1)(b_k-2) \; .$$

Introduce the following elements of L^*:

$$P = \frac{-p_{k-1}}{W-W'} + \frac{q_{k-1}}{W-W'}W' \; ,$$

$$Q = \frac{p_{k+1}}{W+W'} + \frac{q_{k+1}}{W-W'}W' \; .$$

Define

$$P_i = \frac{i}{b_k}P + \frac{b_k-i}{b_k}Q \text{ for } 1 \leq i \leq b_k \; .$$

We have

(1) $P_1 = W_k$, $P_{b_k-1} = W_{k-1}$ (from (1.6));

(2) $P_i \in L^*$, because $\frac{1}{b_k}(P-Q) = \frac{-p_k}{W-W'} + \frac{q_k}{W-W'}W'$ (from (1.6));

(3) $(P_i, V_k) = 1$ (from (1));

(4) $\{P_i, P_{i+1}\}$ form a basis for L^* .

To see the last point, calculate the determinant with respect to the basis $\left\{ \frac{1}{W-W'}, \frac{W'}{W-W'} \right\}$ of L^*:

$$\det\{P_i, P_{i+1}\} = \begin{vmatrix} \frac{i+1}{b_k} & \frac{b_k-i-1}{b_k} \\ \frac{i}{b_k} & \frac{b_k-i}{b_k} \end{vmatrix} \begin{vmatrix} -p_{k-1} & q_{k-1} \\ p_{k+1} & -q_{k+1} \end{vmatrix} = \frac{1}{b_k} \cdot b_k = 1 \; .$$

Define $C_{k,i}^* = \mathbb{R}_{>0}P_i + \mathbb{R}_{\geq 0}P_{i+1}$ for $i = 1, \dots, b_k - 2$, and let

$$N_{k,i} = \#\{\alpha \in L^* \cap C_{k,i}^* \mid \varphi(\alpha) < \ell\} \; ,$$

so that $N_k = \sum N_{k,i}$.

Since $\{P_i, P_{i+1}\}$ forms a base for L^*, and $\varphi(P_i) = \varphi(P_{i+1}) = 1$, we get

$$N_{k,i} = 1 + 2 + \cdots + (\ell - 1) = \tfrac{1}{2}\ell(\ell-1) \; ,$$

and hence

$$N_k = (b_k - 2) \cdot \tfrac{1}{2}\ell(\ell-1) \; .$$

□

2 **Projectivity of** $\overline{D/\Gamma}$

2.1

Let D be a bounded symmetric domain, let Γ be an arithmetic subgroup of $\mathrm{Aut}\,(D)^o$, and let $\{\sigma_\alpha^F\}$ be a Γ-admissible decomposition. Then we have constructed the associated compactification $\overline{D/\Gamma}$ of D/Γ. In this section, we are going to study the relation between $\overline{D/\Gamma}$ and Baily–Borel's compactification $(D/\Gamma)^* = D^*/\Gamma$. The main result is that $\overline{D/\Gamma}$, for certain $\{\sigma_\alpha^F\}$, is the blowing-up of D^*/Γ at a certain sheaf of ideals. Consequently $\overline{D/\Gamma}$ is projective in these cases.

For simplicity, we shall assume in this section that Γ *is neat.* (Things can be worked out without this assumption, but things will be much cleaner if we assume Γ to be neat.)

The relevant decompositions are similar to the projective subdivisions in TE I,† Ch. III, §1. We shall describe them first.

As in Chapter III, Section 7, define

$$|\widetilde{\Sigma}| = \bigsqcup_F C(F) \,, \ |\Sigma| = |\widetilde{\Sigma}|/\Gamma \text{ and } \Sigma_{\mathbb{Z}} \,.$$

Definition 2.1 A Γ-admissible decomposition $\{\sigma_\alpha^F\}$ is *projective* if there exists a continuous convex piecewise-linear function $\varphi : |\Sigma| \longrightarrow \mathbb{R}$ such that

(1) $\varphi(x) > 0$ for $x \neq 0$;

(2) φ is linear on the image of $\overline{\sigma_\alpha^F}$, and $\overline{\sigma_\alpha^F}$ are the maximal polyhedral cones in $\overline{C(F)}$ on which φ is linear ;

(3) φ is integral on $\Sigma_{\mathbb{Z}}$.

The existence of such decompositions follows from the theory of co-cores (see Chapter II, Section 5). Indeed, let $\{\Delta_F\}$ be a system of $\overline{\Gamma}_F$-polyhedral co-cores, one for each rational boundary component F, such that

(1) for $F \subset \overline{F}'$, $\Delta_F \cap \overline{C(F')} = \Delta_{F'}$,

(2) for $F' = \gamma F$ with $\gamma \in \Gamma$, we have $\Delta_{F'} = \gamma \Delta_F$.

From Chapter II, Section 5, we know that the cones over the faces of Δ_F define a Γ-admissible decomposition. Fix a sufficiently divisible integer N. For each F, let φ_F be the unique convex piecewise-linear function on $\overline{C(F)}$ such that it has value N at each face of Δ_F and is linear on the cone over each face. Then $\{\varphi_F\}$ defines a function φ on $|\Sigma|$ with the required properties.

Let us start with a projective Γ-admissible decomposition $\{\sigma_\alpha^F\}$ and $\varphi : |\Sigma| \longrightarrow \mathbb{R}$ satisfying the required properties. Then φ defines a collection of continuous piecewise-linear functions $\{\varphi_F\}$ on $\{\overline{C(F)}\}$ such that φ_F is linear

† Recall this reference from p. x.

on each σ_α^F. We define a dual piecewise-linear function φ_F^* on $C(F)$ such that, in the terminology of Chapter II, Subsection 5.2,

$$\{\varphi_F^*(\lambda) \geq 1\} = \text{the core dual to the co-core } \{\varphi_F(x) \geq 1\} \ .$$

More explicitly, let

$$\{P_{\alpha,i}\} = \text{vertices of } \sigma_\alpha^F \cap \{\varphi_F = 1\} \ ,$$

and define

$$\varphi_F^*(\lambda) = \min_{\alpha,i} \langle \lambda, P_{\alpha,i} \rangle \ , \text{ for } \lambda \in C(F) \ .$$

Moreover, for all top-dimensional σ_α^F, let

$$\lambda_\alpha \in U(F)_{\mathbb{Z}}^* = \text{Hom}(U(F)_{\mathbb{Z}}, \mathbb{Z})$$

be the linear function such that

$$\lambda_\alpha|_{\sigma_\alpha^F} = \varphi_F|_{\sigma_\alpha^F}$$

(this exists by assumptions (2) and (3) on φ). Note that $\varphi_F^*(\lambda_\alpha) = 1$. Now we can define the ideal of the blowing-up as follows.

Let $x \in F/\Gamma(F) \subset D^*/\Gamma$. Then the holomorphic functions around x are described by the Fourier–Jacobi series (for notation, see Subsection 2.2 below) of the following form:

$$f = \sum_{\rho \in \overline{C(F)} \cap U(F)_{\mathbb{Z}}^*} \theta_\rho(u,t) \exp(2\pi i \langle \rho, z \rangle) \ .$$

Define

$$\mathscr{I}_{m,x} = \{ f \in \mathscr{O}_x \mid \theta_\rho \neq 0 \text{ only for } \rho \in U(F)_{\mathbb{Z}}^* \cap C(F) \text{ and } \varphi_F^*(\rho) \geq m \} \ .$$

Since φ_F^* is convex, $\mathscr{I}_{m,x}$ is an ideal. We shall see later that $\{\mathscr{I}_{m,x}\}$ form a coherent sheaf of ideals $\mathscr{I}_m$ concentrated at the boundary. (Note that $\mathscr{I}_m$ depends on the choice of φ as well as on m.) We shall often denote $\mathscr{I}_m$ simply by $\mathscr{I}$.

Our aim is to prove the following theorem.

Theorem 2.2 Let $\{\sigma_\alpha^F\}$ be a projective Γ-admissible decomposition, let $\mathscr{I}_m$ be the sheaf of ideals on D^*/Γ constructed as above (with a suitable m fixed in the course of the proof), and let $\widetilde{(D/\Gamma)}_{\mathscr{I}}$ be the normalization of the blowing-up at $\mathscr{I}$. Then $\widetilde{(D/\Gamma)}_{\mathscr{I}}$ is isomorphic to the compactification $\overline{D/\Gamma}$ associated to $\{\sigma_\alpha^F\}$.

Corollary 2.3 If $\{\sigma_\alpha^F\}$ is projective, then the associated $\overline{D/\Gamma}$ is projective. $\square$

Corollary 2.4 *There are* Γ-*admissible collections* $\{\sigma_\alpha\}$ *of polyhedra such that the associated compactification* $\overline{D/\Gamma}$ *is non-singular and projective.*

Proof Indeed, start with some projective $\{\sigma_\alpha^F\}$ and then apply the refining procedure of Chapter III, Corollary 7.6. $\qquad\square$

2.2

Let F be a rational boundary component of D, let $x' \in F \subset D^*$, let x be the image of x' in D^*/Γ, and let U be a neighborhood of x' in D^* such that

$$U \cap D = \pi_F^{-1}(E) \cap \Phi_F^{-1}(\ell C_0),$$

where E is a relatively compact open neighborhood of x' in F such that

- $\gamma E \cap E = \emptyset$ for id $\neq \gamma \in \Gamma(F)$,
- C_0 is the interior of a core in $C(F)$,
- $\ell > 0$,
- $\gamma U = U$ for $\gamma \in \Gamma_{x'}$.

By Chapter III, Section 6, such U actually form a fundamental system of neighborhoods of x'.

Let f be a holomorphic function on $U \cap D$; such an f defines an element in $\mathcal{O}_x$ if, for all $y \in U \cap D$,

$$f(\gamma y) = f(y) \text{ for all } \gamma \in \Gamma_{x'}.$$

To study this invariance condition we need explicit forms of the $N(F)$-action. As in Chapter III, Section 4, and in Koranyi–Wolf [7], D can be realized as a Siegel domain of the third kind:

$$D \cong \{(t,u,z) \in F \times \mathbb{C}^k \times U(F)_{\mathbb{C}} \mid \operatorname{Im} z - \operatorname{Re} L_t(u,u) \in C(F)\},$$

where L_t is a quasi-hermitian form (a sum of a hermitian form H_t and a symmetric form S_t) depending analytically on t, where $2k = \dim V(F)$, and where $H_t(u,u) \in \overline{C(F)}$, for all u. Recall that, for each $t_0 \in F$, the action of $V(F)$ on the points $(t_0, u, \cdot) \in D$ makes $\mathbb{C}^k$ into a principal homogeneous space over $V(F)$, and hence identifies $V(F)$ with $\mathbb{C}^k$ via mapping x to the second coordinate of $x(t_0, 0, \cdot)$. But this identification *varies with* t_0.

Proposition 2.5

(1) *In the coordinates* (t,u,z), *the action of* $N(F)$ *consists of quasi-linear*

transformations, i.e., every $\gamma \in N(F)$ *acts by:*

$$z \longmapsto Az + a(u,t) \,,$$
$$u \longmapsto B_t u + b_t \,,$$
$$t \longmapsto g(t) \,,$$

where $A \in \mathrm{Aut}\,(C(F))$, B_t *is linear in* u, *and* B_t, b_t, $g(t)$, *and* $a(u,t)$ *are analytic in* t *and* u.

(2) $W(F)$ *consists of the following transformations:*

$$(b,a) = \begin{cases} z & \longmapsto & z + a + 2\mathrm{i}L_t(u,b_t) + \mathrm{i}L_t(b_t,b_t) \,, \\ u & \longmapsto & u + b_t \,, \\ t & \longmapsto & t \,, \end{cases}$$

where $a \in U(F)$, $b \in V(F)$, *and* $b_t \in \mathbb{C}^k$ *is given by the identification of* $V(F)$ *and* $\mathbb{C}^k$, *depending analytically on* t.

(3) *Let* $Z'(F) \subset Z(F)$ *be the subgroup consisting of transformations of the form:*

$$(B,A) = \begin{cases} z & \longmapsto & Az + \mathrm{i}L_t(B_t u, B_t u) - \mathrm{i}AL_t(u,u) \,, \\ u & \longmapsto & B_t u \,, \\ t & \longmapsto & t \,, \end{cases}$$

with A, B_t *satisfying*

$$AH_t(u,u) = H_t(B_t u, B_t u), \quad A \in \mathrm{Aut}\,(C(F)) \,.$$

Then we have the semi-direct product decomposition:

$$Z(F) = Z'(F) \ltimes W(F) \,.$$

For the proof of (1) and (2), see [7] and [9]; for (3), see [6]. We make some remarks about (2) and (3).

(i) By simple computations, one can show

$$(b,a)(b',a') = (b + b', a + a' + 2[b,b']) \,,$$

where $[b,b'] = \mathrm{Im}\,H_t(b_t, b'_t)$. Hence

$$(b,a)(b',a')(b,a)^{-1}(b',a')^{-1} = (0,4[b,b']) \,.$$

(ii) The form that (B,A) is given shows that (B,A) transforms D to D (in the given realization). The condition $AH_t(u,u) = H_t(B_t u, B_t u)$ guarantees that $Z'(F)$ normalizes $W(F)$. By computation, we have that, if $\gamma = (b,a)(B,A)$, then

$$\gamma(b',a')\gamma^{-1} = (Bb', Aa' + 4[b,Bb']) \,.$$

(iii) For every $\beta \in V(F)$, there is a unique function b_t as in (2) such that $(b,0)$ represents the action of β, and all the b_t correspond to some β.

(iv) In the notation of Chapter III, Section 4, $G_\ell(F) \cdot M(F)$ is the identity component of $Z'(F)$.

Let $V(F)_{\mathbb{Z}}$ be the image of $W(F) \cap \Gamma$ in $V(F)$ considered as the quotient $W(F)/U(F)$.

Let $\overline{\Gamma}_F$ be the image of $Z(F) \cap \Gamma$ in $\mathrm{Aut}\,(C(F))$. Note that this is a subgroup of finite index in the $\overline{\Gamma}_F$ defined in Chapter III, Section 5, which was the image of $N(F) \cap \Gamma$ in $\mathrm{Aut}\,(C(F))$.

Lemma 2.6 *The following sequences are exact:*

$$1 \longrightarrow U(F)_{\mathbb{Z}} \longrightarrow W(F) \cap \Gamma \longrightarrow V(F)_{\mathbb{Z}} \longrightarrow 1 \,,$$

$$1 \longrightarrow W(F) \cap \Gamma \longrightarrow Z(F) \cap \Gamma \longrightarrow \overline{\Gamma}_F \longrightarrow 1 \,.$$

Proof The exactness of the first sequence follows from the definition of $V(F)_{\mathbb{Z}}$. To prove that the second sequence is exact, we need to prove

$$\ker\,(Z(F) \cap \Gamma \longrightarrow \mathrm{Aut}\,(C(F))) = W(F) \cap \Gamma \,.$$

Recall from Chapter III, Section 4, that we have

$$Z(F)^o = [G_\ell(F) \cdot M(F)] \ltimes W(F) \,.$$

Therefore $\ker\,(Z(F) \cap \Gamma \longrightarrow \mathrm{Aut}\,(C(F)))\,/W(F) \cap \Gamma$ is contained in the compact factors, and, since it is discrete, it is finite. But

$$\ker\,(Z(F) \cap \Gamma \to \mathrm{Aut}\,(C(F)))\,/W(F) \cap \Gamma \subset (\mathscr{N}(F) \cap \Gamma)/\mathscr{W}(F) \cap \Gamma \,,$$

which is torsion-free, since Γ is neat and $\mathscr{N}(F)$ and $\mathscr{W}(F)$ are both defined over $\mathbb{Q}$. $\qquad\qquad\square$

Lemma 2.7 $[G_\ell(F) \cap \Gamma] \ltimes [W(F) \cap \Gamma]$ *is of finite index in* $Z(F) \cap \Gamma$.

Proof Since G_ℓ, W, and $G_\ell \ltimes W$ are defined over $\mathbb{Q}$, the following are arithmetic subgroups:

$$G_\ell(F) \cap \Gamma \subset G_\ell(F) \,,$$

$$W(F) \cap \Gamma \subset W(F) \,,$$

$$(G_\ell(F) \ltimes W(F)) \cap \Gamma \subset G_\ell(F) \ltimes W(F) \,.$$

From the general theory of arithmetic subgroups, $(G_\ell(F) \cap \Gamma) \ltimes (W(F) \cap \Gamma)$ is also an arithmetic subgroup of $G_\ell(F) \ltimes W(F)$, and hence $(G_\ell(F) \cap \Gamma) \ltimes (W(F) \cap \Gamma)$ is of finite index in $(G_\ell(F) \ltimes W(F)) \cap \Gamma$. But $Z(F)/(G_\ell(F) \ltimes W(F))$ is compact, so $Z(F) \cap \Gamma/(G_\ell(F) \ltimes W(F)) \cap \Gamma$ is also finite. $\qquad\square$

Going back to our invariance condition, if we start with f in $\mathcal{O}_x$, then $f(\gamma y) = f(y)$, for all $\gamma \in \Gamma_x$, $y \in U \cap D$.

The invariance condition by the elements in $U(F)_{\mathbb{Z}}$ implies that f admits the Fourier–Jacobi series expansion:

$$f(t,u,z) = \sum_{\rho \in U(F)_{\mathbb{Z}}^*} \theta_\rho(u,t) \exp(2\pi i \langle \rho, z \rangle) .$$

If $\gamma \in V(F)_{\mathbb{Z}}$ and lifts to $(b,a) \in W(F) \cap \Gamma$, the invariance condition shows

$$\theta_\rho(u + b_t, t) = \theta_\rho(u,t) \exp(2\pi i \langle \rho, -2iL_t(u,b_t) - iL_t(b_t,b_t) - a \rangle) . \qquad (2.1)$$

For simplicity, call $e_{b_t}(u)$ the exponential factor appearing here. (Note that a is uniquely determined modulo $U(F)_{\mathbb{Z}}$.) Define the line bundle $\mathscr{L}_\rho$ on the family of complex tori $(\mathbb{C}^k \times F)/V(F)_{\mathbb{Z}} \xrightarrow{\pi} F$ as $\mathbb{C} \times \mathbb{C}^k \times F$ modulo the following action of $V(F)_{\mathbb{Z}}$:

$$(\alpha, u, t) \longmapsto (e_{v_t}(u)\alpha, v_t + u, t) , \ v \in V(F)_{\mathbb{Z}} .$$

Equation (2.1) just means $\theta_\rho \in \Gamma(\pi^{-1}(E), \mathscr{L}_\rho)$.

It is shown in [9] and [1] that the convergence of the series for f requires that $\rho \in \overline{C(F)}$ whenever $\theta_\rho \neq 0$.

Define

$$g(u) = \exp(-2\pi \langle \rho, S_t(u,u) \rangle) ,$$

$$e'_{v_t}(u) = e_{v_t}(u) g(v_t + u) g(u)^{-1} .$$

By simple computations,

$$e'_{v_t}(u) = \exp(2\pi \langle \rho, 2H_t(u,v) + H_t(v,v) \rangle) \cdot \exp(-2\pi i \langle \rho, a \rangle) .$$

In the notation of [8], this shows that the line bundle $\mathscr{L}_\rho|_{\pi^{-1}(t)}$ is algebraically equivalent to $\mathscr{L}(4\langle \rho, H_t \rangle)$.

Furthermore, for $\rho \in L^* \cap C(F)^*$, since $H_t(u,u) \in \overline{C(F)}$ for all u, we have that $4\langle \rho, H_t \rangle$ is positive-definite. By the remarks following Proposition 2.5, $\operatorname{Im} 4\langle \rho, H_t(u,v) \rangle$ is integral on $V(F)_{\mathbb{Z}} \times V(F)_{\mathbb{Z}}$. In particular, for our λ_α defined in Subsection 2.1, we know $\mathscr{L}_{m\lambda_\alpha}|_{\pi^{-1}(t)}$ is generated by its sections if $m \geq 2$, and hence so is $\mathscr{L}_{m\lambda_\alpha}|_{\pi^{-1}(E)}$.

Now consider $A \in \overline{\Gamma}_F$, and assume it lifts to $(d,c)(B,A)$ in $Z(F) \cap \Gamma$. By Lemma 2.7, for all $A \in \overline{\Gamma}_F$, we can get such liftings while choosing (d,c) in a finite set.

For $\gamma = (d,c)(B,A)$, the invariance condition gives:

$$\theta_{A^*\rho}(u,t) = \theta_\rho(B_t u + d_t, t) \exp(2\pi i \langle \rho, a(u,t) \rangle) , \qquad (2.2)$$

where

$$a(u,t) = iL_t(B_t u, B_t u) - iAL_t(u,u) + 2iL_t(B_t u, d_t) + iL_t(d_t, d_t) + c .$$

Note that by Lemma 2.6, if $A \in \overline{\Gamma}_F$, then (d,c) is uniquely determined modulo $W(F) \cap \Gamma$, and B_t is uniquely determined.

Conversely, start with $\theta_\lambda \in \Gamma(\pi^{-1}(E), \mathscr{L}_\lambda)$ for some neighborhood E of x' in F and define $\theta_{A*\lambda}$ by (2.2) ($A \in \overline{\Gamma}_F$). Then form the sum

$$f\theta_\lambda = \sum_{A \in \overline{\Gamma}_F} \theta_{A*\lambda} \exp(2\pi i \langle A^*\lambda, z \rangle) \, .$$

By Lemma 2.6, and the discussion above, the local ring $\mathscr{O}_x$ is generated by such $f\theta_\lambda$ if we know their convergence. For this, we prove the following proposition.

Proposition 2.8 *For all* $A \in \overline{\Gamma}_F$, $a \in C(F)$, *and* θ_λ *as above, there is an* $M > 0$ *such that*

$$|f\theta_\lambda| \le M \text{ in } \pi_F^{-1}(E) \cap \Phi_F^{-1}(a + C(F)) \, .$$

Proof Since F is $\mathbb{Q}$-rational, $V(F)/V(F)_{\mathbb{Z}}$ is compact. Let $S \subset V(F)$ be a compact fundamental set for $V(F)_{\mathbb{Z}}$. Using the action of $V(F)$ on $\mathbb{C}^k \times F$, let $S_1 \subset \mathbb{C}^k$ be a compact set such that $S \cdot [(0) \times E] \subset S_1 \times E$.

By (2.2), we have

$$|\theta_{A*\lambda}(u,t)| = |\theta_\lambda(B_t u + d_t, t)| \exp(-2\pi \langle \lambda, \operatorname{Im} a(u,t) \rangle) \, .$$

Writing $h_t = \operatorname{Re} L_t$, we find

$$\operatorname{Im} a(u,t) = h_t(B_t u, B_t u) - A h_t(u,u) + 2 h_t(B_t u, d_t) + h_t(d_t, d_t) \, .$$

Decompose $B_t u$ as $u_t + b_t$, with $u_t \in S_1$ and $b \in V(F)_{\mathbb{Z}}$. Then

$$|\theta_\lambda(B_t u + d_t, t)| = |\theta_\lambda(u_t + d_t + b_t, t)|$$
$$= |\theta_\lambda(u_t + d_t, t)| \exp(-2\pi \langle \lambda, a'(u,t) \rangle) \, ,$$

where, by (2.1),

$$a'(u,t) = -2 h_t(u_t + d_t, b_t) - h_t(b_t, b_t)$$
$$= -2 h_t(d_t, b_t) + h_t(u_t, u_t) - h_t(u_t + b_t, u_t + b_t)$$
$$= -2 h_t(d_t, B_t u) + 2 h_t(d_t, u_t) + h_t(u_t, u_t) - h_t(B_t u, B_t u) \, .$$

Since t is in a compact set and the (d,c) are in a finite set, the d_t are in a compact set. Further, $u_t \in S_1$, which is compact. Hence

$$|\theta_{A*\lambda}(u,t) \exp(2\pi i \langle A^*\lambda, z \rangle)| \le C \cdot \exp(-2\pi \langle A^*\lambda, \operatorname{Im} z \rangle) \exp(2\pi \langle \lambda, A h_t(u,u) \rangle)$$
$$= C' \cdot \exp(-2\pi \langle A^*\lambda, \operatorname{Im} z - h_t(u,u) \rangle)$$
$$\le C'' \cdot \exp(-2\pi \langle A^*\lambda, a \rangle) \, ,$$

for suitable constants C, C', C''. Let

$$a_n = \#\{A \in \overline{\Gamma}_F \mid n \le \langle A^*\lambda, a \rangle \le n+1\} \, ;$$

then (because the action of $\overline{\Gamma}_F$ is fixed-point free)

$$a_n \leq \#\{x \in U(F)^*_{\mathbb{Z}} \cap C(F) \mid n \leq (x,a) \leq n+1\},$$

which grows like the volume of $C(F) \cap \{(x,a) = n\}$. Hence $a_n \leq \text{const.} \cdot n^K$ for some constant K. Therefore,

$$\sum \exp(-2\pi\langle A^*\lambda, a\rangle) \leq \sum_{n\geq 0} a_n \exp(-2\pi n) \leq \text{const.} \cdot \sum_{n\geq 0} n^K \exp(-2\pi n) < \infty.$$

$\square$

Proposition 2.9 $\mathcal{O}_x$ is generated by f_{θ_λ} with $\theta_\lambda \in \Gamma(\pi^{-1}(E), \mathscr{L}_\lambda)$ and $\lambda \in \overline{C(F)} \cap U(F)^*_{\mathbb{Z}}$.

Proof This follows from Proposition 2.8 and the fact that ℓC_0 is, modulo $\overline{\Gamma}_F$, contained in a finite union of cylindrical sets $a + C(F)$. $\square$

Proposition 2.10 *Assume* $\text{Int}\,\sigma^F_\alpha \subset C(F)$, *with* σ^F_α *top-dimensional. Fix* $a \in C(F)$, $K > 0$, *and* $\theta_\lambda \in \Gamma(\pi^{-1}(E), \mathscr{L}_\lambda)$, *where* $\varphi^*_F(\lambda) \geq m$. *Then there exists* $M > 0$ *such that*

$$|f_{\theta_\lambda} \exp(\langle -m\lambda_\alpha, z\rangle)| < M$$

in $\pi^{-1}_F(E) \cap \Phi^{-1}_F(\sigma^F_\alpha + a) \cap \{|u| \leq K\}$.

Proof Recall the definition of φ^*_F: we have points $\{P_{i,\alpha}\}$ such that

$$\sigma^F_\alpha = \sum \mathbb{R}_{\geq 0} P_{i,\alpha} \text{ and } \langle \lambda_\alpha, P_{i,\alpha}\rangle = 1,$$
$$\varphi^*_F(\lambda) = \min_{i,\alpha} \langle \lambda, P_{i,\alpha}\rangle.$$

Since $\varphi^*_F(A^*\lambda) = \varphi^*_F(\lambda) \geq m$, we have $A^*\lambda - m\lambda_\alpha \geq 0$ in σ_α, and hence

$$\langle A^*\lambda - m\lambda_\alpha, a\rangle \geq 0 \text{ for all } A \in \overline{\Gamma}_F.$$

The rest of the proof runs parallel to that of Proposition 2.8, where we proved the similar statement for f_{θ_λ}. $\square$

We now prove that the ideals $\mathscr{I}_{m,x}$ defined in Subsection 2.1 form a coherent sheaf of ideals. We first define a sheaf of principal ideals $\mathscr{J}$ on $\overline{D/\Gamma}$ in the following steps.

(1) Define $\mathscr{J}_F$ on $(D/U(F)_{\mathbb{Z}})_{\{\sigma^F_\alpha\}}$: Let $X_\alpha = (D/U(F)_{\mathbb{Z}})_{\sigma^F_\alpha}$, and, for all top-dimensional σ^F_α, let

$$\mathscr{J}_{F,\alpha} = \mathcal{O}_{X_\alpha} \mathfrak{x}^{\lambda_\alpha}, \text{ where } \mathfrak{x}^{\lambda_\alpha} = \exp(2\pi i\langle \lambda_\alpha, z\rangle).$$

The $\{\mathscr{J}_{F,\alpha}\}$ define an ideal $\mathscr{J}_F$ on $(D/U(F)_{\mathbb{Z}})_{\{\sigma_\alpha^F\}}$ since, if $\sigma_\alpha \cap \sigma_\beta = \sigma_\gamma$, then on X_γ we have $\mathscr{J}_{F,\gamma} = \mathscr{O}_{X_\gamma} \mathfrak{X}^{\lambda_\alpha} = \mathscr{O}_{X_\gamma} \mathfrak{X}^{\lambda_\beta}$.

(2) If $F \subset \overline{F'}$, then $\overline{C(F)} \supset \overline{C(F')}$, and we have the glueing map:

$$\alpha_{F',F} : (D/U(F')_{\mathbb{Z}})_{\{\sigma_\alpha^{F'}\}} \longrightarrow (D/U(F)_{\mathbb{Z}})_{\{\sigma_\alpha^F\}} .$$

The $\{\sigma_\alpha^{F'}\}$ are the cones $\sigma_\alpha^F \cap \overline{C(F')}$. We get $\alpha_{F',F}$ by dividing out $U(F)_{\mathbb{Z}}$, plus an open immersion. If $\sigma_\alpha^{F'} = \sigma_\alpha^F \cap \overline{C(F')}$, then $\lambda_\alpha = \lambda'_\alpha$ on $\sigma_\alpha^{F'}$, hence

$$\alpha_{F',F}^*(\mathscr{J}_F) = \mathscr{J}_{F'} .$$

(3) Similarly, if $F' = \gamma F$ for some $\gamma \in \Gamma$, then we have

$$\alpha_{F',F} : (D/U(F')_{\mathbb{Z}})_{\{\sigma_\alpha^{F'}\}} \longrightarrow (D/U(F)_{\mathbb{Z}})_{\{\sigma_\alpha^F\}} ,$$

with $\{\sigma_\alpha^{F'}\} = \{\gamma \sigma_\alpha^F\}$. Now, $\{\lambda'_\alpha\}$ is just $\{\gamma^* \lambda_\alpha\}$, hence

$$\alpha_{F',F}^*(\mathscr{J}_F) = \mathscr{J}_{F'} .$$

(4) As in Chapter III, Section 6, define

$$\widetilde{D/\Gamma} = \bigsqcup_F (D/U(F)_{\mathbb{Z}})_{\{\sigma_\alpha^F\}} .$$

Let ι_F be the injection $(D/U(F)_{\mathbb{Z}})_{\{\sigma_\alpha^F\}} \longrightarrow \widetilde{D/\Gamma}$ and define $\mathscr{J}$ on $\widetilde{D/\Gamma}$ as $\bigoplus_F \iota_{F,*}\mathscr{J}_F$.

(5) Let p be the map $\widetilde{D/\Gamma} \longrightarrow \overline{D/\Gamma}$, and define $\mathscr{J}$ by

$$\Gamma(U, \mathscr{J}) = \left\{ s \in \Gamma(p^{-1}(U), \widetilde{\mathscr{J}}) \ \middle| \ \begin{array}{l} \alpha_{F',F}^* \iota_F^*(s) = \iota_{F'}^*(s) \text{ if } F \subset \overline{F'} \\ \text{or } F' = \gamma F \text{ for some } \gamma \in \Gamma \end{array} \right\} .$$

It is easy to check that in each step we get a principal sheaf of ideals and that $\mathscr{J}$ is locally generated by $\mathfrak{X}^{\lambda_\alpha}$.

Proposition 2.11 *Let* $f : \overline{D/\Gamma} \longrightarrow D^*/\Gamma$ *be the map defined in Chapter III, Section 5, and let* $\mathscr{I}_m = f_* \mathscr{J}^m$. *Then*

(1) *$\mathscr{I}_m$ is a coherent sheaf of ideals.*

(2) *Let* $x \in F/\Gamma(F) \subset D^*/\Gamma$. *Then* $\mathscr{I}_{m,x}$ *is the ideal of holomorphic functions around x such that in their Fourier–Jacobi series expansion,*

$$\sum \theta_\rho \exp(2\pi i \langle \rho, z \rangle) ,$$

the coefficient $\theta_\rho \neq 0$ *only for* $\rho \in U(F)_{\mathbb{Z}}^* \cap C(F)$ *and* $\varphi_F^*(\rho) \geq m$.

Proof (1) follows from Grauert's coherency theorem, since f is obviously proper.

To prove (2), let V be an open neighborhood of x in D^*/Γ; let U be a connected component of the inverse image of V in D^*. We may choose U as at the beginning of this section:

$$U \cap D = \pi_F^{-1}(E) \cap \Phi_F^{-1}(\ell C_0) .$$

Let $U_\alpha = U \cap \Phi_F^{-1}(\sigma_\alpha^F)$. Then

$$\Gamma(V, \mathscr{I}_m) = \{ f \in \mathscr{O}_V \mid f \cdot \mathfrak{X}^{-m\lambda_\alpha} \text{ is holomorphic in } U_\alpha \text{ for all } \alpha \} .$$

If $f \in \Gamma(V, \mathscr{I}_m)$, consider the Fourier expansion of f:

$$f = \sum \theta_\rho \exp(2\pi i \langle \rho, z \rangle) .$$

As above, let $P_{i,\alpha}$ be the vertices of σ_α^F, and let $D_{i,\alpha}$ be the corresponding codimension-one strata of $U_\alpha \setminus U_\alpha \cap D$. Then a term

$$\theta_\rho \exp(2\pi i \langle \rho - m\lambda_\alpha, z \rangle)$$

of $f \cdot \mathfrak{X}^{-m\lambda_\alpha}$ vanishes on $D_{i,\alpha}$ to order $\langle \rho - m\lambda_\alpha, P_{i,\alpha} \rangle$ (or has a pole on $D_{i,\alpha}$ if $\langle \rho - m\lambda_\alpha, P_{i,\alpha} \rangle < 0$). Thus $f \in \Gamma(V, \mathscr{I}_m)$ implies that $\theta_\rho \neq 0$ only when

$$\langle \rho - m\lambda_\alpha, P_{i,\alpha} \rangle \geq 0 \text{ for all } i, \alpha .$$

But this means that

$$\varphi_F^*(\rho) = \min \langle \rho, P_{i,\alpha} \rangle \geq \min \langle m\lambda_\alpha, P_{i,\alpha} \rangle = m .$$

Conversely, if $\varphi_F^*(\rho) \geq m$ whenever $\theta_\rho \neq 0$, then, by Proposition 2.10, the function $f \cdot \mathfrak{X}^{-m\lambda_\alpha}$ is bounded on $U \cap D \cap \Phi_F^{-1}(\sigma_\alpha + a)$, for all α and a, hence extends holomorphically to U_α.   □

2.3

We will now give the proof of Theorem 2.2. We shall in fact prove slightly more: namely, in the notation of Proposition 2.11, we shall prove that, for suitable m,

$$\overline{D/\Gamma} = \text{normalization of blow-up of } D^*/\Gamma \text{ along } \mathscr{I}_m ,$$
$$\mathscr{J}^m = f^* \mathscr{I}_m .$$

Step I We first remark that it suffices to prove the theorem for some normal subgroup Γ' of finite index in Γ.

Write $X = \overline{D/\Gamma}$ and $H = \Gamma/\Gamma'$, and let $X' = \overline{D/\Gamma'}$ be the compactification corresponding to the same $\{\sigma_\alpha^F\}$. By uniqueness of $\overline{D/\Gamma}$, we have

$$X \cong X'/H .$$

Write $S = D^*/\Gamma$ and $S' = D^*/\Gamma'$. The group H acts on S' and defines a map $h : S' \longrightarrow S$ which induces an isomorphism $S'/H \cong S$.

Lemma 2.12 $(h_* \mathscr{I}'_m)^H = \mathscr{I}_m.$

Proof Consider the following commutative diagram:

$$
\begin{array}{ccc}
X' & \xrightarrow{\;g\;} & X \\
\downarrow{\scriptstyle f'} & & \downarrow{\scriptstyle f} \\
S' & \xrightarrow{\;h\;} & S
\end{array}
$$

We have $\mathscr{J}$ and $\mathscr{J}'$ on X and X', both are generated locally by $\mathfrak{x}^{\lambda_\alpha}$, and $f_* \mathscr{J}^m = \mathscr{I}_m$ and $f'_* \mathscr{J}'^m = \mathscr{I}'_m$. Moreover $\mathscr{J}$ and $\mathscr{J}'$ are related by $\mathscr{J} = (g_* \mathscr{J}')^H$; hence

$$(h_* \mathscr{I}'_m)^H = (h_* f'_* \mathscr{J}'^m)^H = (f_* g_* \mathscr{J}'^m)^H = f_*(g_* \mathscr{J}'^m)^H = f_* \mathscr{J}^m = \mathscr{I}_m .$$

$\square$

Lemma 2.13 *If* $f'^* \mathscr{I}'_m = \mathscr{J}'^m$, *then* $\mathscr{I}'_{mk}$ *is the integral closure* $\widehat{\mathscr{I}'^k_m}$ *of* $\mathscr{I}'^k_m$, *for all* $k \geq 1$.

Proof Since f' is proper, whenever $f'^* \mathscr{K}$ is a sheaf of principal ideals, then $f'_* f'^* \mathscr{K} = \mathscr{K}$. Therefore

$$\mathscr{I}'_{mk} = f'_*(\mathscr{J}'^{mk}) = f'_* f'^*(\mathscr{I}'^k_m) = \widehat{\mathscr{I}'^k_m} .$$

$\square$

Introduce the notation $\widetilde{X}_\mathscr{I}$ for the normalization of the blow-up of a variety X along a coherent sheaf of ideals $\mathscr{I}$. It is well known that

(a) $\widetilde{X}_\mathscr{I} \cong \widetilde{X}_{\mathscr{I}^n}$ for all $n \geq 1$,
(b) $\widetilde{X}_\mathscr{I} \cong \widetilde{X}_{\widehat{\mathscr{I}}}$ with $\widehat{\mathscr{I}}$ the integral closure of $\mathscr{I}$.

We will apply the general fact:

Lemma 2.14 *Let X be a normal quasi-projective variety, let H be a finite group acting on X, and let $\mathscr{I}$ be an H-invariant coherent sheaf of ideals on X. If n denotes the order of H, and $\pi : X \longrightarrow X/H$ the canonical map and $\mathscr{J} = \pi_*(\mathscr{I}^n)^H$, then*

$$\tilde{X}_{\mathscr{I}}/H \cong \widetilde{(X/H)}_{\mathscr{J}} \ .$$

Moreover, $\mathscr{J}$ induces the ideal $\mathscr{I}^n \cdot \mathscr{O}_{\tilde{X}_{\mathscr{I}}}$ on $\tilde{X}_I$.

Proof It is easy to check that

$$\tilde{X}_{\pi^* \mathscr{J}}/H \cong \widetilde{(X/H)}_{\mathscr{J}} \ ,$$

and hence the lemma follows if we prove that $\widehat{\mathscr{I}^n} = \widehat{\pi^* \mathscr{J}}$. To do this, it suffices to prove that $\mathscr{I}^n$ and $\pi^* \mathscr{J}$ generate the same sheaf of ideals on $\hat{X}_{\mathscr{I}}$. The difficult point here is to check that $\mathscr{J}$ generates the full sheaf $(\mathscr{I} \cdot \mathscr{O}_{\tilde{X}_{\mathscr{I}}})^n$. Take any point $\tilde{x} \in \tilde{X}_{\mathscr{I}}$, let x be its image in X, and let $U \subset X$ be an H-invariant affine neighborhood of x. It is easy to see that there is an $f \in \Gamma(U, \mathscr{I})$ that generates the principal ideal sheaf $\mathscr{I} \cdot \mathscr{O}_{\tilde{X}_{\mathscr{I}}}$ at each of the points $\sigma \tilde{x}$, $\sigma \in H$. Then $f' = \prod_{\sigma \in H}(\sigma f)$ is a section of $\mathscr{J}$ on U/H which generates $(\mathscr{I} \cdot \mathscr{O}_{\tilde{X}_{\mathscr{I}}})^n$ at $\tilde{x}$. □

Corollary 2.15 *The same thing holds if we let $\mathscr{J}$ equal $\pi_*(\widehat{\mathscr{I}^n})^H$.*

Proof The same proof works in fact. □

Now, assuming the theorem for Γ', we have an m such that

$$X' \cong \tilde{S}'_{\mathscr{I}'_m} \ , \quad f'^* \mathscr{I}'_m = \mathscr{J}'^m \ .$$

Therefore, by Lemmas 2.12 and 2.13,

$$(h_* \widehat{(\mathscr{I}'_m)^k})^H = (h_* \mathscr{I}'_{mk})^H = \mathscr{I}_{mk} \ ,$$

and, by Lemma 2.14,

$$X \cong X'/H \cong \tilde{S}'_{\mathscr{I}'_m}/H \cong \widetilde{(S'/H)}_{\mathscr{I}_{mk}} = \tilde{S}_{\mathscr{I}_{mk}} \ ,$$

and $\mathscr{I}_{mk}$ induces the ideal $\mathscr{J}'^{mk}$ on X', hence $\mathscr{I}_{mk}$ induces the ideal $\mathscr{J}^{mk}$ on X.

Step II Since, by Step I, it suffices to prove the theorem for some normal subgroup of finite index of Γ, we can make the following assumption about Γ.

For each σ_α^F, let $\tilde{\sigma}_\alpha^F = \sigma_\alpha^F \setminus \sigma_\alpha^F \cap \partial C(F)$. (When there is no confusion, we shall drop the superscript or subscript F.)

Assumption $\langle \lambda_\alpha, x \rangle < \langle A^* \lambda_\alpha, x \rangle$ for all $x \in \overset{\circ}{\sigma}_\alpha$, id $\neq A \in \overline{\Gamma}_F$.

This assumption is justified for the following reason. There are only finitely many $A \in \overline{\Gamma}_F$ such that $A\sigma_\alpha \cap \sigma_\alpha \cap C(F) \neq \emptyset$; since there are only finitely many σ_α modulo $\overline{\Gamma}_F$, we may take a subgroup Γ' of finite index (e.g., a suitable congruence subgroup) such that all such A are not in $\overline{\Gamma}'_F$ for all F.

With the above assumption, we have:

Proposition 2.16

(1) *Let* $x_n = (t_n, u_n, z_n)$ *be a sequence in* D, *with* $\lim(t_n, u_n) = (\bar{t}, \bar{u})$, $\lim \operatorname{Re} z_n = \bar{x}$ *and* $\lim \operatorname{Im} z_n = \bar{y} + \infty \cdot \sigma_\gamma$, *where* σ_γ *is a face of the top-dimensional* σ_α *and* $\operatorname{Int} \sigma_\gamma \subset C(F)$. *Then*

$$\lim_{n \to \infty} f_{\theta_{m\lambda_\alpha}}(x_n) \mathfrak{X}^{-m\lambda_\alpha}(z_n) = \theta_{m\lambda_\alpha}(\bar{t}, \bar{u}) \ .$$

(2) *Assume further that* σ_γ *is a face of the top-dimensional* σ_α *and* σ_β, *with* $\theta_{m\lambda_\alpha}(\bar{t}, \bar{u}) \neq 0$, *and let* $\bar{z} = \bar{x} + i\bar{y}$. *Then*

$$\lim_{n \to \infty} \frac{f_{\theta_{m\lambda_\beta}}(x_n)}{f_{\theta_{m\lambda_\alpha}}(x_n)} = \mathfrak{X}^{m(\lambda_\beta - \lambda_\alpha)}(\bar{z}) \cdot \frac{\theta_{m\lambda_\beta}(\bar{t}, \bar{u})}{\theta_{m\lambda_\alpha}(\bar{t}, \bar{u})} \ .$$

Proof For the proof of (1), by the same argument as in Proposition 2.8,

$$|\theta_{mA^*\lambda_\alpha} \mathfrak{X}^{m(A^*\lambda_\alpha - \lambda_\alpha)}(z_n)| \leq \text{const.} \cdot \exp(-2\pi m \langle A^*\lambda_\alpha - \lambda_\alpha, \operatorname{Im} z_n \rangle) \ .$$

By the assumption on Γ, for $A \neq$ id, we have that $A^*\lambda_\alpha - \lambda_\alpha$ is positive on σ_γ, and $\langle A^*\lambda_\alpha - \lambda_\alpha, \operatorname{Im} z_n \rangle \longrightarrow \infty$ uniformly as $n \longrightarrow \infty$; therefore, except for $A = $ id, every term of $f_{\theta_{m\lambda_\alpha}} \cdot \mathfrak{X}^{-m\lambda_\alpha}(z_n)$ tends to zero uniformly.

(2) follows from (1), since $\operatorname{Im} z_n = \bar{y} + \varepsilon_n + w_n$ with $\varepsilon_n \longrightarrow 0$, $w_n \in \sigma_\gamma$; hence

$$\lim \mathfrak{X}^{\lambda_\beta - \lambda_\alpha}(z_n) = \lim \exp(-2\pi i \langle \lambda_\beta - \lambda_\alpha, \bar{z} + \varepsilon_n + w_n \rangle) = \mathfrak{X}^{\lambda_\beta - \lambda_\alpha}(\bar{z}) \ .$$

$\square$

For later use, we may put (2) in a slightly more general form. Define $\hat{\sigma}_\alpha = \{\lambda \in U(F)^* \mid \lambda \geq 0 \text{ on } \sigma_\alpha\}$.

Since $\{\sigma_\alpha\}$ is the biggest decomposition such that λ_α is linear on each σ_α, it follows that $\hat{\sigma}_\alpha$ is generated by the $\lambda_\beta - \lambda_\alpha$, where β runs through the top-dimensional simplices such that $\sigma_\beta \cap \sigma_\alpha$ has codimension one.

Let $I_\gamma = \{\lambda \in \hat{\sigma}_\alpha \cap U(F)^*_{\mathbb{Z}} \mid \lambda \equiv 0 \text{ on } \sigma_\gamma\}$. Then I_γ is generated by the $\lambda_{\beta_i} - \lambda_\alpha$ with $\lambda_{\beta_i} \equiv \lambda_\alpha$ on σ_γ, i.e., every $\lambda \in I_\gamma$ is of the form $\lambda = \sum a_i(\lambda_{\beta_i} - \lambda_\alpha)$, where the a_i are positive rational numbers.

There exists $k \in \mathbb{Z}_{\geq 0}$ such that $\lambda + k\lambda_\alpha \in C(F)$, and is a positive linear combination of the λ_{β_i} and λ_α; hence

$$\langle \lambda + k\lambda_\alpha, x \rangle < \langle \lambda + k\lambda_\alpha, Ax \rangle \ , \ \text{for all id} \neq A \in \overline{\Gamma}_F \ , \ x \in \text{Int} \, \sigma_\gamma \ .$$

Proposition 2.17 *With the same notation as in Proposition 2.16, and for λ, k as above, if $\theta_{k\lambda_\alpha}(\bar{t}, \bar{u}) \neq 0$, then*

$$\lim_{n \to \infty} \frac{f_{\theta_{\lambda + k\lambda_\alpha}}(x_n)}{f_{\theta_{k\lambda_\alpha}}(x_n)} = \frac{\theta_{\lambda + k\lambda_\alpha}(\bar{t}, \bar{u})}{\theta_{k\lambda_\alpha}(\bar{t}, \bar{u})} \mathfrak{X}^\lambda(\bar{z}) \ .$$

□

Step III Recall from Chapter III, Section 5 the following commutative diagram:

$$
\begin{array}{ccc}
D/U(F)_{\mathbb{Z}} & \longhookrightarrow & (D/U(F)_{\mathbb{Z}})_{\{\sigma_\alpha^F\}} \\
\downarrow & & \Big| \ \pi_F \\
& & \downarrow \\
D/\Gamma & \longhookrightarrow & \overline{D/\Gamma}
\end{array}
$$

The map π_F is étale, $\bigsqcup \pi_F$ is surjective, and $\overline{D/\Gamma}$ is the unique compact analytic space with these properties. Consider the map

$$f : \overline{D/\Gamma} \longrightarrow D^*/\Gamma \ .$$

By Proposition 2.11, $f^* \mathscr{I}_m$ is a subsheaf of $\mathscr{J}^m$. By Proposition 2.16, (1), if $m \geq 2$, then $\mathscr{I}_m$ has a section at each point which, locally on $\overline{D/\Gamma}$, is a unit times $\mathfrak{X}^{m\lambda_\alpha}$: this is because, if $m \geq 2$, then $\mathscr{L}_{m\lambda_\alpha}|_{\pi^{-1}(E)}$ is generated by its sections, so we can find $\theta_{m\lambda_\alpha}$ with $\theta_{m\lambda_\alpha}(\bar{t}, \bar{u}) \neq 0$. Therefore, if $m \geq 2$, we have $f^* \mathscr{I}_m = \mathscr{J}^m$, hence $f^* \mathscr{I}_m$ is principal. Since $\overline{D/\Gamma}$ is normal, by the universal property of $Y = \widetilde{(D/\Gamma)}_{\mathscr{I}}$, there is an analytic morphism ψ such that the following diagram is commutative:

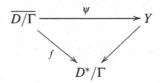

We shall prove that ψ is a local isomorphism, i.e.,

$$\psi^* : \mathscr{O}_y \overset{\sim}{\longrightarrow} \mathscr{O}_x \ , \ \text{for all } x \in D/\Gamma \ , \ \text{where } y = \psi(x) \ .$$

Then Y will have the property characterizing $\overline{D/\Gamma}$, and hence ψ will be an isomorphism.

Step IV Since ψ induces the identity map on D/Γ, it is clear that ψ^* is injective, so it suffices to prove ψ^* is surjective, and again we only need to check this for each stratum of $(D/U(F)_{\mathbb{Z}})_{\{\sigma_\alpha\}}$.

We first recall the notion of strata of $(D/U(F)_{\mathbb{Z}})_{\sigma_\alpha}$.

For each face σ_γ of σ_α, we have the orbit $\mathbb{O}^\gamma$ of $T(F)$ in $T(F)_{\sigma_\alpha}$, cf. TE I, Ch. I; see also Chapter I, Section 1.

Let S_γ be the subset of $(D/U(F)_{\mathbb{Z}})_{\sigma_\alpha}$ given by

$$S_\gamma = (D(F)/U(F)_{\mathbb{Z}}) \times^{T(F)} \mathbb{O}^\gamma .$$

Then $(D/U(F)_{\mathbb{Z}})_{\sigma_\alpha}$ can be broken up into $\bigsqcup S_\gamma$. For $x \in S_\gamma$, $\mathscr{O}_{S_\gamma,x}$ is generated by t, u, and by $\mathfrak{X}^\lambda$ with $\lambda \in I_\gamma = \{\lambda \in U(F)_{\mathbb{Z}}^* \mid \lambda \equiv 0 \text{ on } \sigma_\gamma\}$.

We have a sequence of maps:

$$S_\gamma \overset{\iota_\gamma}{\hookrightarrow} (D/U(F)_{\mathbb{Z}})_{\{\sigma_\alpha\}} \overset{\pi_F}{\longrightarrow} \overline{D/\Gamma} \overset{\psi}{\longrightarrow} Y .$$

Claim *To prove* ψ^* *is surjective, it is sufficient to prove that* $(\psi \circ \pi_F \circ \iota_\gamma)^*$ *is surjective for all* F, α, γ.

Proof of the claim Indeed, assume ψ^* is not surjective at $y \in Y$. Then there is a curve in $\overline{D/\Gamma}$ mapping to y under ψ. (By Zariski's Main Theorem, if $\psi^{-1}(y)$ is finite, then ψ is a local isomorphism at each point of $\psi^{-1}(y)$.) Hence there is a curve in $(D/U(F)_{\mathbb{Z}})_{\{\sigma_\alpha\}}$ mapping to y under $\psi \circ \pi_F$ for some F. It follows that, for some stratum S_γ in $(D/U(F)_{\mathbb{Z}})_{\sigma_\alpha}$, there is a curve in S_γ mapping to y under $\psi \circ \pi_F \circ \iota_\gamma$.

If we can prove that $(\psi \circ \pi_F \circ \iota_\gamma)^*$ is surjective at y for all F, α, γ, then such a curve cannot occur in any stratum S_γ, and hence ψ^* is surjective. $\qquad \square$

By this reduction step, we only need to consider the situation

$$\psi : S_\gamma \longrightarrow Y ,$$

with $x \in S_\gamma$, $\psi(x) = y \in Y$, and to prove that $\psi^* \mathscr{O}_{Y,y} = \mathscr{O}_{S_\gamma,x}$.

Note that, if S_γ is a stratum in $(D/U(F)_{\mathbb{Z}})_{\{\sigma_\alpha\}}$, we may choose F to be the associated boundary component of x, i.e., $\text{Int}\,\sigma_\gamma \subset C(F)$.

The local ring $\mathscr{O}_{S_\gamma,x}$ is generated by t, u, and $\mathfrak{X}^\lambda$ with $\lambda \in I_\gamma$. Let $t(x) = \bar{t}$, $u(x) = \bar{u}$.

We now assume $m \geq 3$, so that $\mathscr{L}_{m\lambda_\alpha}|_{\pi^{-1}(\bar{t})}$ is very ample and we can choose $\theta_{m\lambda_\alpha}^{(i)} \in \Gamma(\pi^{-1}(\bar{t}), \mathscr{L}_{m\lambda_\alpha})$ with $0 \leq i \leq \dim V(F)$ so that

$$u \longmapsto \frac{\theta_{m\lambda_\alpha}^{(i)}(\bar{t}, u)}{\theta_{m\lambda_\alpha}^{(0)}(\bar{t}, u)}$$

is a local isomorphism at $\bar{u}$. Therefore, $\mathcal{O}_{S_\gamma,x}$ is generated by t, $\dfrac{\theta^{(i)}_{m\lambda\alpha}}{\theta^{(0)}_{m\lambda\alpha}}$ and $\mathfrak{X}^\lambda$ for $\lambda \in I_\gamma$.

Applying Proposition 2.16, (2) for $f_{\theta^{(i)}_{m\lambda\alpha}}$, we know that all the θ's are in $\psi^* \mathcal{O}_{Y,y}$. Then applying Proposition 2.17, all $\mathfrak{X}^\lambda$ with $\lambda \in I_\gamma$ are in $\psi^* \mathcal{O}_{Y,y}$. Hence ψ^* is surjective, and the proof of Theorem 2.2 is complete.

References

[1] W. L. Baily, Fourier–Jacobi series, in *Proc. Symp. Pure Math IX*. Providence, RI: American Mathematical Society, 1966, pp. 296–300,

[2] W. L. Baily and A. Borel, Compactification of arithmetic quotients of bounded symmetric domains, *Ann. of Math.* **84** (1966), 442–528.

[3] A. Borel and Harish-Chandra, Arithmetic subgroups of algebraic groups, *Ann. of Math.* **75** (1962), 485–535.

[4] F. Hirzebruch, *Hilbert Modular Surfaces*. Monographie No. 21 de L'Enseignement Mathematique, 1973.

[5] I. Igusa, A desingularization problem in the theory of Siegel modular functions, *Math. Ann.* **168** (1967), 228–260.

[6] I. Igusa, On the theory of compactifications, in *AMS Summer Institute on Algebraic Geometry, Woods Hole*, (1964), Lecture Note.

[7] A. Korányi and J. Wolf, Generalized Cayley transformations of bounded symmetric domains, *Am. J. Math.* **87** (1965), 899–939.

[8] D. Mumford, *Abelian Varieties*. Oxford: Oxford University Press, 1970.

[9] I. Piatetskii-Shapiro, *Geometry of Classical Domains and Theory of Automorphic Functions*. New York: Gordon and Breach, 1969.

Supplementary Bibliography

The following is a list of references to the more recent literature in the subject.

Survey Papers and General Expositions

[1] Ching-Li Chai. *Compactification of Siegel Moduli Schemes*, volume 107 of *London Mathematical Society Lecture Note Series*. Cambridge University Press, Cambridge, 1985.

[2] Ching-Li Chai. Arithmetic compactification of the Siegel moduli space. In *Theta functions—Bowdoin 1987, Part 2 (Brunswick, ME, 1987)*, volume 49 of *Proc. Sympos. Pure Math.*, pages 19–44. Amer. Math. Soc., Providence, RI, 1989

[3] Gerd Faltings and Ching-Li Chai. *Degeneration of Abelian Varieties*, volume 22 of *Ergebnisse der Mathematik und ihrer Grenzgebiete (Series 3)*. Springer-Verlag, Berlin, 1990. With an appendix by David Mumford.

[4] Mark Goresky. Compactifications and cohomology of modular varieties. In *Harmonic Analysis, the Trace Formula, and Shimura Varieties*, volume 4 of *Clay Math. Proc.*, pages 551–582. Amer. Math. Soc., Providence, RI, 2005.

[5] K. Hulek and G. K. Sankaran. The geometry of Siegel modular varieties. In *Higher Dimensional Birational Geometry (Kyoto, 1997)*, volume 35 of *Adv. Stud. Pure Math.*, pages 89–156. Math. Soc. Japan, Tokyo, 2002.

[6] Klaus Hulek, Constantin Kahn, and Steven H. Weintraub. *Moduli Spaces of Abelian Surfaces: Compactification, Degenerations, and Theta Functions*, volume 12 of *de Gruyter Expositions in Mathematics*. Walter de Gruyter & Co., Berlin, 1993.

[7] Lizhen Ji. Buildings and their applications in geometry and topology. *Asian J. Math.*, 10(1):11–80, 2006.

[8] Lizhen Ji. *Arithmetic Groups and their Generalizations: What, Why, and How*, volume 43 of *AMS/IP Studies in Advanced Mathematics*. American Mathematical Society, Providence, RI, 2008.

[9] Yukihiko Namikawa. *Toroidal Compactification of Siegel Spaces*, volume 812 of *Lecture Notes in Mathematics*. Springer-Verlag, Berlin, 1980.

[10] Ichirô Satake. *Algebraic Structures of Symmetric Domains*, volume 4 of *Kanô Memorial Lectures*. Iwanami Shoten, Tokyo, 1980.

[11] Ichirô Satake. *Compactifications, Old and New. Sugaku Expositions*, 14(2):175–189, 2001. [Translation of Sūgaku **51** (1999), no. 2, 129–141.]

[12] Gerard van der Geer. *Hilbert Modular Surfaces*, volume 16 of *Ergebnisse der Mathematik und ihrer Grenzgebiete (Series 3)*. Springer-Verlag, Berlin, 1988.

[13] Gerard van der Geer and Frans Oort. Moduli of abelian varieties: a short introduction and survey. In *Moduli of Curves and Abelian Varieties*, Aspects Math., E33, pages 1–21. Vieweg, Braunschweig, 1999.

Geometric applications and classification problems

[1] Alessandro Arsie. Very ampleness of multiples of principal polarization on degenerate abelian surfaces. *Rev. Mat. Complut.*, 18(1):119–141, 2005.

[2] C. Erdenberger. The Kodaira dimension of certain moduli spaces of abelian surfaces. *Math. Nachr.*, 274/275:32–39, 2004.

[3] Eberhard Freitag. Die Kodairadimension von Körpern automorpher Funktionen. *J. Reine Angew. Math.*, 296:162–170, 1977.

[4] V. Gritsenko, K. Hulek, and G. K. Sankaran. The Hirzebruch-Mumford volume for the orthogonal group and applications. *Doc. Math.*, 12:215–241 (electronic), 2007.

[5] V. Gritsenko, K. Hulek, and G. K. Sankaran. The Kodaira dimension of the moduli of $K3$ surfaces. *Invent. Math.*, 169(3):519–567, 2007.

[6] J. William Hoffman and Steven H. Weintraub. Cohomology of the Siegel modular group of degree two and level four, *Mem. Amer. Math. Soc.*, 133(631):ix,59–75, 1998.

[7] J. William Hoffman and Steven H. Weintraub. The Siegel modular variety of degree two and level three. *Trans. Amer. Math. Soc.*, 353(8):3267–3305 (electronic), 2001.

[8] Rolf-Peter Holzapfel. *Ball and Surface Arithmetics*. Aspects of Mathematics, E29. Friedr. Vieweg & Sohn, Braunschweig, 1998.

[9] K. Hulek and G. K. Sankaran. The Kodaira dimension of certain moduli spaces of abelian surfaces. *Compositio Math.*, 90(1):1–35, 1994.

[10] K. Hulek and G. K. Sankaran. The nef cone of toroidal compactifications of $\mathscr{A}_4$. *Proc. London Math. Soc. (3)*, 88(3):659–704, 2004.

[11] Klaus Hulek. Nef divisors on moduli spaces of abelian varieties. In *Complex Analysis and Algebraic Geometry*, pages 255–274. de Gruyter, Berlin, 2000.

[12] Klaus Hulek. Igusa's modular form and the classification of Siegel modular threefolds. In *Moduli of abelian varieties (Texel Island, 1999)*, volume 195 of *Progr. Math.*, pages 217–229. Birkhäuser, Basel, 2001.

[13] Jun-Muk Hwang. On the volumes of complex hyperbolic manifolds with cusps. *Internat. J. Math.*, 15(6):567–572, 2004.

[14] Jun-ichi Igusa. A desingularization problem in the theory of Siegel modular functions. *Math. Ann.*, 168:228–260, 1967.

[15] Friedrich W. Knöller. Über die Plurigeschlechter Hilbertscher Modulmannigfaltigkeiten. *Math. Ann.*, 264(4):413–422, 1983.

[16] Shigeyuki Kondō. On the Kodaira dimension of the moduli space of $K3$ surfaces, *Compositio Math.*, 89(3):251–299, 1993.

[17] David Mumford. On the Kodaira dimension of the Siegel modular variety. In *Algebraic Geometry—Open Problems (Ravello, 1982)*, volume 997 of *Lecture Notes in Math.*, pages 348–375, Springer, Berlin, 1983.

[18] E. Oeljeklaus and C. Schmerling. Hyperbolicity properties of quotient surfaces by freely operating arithmetic lattices. *Ann. Inst. Fourier (Grenoble)*, 50(1):197–210, 2000.

[19] G. K. Sankaran. Fundamental group of locally symmetric varieties. *Manuscripta Math.*, 90(1):39–48, 1996.

[20] G. K. Sankaran. Moduli of polarised abelian surfaces. *Math. Nachr.*, 188:321–340, 1997.

[21] G. K. Sankaran and J. G. Spandaw. The moduli space of bilevel-6 abelian surfaces. *Nagoya Math. J.*, 168:113–125, 2002.

[22] Eric Schellhammer. The Kodaira dimension of Siegel modular varieties of genus 3 or higher. *Boll. Unione Mat. Ital. Sez. B Artic. Ric. Mat. (8)*, 9(3):749–776, 2006.

[23] N. I. Shepherd-Barron. Perfect forms and the moduli space of abelian varieties. *Invent. Math.*, 163(1):25–45, 2006.

[24] Yung-Sheng Tai. On the Kodaira dimension of the moduli space of abelian varieties. *Invent. Math.*, 68(3):425–439, 1982.

[25] Yung-Sheng Tai. On the Kodaira dimension of moduli spaces of abelian varieties with non-principal polarizations. In *Abelian varieties (Egloffstein, 1993)*, pages 293–302. de Gruyter, Berlin, 1995.

[26] Wing-Keung To. Total geodesy of proper holomorphic immersions between complex hyperbolic space forms of finite volume. *Math. Ann.*, 297(1):59–84, 1993.

[27] R. Weissauer. Untervarietäten der Siegelschen Modulmannigfaltigkeiten von allgemeinem Typ, *Math. Ann.*, 275(2):207–220, 1986.

[28] R. Weissauer. The Picard group of Siegel modular threefolds. *J. Reine Angew. Math.*, 430: 179–211, 1992. With an erratum by the author: "Differential forms attached to subgroups of the Siegel modular group of degree two" [*J. Reine Angew. Math.* 391: 100–156, 1988; MR0961166 (89i:32074)].

[29] Jörg Zintl. Invariants of moduli spaces of abelian surfaces. *Internat. J. Math.*, 11(1):113–131, 2000.

Cohomological applications

[1] Jonas Bergström, Carel Faber, and Gerard van der Geer. Siegel modular forms of genus 2 and level 2: cohomological computations and conjectures. *Int. Math. Res. Not. IMRN*, Art. ID rnn 100, 20, 2008.

[2] Jan H. Bruinier, José I. Burgos Gil, and Ulf Kühn. Borcherds products and arithmetic intersection theory on Hilbert modular surfaces. , *Duke Math. J.*, 139(1):1–88, 2007.

[3] José I. Burgos and Jörg Wildeshaus. Hodge modules on Shimura varieties and their higher direct images in the Baily-Borel compactification, *Ann. Sci. École Norm. Sup. (4)*, 37(3):363–413, 2004.

[4] W. Casselman. Introduction to the L^2-cohomology of arithmetic quotients of bounded symmetric domains. In *Complex Analytic Singularities*, volume 8 of *Adv. Stud. Pure Math.*, pages 69–93. North-Holland, Amsterdam, 1987.

[5] Torsten Ekedahl and Gerard van der Geer. Cycles representing the top Chern class of the Hodge bundle on the moduli space of abelian varieties. *Duke Math. J.*, 129(1):187–199, 2005.

[6] Carel Faber and Gerard van der Geer. Sur la cohomologie des systèmes locaux sur les espaces de modules des courbes de genre 2 et des surfaces abéliennes. I. *C. R. Math. Acad. Sci. Paris*, 338(5):381–384, 2004.

[7] Carel Faber and Gerard van der Geer. Sur la cohomologie des systèmes locaux sur les espaces de modules des courbes de genre 2 et des surfaces abéliennes. II. *C. R. Math. Acad. Sci. Paris*, 338(6):467–470, 2004.

[8] Gerd Faltings. On the cohomology of locally symmetric Hermitian spaces. In *Paul Dubreil and Marie-Paule Malliavin algebra seminar, 35th year (Paris, 1982)*, volume 1029 of *Lecture Notes in Math.*, pages 55–98. Springer, Berlin, 1983.

[9] M. Goresky, G. Harder, and R. MacPherson. Weighted cohomology. *Invent. Math.*, 116(1-3):139–213, 1994.

[10] M. Goresky, G. Harder, R. MacPherson, and A. Nair. Local intersection cohomology of Baily–Borel compactifications. *Compositio Math.*, 134(3):243–268, 2002.

[11] Mark Goresky and Robert MacPherson. The topological trace formula. *J. Reine Angew. Math.*, 560:77–150, 2003.

[12] Mark Goresky and William Pardon. Chern classes of automorphic vector bundles. Invent. Math., 147(3):561–612, 2002.

[13] Michael Harris. Automorphic forms and the cohomology of vector bundles on Shimura varieties. In *Automorphic forms, Shimura varieties, and L-functions, Vol. II (Ann Arbor, MI, 1988)*, volume 11 of *Perspect. Math.*, pages 41–91. Academic Press, Boston, MA, 1990.

[14] Michael Harris. Automorphic forms of $\overline{\partial}$-cohomology type as coherent cohomology classes. *J. Differential Geom.*, 32(1):1–63, 1990.

[15] Michael Harris and Steven Zucker. Boundary cohomology of Shimura varieties. I. Coherent cohomology on toroidal compactifications. *Ann. Sci. École Norm. Sup. (4)*, 27(3):249–344, 1994.

[16] Michael Harris and Steven Zucker. Boundary cohomology of Shimura varieties. II. Hodge theory at the boundary. *Invent. Math.*, 116(1-3):243–308, 1994.

[17] Michael Harris and Steven Zucker. Boundary cohomology of Shimura varieties. III. Coherent cohomology on higher-rank boundary strata and applications to Hodge theory. *Mém. Soc. Math. Fr. (N.S.)*, 85:vi+116, 2001.

[18] J. William Hoffman and Steven H. Weintraub. Cohomology of the bound-

ary of Siegel modular varieties of degree two, with applications. *Fund. Math.*, 178(1):1–47, 2003.

[20] E. Looijenga and M. Rapoport. Weights in the local cohomology of a Baily–Borel compactification. In *Complex Geometry and Lie Theory (Sundance, UT, 1989)*, volume 53 of *Proc. Sympos. Pure Math.*, pages 223–260. Amer. Math. Soc., Providence, RI, 1991.

[20] Eduard Looijenga. L^2-cohomology of locally symmetric varieties. *Compositio Math.*, 67(1):3–20, 1988.

[21] David Mumford. Hirzebruch's proportionality theorem in the noncompact case. *Invent. Math.*, 42:239–272, 1977.

[22] Richard Pink. On the calculation of local terms in the Lefschetz–Verdier trace formula and its application to a conjecture of Deligne. *Ann. of Math. (2)*, 135(3):483–525, 1992.

[23] Leslie Saper. On the cohomology of locally symmetric spaces and of their compactifications. In *Current Developments in Mathematics, 2002*, pages 219–289. Int. Press, Somerville, MA, 2003.

[24] Leslie Saper. L^2-cohomology of locally symmetric spaces. I. *Pure Appl. Math. Q.*,1(4, part 3):889–937, 2005.

[25] Leslie Saper. $\mathscr{L}$-modules and the conjecture of Rapoport and Goresky–MacPherson. In *Automorphic forms. I, Astérisque*, 298:319–334, 2005.

[26] Leslie Saper and Mark Stern. L_2-cohomology of arithmetic varieties. *Ann. of Math. (2)*, 132(1):1–69, 1990.

[27] Gerard van der Geer. The Chow ring of the moduli space of abelian three-folds. *J. Algebraic Geom.*, 7(4):753–770, 1998.

[28] Steven Zucker. On the reductive Borel-Serre compactification: L^p-cohomology of arithmetic groups (for large p). *Amer. J. Math.*, 123(5):951–984, 2001.

Papers with an arithmetic flavor or functor descriptions

[1] Valery Alexeev. On extra components in the functorial compactification of A_g. In *Moduli of Abelian Varieties (Texel Island, 1999)*, volume 195 of *Progr. Math.*, pages 1–9. Birkhäuser, Basel, 2001.

[2] Valery Alexeev. Complete moduli in the presence of semiabelian group action. *Ann. of Math. (2)*, 155(3):611–708, 2002.

[3] Valery Alexeev and Iku Nakamura. On Mumford's construction of degenerating abelian varieties. *Tohoku Math. J. (2)*, 51(3):399–420, 1999.

[4] Michel Brion. Compactification de l'espace des modules des variétés abéliennes principalement polarisées (d'après V. Alexeev). In *Séminaire Bourbaki. Vol. 2005/2006. Astérisque*, (311):Exp. No. 952, vii, 1–31, 2007.

[5] J. I. Burgos Gil, J. Kramer, and U. Kühn. Arithmetic characteristic classes of automorphic vector bundles. *Doc. Math.*, 10:619–716 (electronic), 2005.

[6] Alexander Caspar. Realisations of Kummer–Chern–Eisenstein-motives. *Manuscripta Math.*, 122(1):23–57, 2007.

[7] C.-L. Chai. Arithmetic minimal compactification of the Hilbert-Blumenthal moduli spaces. *Ann. of Math. (2)*, 131(3):541–554, 1990.

[8] David A. Cox. The functor of a smooth toric variety. *Tohoku Math. J. (2)*, 47(2):251–262, 1995.

[9] Mladen Dimitrov. Compactifications arithmétiques des variétés de Hilbert et formes modulaires de Hilbert pour $\Gamma_1(\mathfrak{c}, \mathfrak{n})$. In *Geometric Aspects of Dwork Theory. Vol. I, II*, pages 527–554. Walter de Gruyter GmbH & Co. KG, Berlin, 2004.

[10] Michael Harris. Functorial properties of toroidal compactifications of locally symmetric varieties. *Proc. London Math. Soc. (3)*, 59(1):1–22, 1989.

[11] Takeshi Kajiwara, Kazuya Kato, and Chikara Nakayama. Logarithmic abelian varieties. I. Complex analytic theory. *J. Math. Sci. Univ. Tokyo*, 15(1):69–193, 2008.

[12] Robert E. Kottwitz and Michael Rapoport. Contribution of the points at the boundary. In *The Zeta Functions of Picard Modular Surfaces*, pages 111–150. Univ. Montréal, Montreal, QC, 1992.

[13] Michael J. Larsen. Arithmetic compactification of some Shimura surfaces. In *The Zeta Functions of Picard Modular Surfaces*, pages 31–45. Univ. Montréal, Montreal, QC, 1992.

[14] A. Miller, S. Müller-Stach, S. Wortmann, Y.-H. Yang, and K. Zuo. Chow–Künneth decomposition for universal families over Picard modular surfaces. In *Algebraic Cycles and Motives. Vol. 2*, volume 344 of *London Math. Soc. Lecture Note Ser.*, pages 241–276. Cambridge University Press, Cambridge, 2007.

[15] Abdellah Mokrane and Jacques Tilouine. Cohomology of Siegel varieties with *p*-adic integral coefficients and applications. In *Cohomology of Siegel varieties. Astérisque*, (280):1–95, 2002.

[16] Sophie Morel. Complexes pondérés sur les compactifications de Baily–Borel: le cas des variétés de Siegel. *J. Amer. Math. Soc.*, 21(1):23–61 (electronic), 2008.

[17] Iku Nakamura. On moduli of stable quasi abelian varieties. *Nagoya Math. J.*, 58:149–214, 1975.

[18] Yukihiko Namikawa. A new compactification of the Siegel space and degeneration of Abelian varieties. I, II. *Math. Ann.*, 221(2, 3):97–141, 201–241, 1976.

[19] Martin C. Olsson. Semistable degenerations and period spaces for polarized $K3$ surfaces. *Duke Math. J.*, 125(1):121–203, 2004.

[20] Martin C. Olsson. *Compactifying Moduli Spaces for Abelian Varieties*, volume 1958 of *Lecture Notes in Mathematics*. Springer-Verlag, Berlin, 2008.

[21] Richard Pink. *Arithmetical Compactification of Mixed Shimura Varieties*. Bonner Mathematische Schriften [Bonn Mathematical Publications], 209. Universität Bonn Mathematisches Institut, Bonn, 1990. Dissertation, Rheinische Friedrich-Wilhelms-Universität Bonn, Bonn, 1989.

[22] Richard Pink. On l-adic sheaves on Shimura varieties and their higher direct images in the Baily–Borel compactification. *Math. Ann.*, 292(2):197–240, 1992.

[23] Michael Rapoport. Compactifications de l'espace de modules de Hilbert-Blumenthal. *Compositio Math.*, 36(3):255–335, 1978.

[24] Michael Rapoport. On the shape of the contribution of a fixed point on the boundary: the case of **Q**-rank one. In *The Zeta Functions of Picard Modular Surfaces*, pages 479–491. Univ. Montréal, Montreal, QC, 1992. With an appendix by Leslie Saper and Mark A. Stern.

[25] Eric Urban. Sur les représentations p-adiques associées aux représentations cuspidales de $GSp_{4/\mathbb{Q}}$. In *Formes automorphes. II. Le cas du groupe $GSp(4)$*. *Astérisque*, (302):151–176, 2005.

[26] Jörg Wildeshaus. On the boundary motive of a Shimura variety. *Compos. Math.*, 143(4):959–985, 2007.

Comparison with other compactifications

[1] Armand Borel and Lizhen Ji. Compactifications of symmetric and locally symmetric spaces. *Math. Res. Lett.*, 9(5-6):725–739, 2002.

[2] Armand Borel and Lizhen Ji. Compactifications of locally symmetric spaces. *J. Differential Geom.*, 73(2):263–317, 2006.

[3] Armand Borel and Lizhen Ji. Compactifications of symmetric spaces. *J. Differential Geom.*, 75(1):1–56, 2007.

[4] Mark Goresky and Yung-Sheng Tai. Toroidal and reductive Borel–Serre compactifications of locally symmetric spaces. *Amer. J. Math.*, 121(5):1095–1151, 1999.

[5] L. Ji. The greatest common quotient of Borel–Serre and the toroidal compactifications of locally symmetric spaces. *Geom. Funct. Anal.*, 8(6):978–1015, 1998.

[6] L. Ji and R. MacPherson. Geometry of compactifications of locally symmetric spaces. *Ann. Inst. Fourier (Grenoble)*, 52(2):457–559, 2002.

[7] Lizhen Ji. Metric compactifications of locally symmetric spaces. *Internat. J. Math.*, 9(4):465–491, 1998.

[8] Adam Korányi. Remarks on the Satake compactifications. *Pure Appl. Math. Q.*, 1(4, part 3):851–866, 2005.

[9] Eduard Looijenga. New compactifications of locally symmetric varieties. In *Proceedings of the 1984 Vancouver Conference in Algebraic Geometry*, volume 6 of *CMS Conf. Proc.*, pages 341–364, Providence, RI, 1986. Amer. Math. Soc.

[10] Eduard Looijenga. Compactifications defined by arrangements. I. The ball quotient case. *Duke Math. J.*, 118(1):151–187, 2003.

[11] Eduard Looijenga. Compactifications defined by arrangements. II. Locally symmetric varieties of type IV. *Duke Math. J.*, 119(3):527–588, 2003.

[12] Steven Zucker. On the reductive Borel-Serre compactification. III. Mixed Hodge structures. *Asian J. Math.*, 8(4):881–911, 2004.

[13] Steven Zucker. Excentric compactifications. *Q. J. Pure Appl. Math.*, 1(1):222–226, 2005.

[14] Steven Zucker. On the reductive Borel-Serre compactification. II. Excentric quotients and least common modifications. *Amer. J. Math.*, 130(4):859–912, 2008.

Explicit resolutions

[1] Fritz Ehlers. Eine Klasse komplexer Mannigfaltigkeiten und die Auflösung einiger isolierter Singularitäten. *Math. Ann.*, 218(2):127–156, 1975.

[2] C. Erdenberger, S. Grushevsky, and K. Hulek. Intersection theory of toroidal compactifications of $\mathscr{A}_4$. *Bull. London Math. Soc.*, 38(3):396–400, 2006.

[3] H. G. Grundman. Explicit resolutions of cubic cusp singularities. *Math. Comp.*, 69(230):815–825, 2000.

[4] K. Hulek and G. K. Sankaran. The Kodaira dimension of certain moduli spaces of abelian surfaces. *Compositio Math.*, 90(1):1–35, 1994.

[5] K. Hulek and G. K. Sankaran. The fundamental group of some Siegel modular threefolds. In *Abelian Varieties (Egloffstein, 1993)*, pages 141–150. de Gruyter, Berlin, 1995.

[6] K. Hulek and G. K. Sankaran. The nef cone of toroidal compactifications of $\mathscr{A}_4$. *Proc. London Math. Soc. (3)*, 88(3):659–704, 2004.

[7] Ichirō Satake. On the blowing-ups of Hilbert modular surfaces. *J. Fac. Sci. Univ. Tokyo Sect. IA Math.*, 24(1):221–229, 1977.

[8] Ichirō Satake. On numerical invariants of arithmetic varieties of **Q**-rank one. In *Automorphic Forms of Several Variables (Katata, 1983)*, volume 46 of *Progr. Math.*, pages 353–369. Birkhäuser Boston, Boston, MA, 1984.

[9] E. Thomas and A. T. Vasquez. On the resolution of cusp singularities and the Shintani decomposition in totally real cubic number fields. *Math. Ann.*, 247(1):1–20, 1980.

[10] G. van der Geer. On the geometry of a Siegel modular threefold. *Math. Ann.*, 260(3):317–350, 1982.

[11] G. van der Geer and A. van de Ven. On the minimality of certain Hilbert modular surfaces. In *Complex Analysis and Algebraic Geometry*, pages 137–150. Iwanami Shoten, Tokyo, 1977.

Higher weight Hodge structures

[1] James A. Carlson, Eduardo H. Cattani, and Aroldo G. Kaplan. Mixed Hodge structures and compactifications of Siegel's space (preliminary report). In *Journées de Géometrie Algébrique d'Angers, Juillet 1979/Algebraic Geometry, Angers, 1979*, pages 77–105. Sijthoff & Noordhoff, Alphen aan den Rijn, 1980.

[2] Eduardo Cattani, Aroldo Kaplan, and Wilfried Schmid. Variations of polarized Hodge structure: asymptotics and monodromy. In *Hodge theory (Sant Cugat, 1985)*, volume 1246 of *Lecture Notes in Math.*, pages 16–31. Springer, Berlin, 1987.

[3] Eduardo H. Cattani and Aroldo G. Kaplan. Extension of period mappings for Hodge structures of weight two. *Duke Math. J.*, 44(1):1–43, 1977.

[4] Phillip Griffiths, editor. *Topics in Transcendental Algebraic Geometry*, volume 106 of *Annals of Mathematics Studies*, Princeton, NJ, 1984. Princeton University Press.

[5] Kazuya Kato and Sampei Usui. Logarithmic Hodge structures and classifying spaces. In *The Arithmetic and Geometry of Algebraic Cycles (Banff, AB, 1998)*, volume 24 of *CRM Proc. Lecture Notes*, pages 115–130. Amer. Math. Soc., Providence, RI, 2000.

[6] Kazuya Kato and Sampei Usui. Borel–Serre spaces and spaces of SL(2)-orbits. In *Algebraic Geometry 2000, Azumino (Hotaka)*, volume 36 of *Adv. Stud. Pure Math.*, pages 321–382. Math. Soc. Japan, Tokyo, 2002.

[7] Kazuya Kato and Sampei Usui. *Classifying Spaces of Degenerating Polarized Hodge Structures*, volume 169 of *Annals of Mathematics Studies*. Princeton University Press, Princeton, NJ, 2009.

[8] Gregory Pearlstein. SL$_2$-orbits and degenerations of mixed Hodge structure. *J. Differential Geom.*, 74(1):1–67, 2006.

[9] Joseph Steenbrink and Steven Zucker. Variation of mixed Hodge structure. I. *Invent. Math.*, 80(3):489–542, 1985.

[10] Sampei Usui. Complex structures on partial compactifications of arithmetic quotients of classifying spaces of Hodge structures. *Tohoku Math. J. (2)*, 47(3):405–429, 1995.

[11] Sampei Usui. Images of extended period maps. *J. Algebraic Geom.*, 15(4):603–621, 2006.

Reduction theory

[1] Avner Ash. Deformation retracts with lowest possible dimension of arithmetic quotients of self-adjoint homogeneous cones. *Math. Ann.*, 225(1):69–76, 1977.

[2] Avner Ash. On eutactic forms. *Canad. J. Math.*, 29(5):1040–1054, 1977.

[3] Avner Ash. Cohomology of congruence subgroups SL$(n, \mathbf{Z})$. *Math. Ann.*, 249(1):55–73, 1980.

[4] Avner Ash. Small-dimensional classifying spaces for arithmetic subgroups of general linear groups. *Duke Math. J.*, 51(2):459–468, 1984.

[5] Avner Ash, Paul E. Gunnells, and Mark McConnell. Cohomology of congruence subgroups of $SL_4(\mathbb{Z})$. *J. Number Theory*, 94(1):181–212, 2002.

[6] Avner Ash, Paul E. Gunnells, and Mark McConnell. Cohomology of congruence subgroups of $SL(4,\mathbb{Z})$. II. *J. Number Theory*, 128(8):2263–2274, 2008.

[7] E. S. Barnes and M. J. Cohn. On Minkowski reduction of positive quaternary quadratic forms. *Mathematika*, 23(2):156–158, 1976.

[8] E. S. Barnes and M. J. Cohn. On the reduction of positive quaternary quadratic forms. *J. Austral. Math. Soc. Ser. A*, 22(1):54–64, 1976.

[9] Bill Casselman. Stability of lattices and the partition of arithmetic quotients. *Asian J. Math.*, 8(4):607–637, 2004.

[10] W. A. Casselman. Geometric rationality of Satake compactifications. In *Algebraic Groups and Lie Groups*, volume 9 of *Austral. Math. Soc. Lect. Ser.*, pages 81–103. Cambridge University Press, Cambridge, 1997.

[11] Mathieu Dutour Sikirić, Achill Schürmann, and Frank Vallentin. A generalization of Voronoi's reduction theory and its application. *Duke Math. J.*, 142(1):127–164, 2008.

[12] R. M. Erdahl and S. S. Ryshkov. The empty sphere. *Canad. J. Math.*, 39(4):794–824, 1987.

[13] Paul E. Gunnells. Modular symbols for **Q**-rank one groups and Voronoï reduction. *J. Number Theory*, 75(2):198–219, 1999.

[14] Paul E. Gunnells and Mark McConnell. Hecke operators and $\mathbb{Q}$-groups associated to self-adjoint homogeneous cones. *J. Number Theory*, 100(1):46–71, 2003.

[15] Peter Kiernan and Shoshichi Kobayashi. Comments on Satake compactification and Picard theorem, *J. Math. Soc. Japan*, 28(3):577–580, 1976.

[16] Enrico Leuzinger. An exhaustion of locally symmetric spaces by compact submanifolds with corners. *Invent. Math.*, 121(2):389–410, 1995.

[17] Robert MacPherson and Mark McConnell. Explicit reduction theory for Siegel modular threefolds. *Invent. Math.*, 111(3):575–625, 1993.

[18] S. S. Ryshkov and R. M. Erdahl. The empty sphere. II. *Canad. J. Math.*, 40(5):1058–1073, 1988.

[19] Leslie Saper. Tilings and finite energy retractions of locally symmetric spaces. *Comment. Math. Helv.*, 72(2):167–202, 1997.

[20] Leslie Saper. Geometric rationality of equal-rank Satake compactifications. *Math. Res. Lett.*, 11(5-6):653–671, 2004.

[21] Mathieu Dutour Sikirić, Achill Schürmann, and Frank Vallentin. Classification of eight-dimensional perfect forms. *Electron. Res. Announc. Amer. Math. Soc.*, 13:21–32 (electronic), 2007.

[22] C. Soulé. Perfect forms and the Vandiver conjecture. *J. Reine Angew. Math.*, 517:209–221, 1999.

[23] Christophe Soulé. The cohomology of $SL_3(\mathbf{Z})$. *Topology*, 17(1):1–22, 1978.

[24] Dan Yasaki. On the existence of spines for $\mathbb{Q}$-rank 1 groups. *Selecta Math. (N.S.)*, 12(3-4):541–564, 2006.

[25] Dan Yasaki. An explicit spine for the Picard modular group over the Gaussian integers. *J. Number Theory*, 128(1):207–234, 2008.

Index

229

Printed in the United States
By Bookmasters